Introduction to Sustainable Energy Transformation

Introduction to Sustainable Energy Transformation

Henryk Anglart

CRC Press
Taylor & Francis Group
Boca Raton London New York

CRC Press is an imprint of the
Taylor & Francis Group, an **informa** business

First edition published 2022
by CRC Press
6000 Broken Sound Parkway NW, Suite 300, Boca Raton, FL 33487-2742

and by CRC Press
2 Park Square, Milton Park, Abingdon, Oxon, OX14 4RN

Library of Congress Cataloging-in-Publication Data

Names: Anglart, Henryk, author.
Title: Introduction to sustainable energy transformations / Henryk Anglart.
Description: First edition. | Boca Raton : CRC Press, 2022. | Includes
bibliographical references and index.
Identifiers: LCCN 2021022808 | ISBN 9780367478612 (hbk) | ISBN
9780367470807 (pbk) | ISBN 9781003036982 (ebk)
Subjects: LCSH: Renewable energy sources.
Classification: LCC TJ808 .A54 2022 | DDC 333.79/4--dc23
LC record available at https://lccn.loc.gov/2021022808

ISBN: 978-0-367-47861-2 (hbk)
ISBN: 978-0-367-47080-7 (pbk)
ISBN: 978-1-003-03698-2 (ebk)

DOI: 10.1201/9781003036982

Typeset in Times New Roman
by KnowledgeWorks Global Ltd.

Dedication

To my family

Contents

SECTION II   Energy Transformation Systems

SECTION III External Effects

Preface

This text was inspired by the need for instructional material for a graduate-level course on sustainable energy systems. The course was developed and has been taught on a yearly basis in the nuclear engineering curriculum at KTH Royal Institute of Technology in Stockholm, Sweden, since 2016. The course serves as an introduction to sustainable energy systems and is open to students who took their bachelor's degrees in either mechanical engineering, electrical engineering, or physics.

The energy science and technology is overlapping many traditional academic disciplines and includes physics, chemistry, biology, earth science, and many fields of engineering. Once discussing energy systems, a variety of disciplines with interests in environmental sustainability are important and should be considered as well. This is because of the widespread concern about depletion of energy resources and environmental degradation. A thorough treatment of all these aspects of energy systems in a single book would have been a very challenging task. One possibility would be to consult many specialized texts and use in teaching the course. Instead I have chosen to try to combine the most important concepts from these different disciplines into a single, self-consistent text. The advantage of this approach is that the level and the style of the book is more coherent and uniform. The disadvantage is that, due to space limitation, only a selected amount of knowledge in the field is included.

Finding no existing book that matched the level of the course and contained the spectrum of topics that should be included, I began to collect lecture notes for the "Introduction to Sustainable Energy Transformation" in 2016. The lecture notes have been gradually expanded in terms of scope and level of detail into this book. The educational objective of this book is to allow deeper understanding of challenges and opportunities in current and future energy transformation systems. To meet this objective, the text includes an extensive discussion of various trades-off that energy systems must balance including environmental impacts, land and material use, risks, safety, and costs.

This book is organized into three parts, addressing energy forms and resources, energy systems, and external effects of energy transformation by humans. The first part introduces the basic ideas and principles used throughout the rest of the text. After a review of units, notation, and nomenclature, the first chapter (§1) introduces fundamental concepts concerned with the structure of matter and the energy stored in matter. The energy forms, supply, consumption, and reserves are presented in the second chapter (§2), whereas basic elements of sustainability are discussed in the third chapter (§3). To proceed further it is necessary to present various energy forms, such as mechanical and electromagnetic energy (§4), biological and chemical energy (§5), nuclear energy (§6), and thermal energy (§7). The first part is closed with two chapters devoted to fluid flows in energy systems (§8) and heat transfer in energy systems (§9). All chapters in part I are intended to be self-contained introductions to these topics for students with no previous exposure.

Part II describes energy transformation systems with focus on their efficiencies and principles of operation. The introductory chapter (§10) is devoted to the concepts of First-Law efficiency and Second-Law efficiency. The introduction is followed by a chapter (§11) containing descriptions of the thermal power systems, including the condensing power, stationary gas turbines, combined cycle power, and cogeneration systems. Extensive sections are devoted to moving water energy (§12), wind energy (§13), solar energy (§14), and nuclear energy (§15).

In Part III the attention is turned to external effects of energy systems, including the environmental effects (§16), as well as risk, safety and cost of energy systems (§17).

The book has five appendices that include a summary of nomenclature and notation (Appendix A), selected useful constants (Appendix B), atomic data of chemical elements and water-steam property data (Appendix C), and selected mathematical tools (Appendix D). SI units are used throughout the book, as introduced in §1 and described in Appendix E. However, occasionally non-SI units are employed, since in energy science and energy applications there are many specialized units that are in a widespread use. In nuclear or atomic applications such unit as electron volt (eV) is preferred, whereas in macroscopic global applications such units as million tons of oil equivalent (Mtoe) are more useful. The list of most frequently used energy units at different scales and conversion factors between the units are provided in Appendix E.

This book can be used in many ways, depending on the expected learning outcome and the reader's background. As an alternative to proceed linearly, the reader interested in a particular area of energy science and technology can use subsets of the book and focus on a given chapter or a chapter selection. For example, readers primarily interested in qualitative understanding of the sustainability of energy systems, including economics and policy, could use §1 (Fundamental concepts), §2 (Energy forms, reserves, supply, and consumption), §3 (Elements of sustainability), §10 (Efficiency of energy transformation), §16 (Energy and environment), and §17 (Risks, safety, and cost analysis).

Similarly those interested in quantitative understanding of the topics could use §4 (Mechanical and electromagnetic energy), §5 (Biological and chemical energy), §6 (Nuclear Energy), §7 (Thermal energy), §11 (Thermal power), §12 (Moving water energy), §13 (Wind energy), §14 (Solar energy), and §15 (Nuclear energy). Readers not exposed to fluid mechanics and heat transfer could include §8 (Fluid flow in energy systems) and §9 (Heat transfer in energy systems) as well.

Completing of this book would not be possible without support of many people. First and foremost I thank my wife Ewa who behind the scenes supported the long hours of preparation that went into this effort. I would like to acknowledge conversations and input on many aspects of the subject matter of this book with colleagues, fellow teachers, and experts in other fields, in particular Haipeng Li, Dmitry Grishchenko, Pavel Kudinov, Weimin Ma, and Per Petersson. Thanks also are due to Jan Dufek and Torbjörn Bäck for supporting development of the course. Also thanks to all the *Introduction to Sustainable Energy Transformation* students for their feedback and enthusiasm for the subject. I am also grateful to KTH and head of the Physics Department, Pär Olsson, for support. Finally, I would like to acknowledge unwavering and patient support throughout a long process of writing this book from Kirsten Barr at CRC Press, Taylor & Francis Group.

Henryk Anglart

Part I

Energy Forms and Resources

1 Fundamental Concepts

Energy (gr. $\varepsilon\nu\varepsilon\rho\gamma\varepsilon\iota\alpha$: action, drive) is an abstract concept that can be traced back to the Galileo Galilei time. The concept of energy was further developed in 19th century, when several naturally occurring phenomena, such as relationships between work, heat, and motion, could be explained. Thus the notions of energy, work, and power are relatively new and they were not clearly understood until James Joule investigated relationship between electricity and heat in his famous experiment in 1841, in which work was converted into heat by stirring a container of water with a paddle wheel. In 1847 Joule formulated the law of conservation of energy, one of the fundamental laws of nature.

Around 1850s, through the work performed by William Thomson (better known as Lord Kelvin, to distinguish him from the physicist J.J. Thomson) and Classius, the first law of thermodynamics was formulated. This law states that heat and work are two forms of energy in transition and that the total energy of an isolated system is conserved. At that time all experimental observations involving work and heat could be correctly explained on the basis of Newton's laws of motion and the first law of thermodynamics.

One of the consequences of the law formulated by Joule is that, in an isolated system, energy cannot be created nor destroyed, and it can only change its form. Even though, strictly speaking, our planet is not an isolated system, it is a good practice to use a term energy transformation once referring to physical processes whose main outcome is a change of energy form from one form to another.

Some energy transformations can occur spontaneously in nature, such as transformation of solar energy into kinetic energy of wind and into heat. Other transformations require dedicated devices, so called converters, to achieve the required form of energy. For example, a wind mill is needed to transform the kinetic energy of wind into electricity.

During the recent century, an intensive energy usage by a humankind has become one of the main contributors to the development of our civilization. The energy sector is between the most important ones in economies of all countries, irrespective of their level of development. Thus, both access to the primary energy resources, as well as availability of modern energy conversion technologies are nowadays vital ingredients of World's economy and politics.

Persistent improvements of energy conversion technologies significantly contributed to the industrial revolution of 20th century. The improvements have been mainly based on the introduction of new and more efficient energy converters, allowing to efficiently use new energy resources. As a result, an affordable energy has become available for a larger population. Certainly, this type of development must have its price in terms of, e.g., negative environmental effects and depletion of natural resources. Thus, one of the fundamental questions would be: is such development sustainable? If not, what should be done to achieve the sustainability?

In order to answer these questions, it is necessary to thoroughly analyze the main aspects of energy transformations and consumption. In this introductory chapter we prepare a proper "ground" for these type of analyses. To this end we introduce basic concepts, properties, units and nomenclature to describe and discern the involved phenomena. We present here the ways in which energy is stored in matter. The various technologies by which this energy can be harvested will be discussed in the rest of this book.

1.1 UNITS AND NOTATION

In any engineering work it is important to develop a habit to follow certain rules and standards. The standards related to physical units, notation, and nomenclature used in technical applications belong

DOI: 10.1201/9781003036982-1

to the most important ones. Below we shortly present the main standards followed in the present book.

1.1.1 UNITS

The SI (abbreviated from the French Système International) unit system is used throughout this book. The system has seven base units: second, meter, kilogram, ampere, kelvin, mole, and candela. The definition of base units was entirely revised on 20 May 2019 and is now expressed in terms of invariant constants of nature. In addition, the SI system contains 22 derived units for other common physical quantities such as, e.g., force, pressure, energy, power, and many others. The new definitions of base units, as well as of the most common derived units are given in Appendix E.

Since energy has many different forms, an exceptionally large number of dedicated energy units have been used. Even though most of these units do not belong to the SI, they are still used for practical reasons. For example, at micro scales (that is at scales comparable to atoms and molecules) the electron volt is used, whereas for global systems (or so-called macro scales), such units as Mtoe[1] are preferred. The conversion factors between various units, for systems at micro, meso (or "human"), and macro scales are provided in Appendix E.

1.1.2 NOTATION

This book covers many different fields of science and technology that have developed their own specific notation. As a result, there are frequent notation conflicts. For example, in the thermodynamics h is usually used to denote the specific enthalpy, in the convective heat transfer the same symbol is used for the heat transfer coefficient, whereas in physics h is the Planck constant. To avoid any confusion, the commonly accepted notation in each of the discussed application fields is followed, and the meaning of used symbols is explained whenever any ambiguity arises. A list of most important symbols and notations used in the book is provided in Appendix A.

1.1.3 ATOMIC AND NUCLEAR NOMENCLATURE

A composition of any atom is uniquely specified by the number of neutrons N and protons Z in its nucleus. For an electrically neutral atom, the number of electrons equals the number of protons Z, which is called the **atomic number**.

All atoms of the same element have the same atomic number. However, they can have different numbers of neutrons in the nucleus. Variants of the same element, but consisting of atoms with different numbers of neutrons, are called **isotopes**. The elements with the most isotopes are cesium and xenon (with 36 known isotopes each), and hydrogen has the fewest number of isotopes with only three.

The **mass number**, denoted by A and also called **atomic mass number**, is the total number of neutrons and protons in an atomic nucleus, hence $A = N + Z$.

The symbols used to denote isotopes of an element X are $_Z^{A_1}\text{X}, _Z^{A_2}\text{X}, ...$, where $A_1 = Z + N_1, A_2 = Z + N_2, ...$ are mass numbers of the isotopes and $N_1, N_2, ...$ are neutron numbers in their nuclei. The use of both X and Z in the notation is redundant and Z is often omitted, since the symbol X uniquely determines the value of Z.

A particular atom or nucleus with a specific neutron number N and atomic number Z is referred to as a **nuclide**. Nuclides are either stable or radioactive (i.e., they spontaneously change to another nuclide with a different Z and/or N by emitting one or more particles). The latter are also termed **radionuclides**.

---

[1]An acronym that stands for million or mega tonnes of oil equivalent. The unit represents the amount of energy released when burning one mega tonne of crude oil.

1.2 STRUCTURE OF MATTER

In this section we consider some basic features of matter that determine how the matter is participating in the energy transformation processes.

1.2.1 MATTER

Matter is the substance of which all physical objects are composed. This includes physical objects that have non-zero rest mass, such as electrons, protons and neutrons, and objects that does not have any rest mass, such as photons.

Atoms, neutrons and protons will be treated as "fundamental" entities of interest, since this point of view is good enough to describe all energy transformations that are discussed in this book. However, it should be mentioned that our current knowledge groups the fundamental constituents of matter into **fermions** (with spin 1/2) and **bosons** (with spin 1). Fermions, in addition, are divided into **quarks** (with fanciful names u—up, d—down, s—strange, c—charm, t—top and b—bottom), and **leptons** (here belong e^-—electron, μ^-—muon and τ^-—tau particle). Bosons include g—**gluons**, γ—photons, $W^\pm$—boson and Z^0—boson. Each fermion has a corresponding antiparticle with the same mass but of opposite charge. The antiparticles are denoted by the same symbol as the corresponding particle having an over-bar; e.g. $\bar{d}$ is the d—antiquark. An exception here is the electron's antiparticle, **positron**, denoted by e^+.

Baryons are composed of three quarks: (qqq); for example, the neutron is a (ddu) composite, whereas the proton is a (uud) composite.

Mesons are composed of quark-antiquark pair, for example, a π^+ meson is a $(u,\bar{d})$ pair and π^- meson is a $(\bar{u},d)$ pair.

Most of the baryons and mesons are extremely unstable, with the heavier quarks decaying rapidly to less massive ones by the force mitigated by bosons, such as photons or $W^\pm$—boson. For example, a free neutron can decay to a proton when a d quark changes to a u quark by emitting a W^- boson, which almost instantaneously decays to an electron (e^-) and an electron antineutrino $(\bar{\nu}_e)$. This process represents the basic β-decay process, which is usually described as

$$n \rightarrow p + e^- + \bar{\nu}_e. \tag{1.1}$$

We observe four forces of very different strength in the world around us. One of the greatest challenges of modern theoretical physics is to develop a unified theory of all these forces. All these forces participate in transformation of the forms of energy that we use on Earth. Gravity pulls together hydrogen atoms in the Sun, whereas strong nuclear interactions between these hydrogen atoms lead to the release of energy that is radiated from the Sun as electromagnetic waves. Nuclear fission and fusion rely on strong nuclear forces, electromagnetic forces, and the weak interactions.

The strong nuclear force, responsible for holding together quarks by the exchange of massless gluons, is a complicated force and often described by empirical formulae with many fit parameters. The strong interaction phenomena have a characteristic length of order 1 femtometer = 10^{-15} m, abbreviated 1 fm. Within this range, the strong interaction potential between two nucleons at rest is given as,

$$V_{\text{strong}}(r) = -\frac{1}{4\pi\varepsilon_0} \frac{g^2_{\text{strong}}}{r} e^{-r/b}, \tag{1.2}$$

with $g^2_{\text{strong}} \approx 10e^2$, where $e \cong 1.602 \times 10^{-19}$ C is the elementary charge, $b \cong 1.4$ fm is the range of the force, and $\varepsilon_0 \cong 8.854 \times 10^{-12}$ F/m is the vacuum electric permittivity.

The magnitude of the electromagnetic force acting between two charges q_1 and q_2 is given as

$$|\mathbf{F}_{em}| = \frac{1}{4\pi\varepsilon_0} \frac{q_1 q_2}{r^2}, \tag{1.3}$$

where r is the distance between the two charges.

The weak interactions appear at roughly the same distance as the strong interactions. The weak potential is approximated with a similar formula as the strong potential,

$$V_{\text{weak}}(r) \approx \frac{1}{4\pi\varepsilon_0} \frac{g_{\text{weak}}^2}{r} e^{-r/b}, \tag{1.4}$$

where $g_{\text{weak}} \sim e$ and $b \approx 2.5 \times 10^{-3}$ fm.

The gravity force is by far the weakest of the four forces but it is always attractive and proportinal to the mass. The gravity force acting between two masses m_1 and m_2 is given as,

$$|\mathbf{F}_g| = G_N \frac{m_1 m_2}{r^2}, \tag{1.5}$$

where r is the distance between the two masses and $G_N \cong 6.674 \times 10^{-11}$ m$^3 \cdot$kg$^{-1} \cdot$s^{-2} is the Newtonian constant of gravitation.

Example 1.1: Comparison of the Gravity and Electromagnetic Forces

Find the values of the electromagnetic force and the gravity force acting between two protons located at a distance of 10^{-10} m from each other.

Solution

The proton mass is $m_p \cong 1.673 \times 10^{-27}$ kg and its electric charge is $q_p = 1e \cong 1.602 \times 10^{-19}$ C. Thus, using Eq. (1.5), the gravity force attracting the two protons is,

$$|\mathbf{F}_g| = G_N \frac{m_p^2}{r^2} = 6.674 \times 10^{-11} \frac{\text{m}^3}{\text{kg s}^2} \frac{(1.673 \times 10^{-27} \text{ kg})^2}{(1 \times 10^{-10} \text{ m})^2} \cong 1.866 \times 10^{-44} \text{ N}, \tag{1.6}$$

and the repulsive electromagnetic force is

$$|\mathbf{F}_{em}| = \frac{1}{4\pi\varepsilon_0} \frac{e^2}{r^2} = \frac{1}{4 \times \pi \times 8.854 \times 10^{-12} \frac{\text{F}}{\text{m}}} \frac{(1.602 \times 10^{-19} \text{ C})^2}{(1 \times 10^{-10} \text{ m})^2} \cong 2.307 \times 10^{-8} \text{ N}. \tag{1.7}$$

The electromagnetic force exceeds the gravitational force by 36 orders of magnitude.

In the study of energy transformations in engineering systems, we can view the electron, neutron, and proton as fundamental indivisible particles. The composite nature of nucleons manifest itself under gigantic energies encountered in the nature only during the first few seconds after the creation of the universe or in high-energy particle accelerators. We will deal with energy scales that are sufficient to rearrange or remove the electrons in an atom (expressed in eV) or the neutrons and protons in an nucleus (expressed in MeV). More comprehensive introduction to modern physics can be found in many books, for example [64]. Reference [53] provides an excellent introduction to energy physics.

Substance

Substance is matter with non-zero rest mass and includes macroscopic bodies composed of atoms and molecules, single atoms and single elementary particles (quarks and leptons) that have non-zero rest masses.

Atomic and Molecular Weights

A special mass unit is introduced to describe masses in the microscopic world. The atomic mass unit (abbreviated as amu or just u, and also called dalton) is defined to be equal to one-twelfth of the mass of the neutral ground-state atom of ^{12}C. Thus, knowing the mass of ^{12}C atom, we have 1 u = 1.660 5387 × 10^{-27} kg. Since atom of ^{12}C contains 12 nucleons of approximately the same mass, 1 u is of the same order of magnitude as the mass of one nucleon. In reality the neutron has a slightly greater mass than the proton: $m_n = 1.008\ 664\ 915\ 6$ u and $m_p = 1.007\ 276\ 466\ 9$ u.

The atomic weight $\mathscr{A}$ of an atom is the ratio of atom's mass to that of 1 u. Similarly, the molecular weight $\mathscr{M}$ of a molecule is the ratio of its mass to that of 1 u. These definitions indicate that molecular and atomic weights, as ratios of masses, are dimensionless numbers.

The definition of the atomic weight indicates that it is approximately equal to the atomic mass number A. Thus for approximate calculations $\mathscr{A} \approx A$.

The majority of naturally occurring elements are composed of two or more isotopes. For such elements the elemental atomic weight, usually provided in the periodic table of the elements, is determined by the isotopic abundance of each constituent isotope γ_i (defined as the fraction of atoms of i-th isotope in a given element) and isotope's atomic weight $\mathscr{A}_i$,

$$\mathscr{A} = \sum_i \frac{\gamma_i(\%)}{100} \mathscr{A}_i. \tag{1.8}$$

Example 1.2: Atomic Weight of the Natural Uranium

Calculate the atomic weight of the natural uranium assuming that it consists of the following three isotopes: ^{234}U, ^{235}U and ^{238}U with atomic weights 234.0409456, 235.0439231, 238.0507826 and isotopic abundances 0.0055%, 0.720%, and 99.22745%, respectively.

Solution

The atomic weight of natural uranium is found as $\mathscr{A}_U = (\gamma_{234}\mathscr{A}_{234} + \gamma_{235}\mathscr{A}_{235} + \gamma_{238}\mathscr{A}_{238})/100 =$ $(0.0055 \times 234.0409456 + 0.720 \times 235.0439231 + 99.2745 \times 238.0507826)/100 = 238.0289....$

Avogadro's Number and Avogadro's Constant

In chemistry and physics, **Avogadro's number** is a dimensionless number equal to a number of constituent particles (atoms or molecules) in one mole of any substance, equal to $N_0 = 6.022\ 140\ 76 \times 10^{23}$. It equals the number of atoms in 12 grams of ^{12}C and is usually approximated to $N_0 \cong 6.022 \times 10^{23}$ in hand calculations.

Avogadro's constant, usually denoted by N_A, is numerically equal to Avogadro's number, but it has a physical unit of the reciprocal of mole: mol^{-1}. Thus, Avogadro's constant multiplied by the amount of substance in a sample measured in moles, gives the number of constituent particles in that sample. This particular feature determines the importance of Avogadro's constant, since it can be interpreted as a link between the atomic and the macroscopic world. For instance, if we know the number of moles n of any species containing atoms with mass M_a, the total mass of the species is found as $m = nN_A M_a$.

1.2.2 THE ATOM

An atom is the smallest particle of a chemical element. However, it is still built from subatomic particles such as neutrons, protons, and electrons.

Structure of an Atom

The basic structure of an atom includes a tiny, relatively massive nucleus, containing at least one proton and neutrons. The number of protons determines the atomic number Z, whereas the total number of protons and neutrons (called collectively **nucleons**)—the mass number A. Neutrons have no electric charge, but protons are positively charged. The outermost regions of the atom are called electron shells and contain the electrons, which are negatively charged. A neutral atom has the same number of electrons and protons.

Size of an Atom

The radii of isolated neutral atoms range between 30 and 300 pm (0.3 to 3 Å).

The size of an atom can be estimated from Avogadro's constant along with the molar mass and bulk density of a solid material,

$$D_{atom} \approx \sqrt[3]{\frac{M}{\rho \cdot N_A}}, \tag{1.9}$$

where M is the molar mass, ρ is the mass density and N_A is Avogadro's constant.

The diameter of the nucleus is in the range from 1.75×10^{-15} m for hydrogen to about 15×10^{-15} m for the heaviest atoms, such as of uranium.

The **radius of a stable nucleus** is roughly proportional to the cube root of the mass number A of the nucleus. This gives a useful approximation of the nucleus radius as

$$R = 1.25 \times 10^{-15} A^{1/3} \text{ m} = 1.25 A^{1/3} \text{ fm}, \tag{1.10}$$

where 1 fm = 1 fermi = 10^{-15} m.

Mass of an Atom

Since $\mathscr{A}$ grams of any isotope contains N_A atoms, the mass of one atom is found as

$$M_a = \frac{\mathscr{A}}{N_A} \approx \frac{A}{N_A} \text{ g/atom}. \tag{1.11}$$

In SI units we have

$$M_a = \frac{10^{-3}\mathscr{A}}{N_A} \approx \frac{10^{-3}A}{N_A} \text{ kg/atom}. \tag{1.12}$$

1.2.3 SOURCES OF NUCLEAR AND ATOMIC INFORMATION

There are many websites and tools that provide nuclear and atomic data. For example, the databases provided by the International Atomic Energy Agency (IAEA) can be accessed at https://www-nds.iaea.org. The following databases are available:

- EXFOR—experimental nuclear reaction data.
- ENDF—evaluated nuclear reaction libraries.
- Interactive chart of nuclides.
- ENSDF—evaluated nuclear structure and decay data.
- AMDC—atomic mass data center.
- NUBASE—experimentally known nuclear properties.

Some of the above-mentioned databases, such as for example AMDC and NUBASE, are available as applications for smartphones and tablets. A list of chemical elements and their representative isotope compositions is provided in Appendix C.

1.3 ENERGY IN MATTER

Use and transformation of energy are strictly related to interplay between internal energy stored in matter and other forms of energy. In this section we study the places and forms in which the internal energy is stored in matter, and how the energy is transformed in chemical and nuclear reactions. Our goal is to provide a necessary introduction here; however, all these topics will be covered in more depth throughout the rest of the book.

Energy can be related to such physical quantities as temperature, time, and length. For any system at a temperature high enough that quantum effects are not important and low enough that atomic excitations are not relevant, the molecule kinetic energy E is related to the temperature through **Boltzmann's constant** k_B as follows

$$E \sim k_B T, \tag{1.13}$$

where $k_B = 1.380\,649 \times 10^{-23}$ J/K (exact value) is one of seven defining constants of the SI. This relationship says that system at temperature T has a characteristic energy scale of order $k_B T$. Quantum mechanics connects energy and frequency (time) of an electromagnetic wave via **Planck's formula**

$$E = h\nu = \hbar\omega, \tag{1.14}$$

where $\omega = 2\pi\nu$ is the angular frequency, $\hbar = h/2\pi$, and $h = 6.626\,070\,15 \times 10^{-34}$ J·s (exact value) is **Planck's constant** and is another defining constant of the SI. This equation is particularly relevant when electromagnetic energy is radiated. It states that the amount of energy E contained in electromagnetic radiation of given frequency ν is quantized in units proportional to the frequency. Further, we can connect energy with wave length (distance) λ as

$$E = \frac{hc}{\lambda}, \tag{1.15}$$

since $\nu = c/\lambda$. Here $c = 299\,792\,458$ m/s (exact value) is the speed of light in vacuum and is yet another defining constant of the SI.

1.3.1 THE EQUIVALENCE OF MASS AND ENERGY

The equivalence of mass and energy was first discovered by Albert Einstein, and described in his famous Short Communication published in *Annalen der Physik* **18**, 639-641 (1905) with the title "Ist die Trägheit eines Körpers von seinem Energieinhalt abhängig?"[2]. Interestingly, this short publication does not contain the famous equation $E_0 = mc^2$, and it is concluded with a rather modest statement: "If the theory agrees with the facts, then radiation transmits inertia between emitting and absorbing bodies".

The equation $E_0 = mc^2$ represents mass-energy equivalence and is, arguably, the most famous equation in 20th century physics. In this equation E_0 is the rest energy of a physical system or an amount of energy as measured in the rest frame of an object, m is the mass (i.e., the rest-mass) of a physical system and c is the speed of light. To prove the mass-energy equivalence, let us follow the derivation provided by Einstein.

According to the relativity theory, the frequency measured by observers in a stationary system S and a system S' that moves with a constant velocity V against S are related with the following equation[3],

$$\nu' = \nu \left(\frac{1 - V/c}{1 + V/c} \right)^{1/2}, \tag{1.16}$$

---

[2]Does the Inertia of a Body Depend Upon Its Energy Content?

[3]Known as the relativistic longitudinal Doppler effect.

where v and v' is a frequency measured by a stationary observer in S and S', respectively. In quantum mechanics, a light impulse can be treated as a number of photons with energy hv (in S system), where h is the Planck constant. When we observe the same photons in system S', the number of photons is the same, but the energy of each photon becomes hv'. Thus, we have the following relationship between the energy of the light impulse in systems S and S',

$$\varepsilon' = \varepsilon \left(\frac{1 - V/c}{1 + V/c} \right)^{1/2}. \tag{1.17}$$

Let us consider a certain stationary body in S, possessing energy E_0 in that system, and corresponding energy E_0' in S'. Next, let us assume that the body emits a light impulse with energy $\frac{1}{2}\varepsilon$ in the positive direction of axis x and a similar impulse with energy $\frac{1}{2}\varepsilon$ in the negative direction of axis x. After that the body remains stationary in S.

Let E_1 and E_1' be an energy of the body after the emission of the two light impulses in system S and S', respectively. The energy conservation law applied in both systems yields,

$$E_0 = E_1 + \frac{1}{2}\varepsilon + \frac{1}{2}\varepsilon,$$

$$E_0' = E_1' + \frac{1}{2}\varepsilon \left(\frac{1 - V/c}{1 + V/c} \right)^{1/2} + \frac{1}{2}\varepsilon \left(\frac{1 + V/c}{1 - V/c} \right)^{1/2} = E_1' + \frac{\varepsilon}{(1 - (V/c)^2)^{1/2}}.$$

Subtracting the two equations from each other yields,

$$E_0 - E_0' = E_1 - E_1' + \varepsilon - \frac{\varepsilon}{(1 - (V/c)^2)^{1/2}}.$$

The energy difference $E_0' - E_0$ is the kinetic energy of the body in S' before the emission of the light pulses. This is so, since the body is stationary in S and its kinetic energy is zero. Correspondingly, the energy difference $E_1' - E_1$ is the kinetic energy of the body in S' after the emission of the light pulses. We can find a change of the kinetic energy as,

$$\Delta E_k \equiv E_{k0} - E_{k1} = \varepsilon \left(\frac{1}{(1 - (V/c)^2)^{1/2}} - 1 \right).$$

As we can see, the kinetic energy of the body decreases as a result of the emission of light pulses. The kinetic energy is given in general as $E_k = \frac{1}{2}MV^2$ and it can change either due to the velocity V change or the mass M change. Assuming that $V \ll c$, we obtain the following approximation of the kinetic energy decrease,

$$\Delta E_k \approx \frac{1}{2}\varepsilon \left(\frac{V}{c} \right)^2 = \frac{1}{2}\frac{\varepsilon}{c^2}V^2.$$

Since V is constant and the same before and after the light impulse emissions, the kinetic energy reduction must be caused by the body mass reduction by factor $\Delta M = \varepsilon/c^2$. Thus the relationship: $\varepsilon = \Delta M c^2$.

1.3.2   INTERNAL ENERGY

Internal energy is located in various bindings present in matter, such as neutron and proton bindings in nuclei, electron bindings in atoms, atom bindings in molecules, and finally bindings between molecules. Additional internal energy is located in motions of molecules. Since there are many physical processes involved, internal energy is difficult to calculate in absolute terms. Usually we are interested in energy differences between different states, and only relevant physical processes need to be taken into account, which facilitates calculations. We discuss all above mentioned physical processes below.

Thermal Energy

A **thermal energy** represents one of the components of an **internal energy**, and refers to the energy contained in relative motions of particles in a macroscopic system. It shouldn't be confused with **heat**, which is the thermal energy in transition from one system to another. A **temperature** is a relative measure of the amount of thermal energy of the system. We expect that in general the thermal energy increases with an increasing temperature and it approaches zero when the temperature approaches the absolute zero. Actually the **absolute zero** is defined to be a state from which no further energy can be removed without changing the nature of the constituents or their chemical or nuclear bindings.

The relationship between internal energy and temperature is very complex and depends on the type of matter and its constituents. In particular, the particles (either atoms or molecules) can move in different ways. We call these different motion modes **degrees of freedom**.

A monoatomic molecule (such as an atom of helium gas) is free to move in three dimensions, so it has three translational kinetic degrees of freedom. A diatomic molecule (e.g. O_2 or CO) can, in addition, rotate about two independent axis perpendicular to the line joining the atoms. A rotation about the axis joining the atoms is assumed to be negligible from the energy-location point of view. Finally, the atoms in diatomic molecule can vibrate like a spring along the axis joining them, adding two additional degrees of freedom: one kinetic and one potential. Thus such molecules have in total seven degrees of freedom: three translational kinetic, two rotational kinetic, one vibrational kinetic and one vibrational potential. Similarly we can deduce that triatomic linear molecules (e.g. CO_2) have 11 degrees of freedom (three translational kinetic, two rotational kinetic, three vibrational kinetic and three vibrational potential), triatomic nonlinear molecules (e.g. H_2O) have 12 degrees of freedom (three translational kinetic, three rotational kinetic, three vibrational kinetic, and three vibrational potential), and finally pure crystalline solid atoms (e.g. Au) have 6 degrees of freedom (zero translational kinetic, zero rotational kinetic, three vibrational kinetic, and three vibrational potential).

The **equipartition theorem** states that in any system in thermal equilibrium at a temperature high enough that quantum effects are not important, each degree of freedom contributes $\frac{1}{2}k_BT$ to the internal energy. At low temperatures, quantum mechanical effects freeze out rotational or vibrational degrees of freedom of gas molecules.

There are two relatively simple and useful models of ideal systems that provide expressions for thermal energy: a **monoatomic ideal gas** model and a model of **ideal monoatomic crystalline solid**.

The thermal energy of N atoms or n moles of a monoatomic ideal gas at temperature T is

$$\mathscr{U} = \frac{3}{2}Nk_BT = \frac{3}{2}nRT \tag{1.18}$$

where $\mathscr{U}$ is the thermal energy, $R = N_Ak_B \simeq 8.314$ J/mol K is the universal gas constant, N_A is Avogadro's constant, and k_B is Boltzmann's constant.

In ideal monoatomic crystalline solid the atoms are held in a regular array by strong forces. The atoms have no translational or rotational kinetic degrees of freedom, but each can vibrate with kinetic and potential energy in three directions. When quantum effects can be neglected, solid's internal energy is

$$\mathscr{U} = 6 \times \frac{1}{2}Nk_BT = 3nRT, \tag{1.19}$$

which is known as the **law of Dulong and Petit**. This law is remarkably accurate for pure elements like sulphur, copper, aluminium, gold and iron at room temperature.

It should be noted that we treat here thermal energy and internal energy as the same quantities. This is actually a common practice in many practical applications, when other components of the internal energy can be neglected. As long as we are interested in internal energy differences, this practice is leading to correct results. However, to avoid confusion, whenever necessary from the clarity point of view, we will use the symbol $\mathcal{U}$ for the thermal energy and E_I for the internal energy.

Melting and Vaporization

Melting and vaporization are examples of phase transition. At certain conditions (e.g. combination of temperature and pressure) the removal or addition of heat to a given material has the effect of changing the relative fractions of the material in different phases, while keeping the material temperature constant. Thus, solids melt (or fuse, as it is called by chemists) and turn into liquid phase when adding heat. With further addition of heat the liquid temperature increases until it reaches the vaporization (or saturation) temperature, at which the liquid evaporates and turns into vapor with additional heat added.

The amount of energy needed to turn a solid to a liquid or a liquid to a vapor is called the latent heat of fusion (melting) or vaporization (evaporation). The physics of fusion and evaporation is quite complex, but expressed in simple terms, the source of the latent heat is in intermolecular binding energy. If the phase change takes place at constant pressure, the added thermal energy equals the increase of enthalpy of the substance: $\Delta I = \Delta\mathcal{U} + p\Delta V$. This expression can be written in terms of enthalpies, internal energies, and volumes per unit mass, known as specific enthalpies, specific internal energies, and specific volumes, respectively. Thus, the specific enthalpy of vaporization is $\Delta i = \Delta u + p\Delta v$. Water has very large enthalpy of vaporization, which makes boiling water a very effective way to store energy. For example, the specific enthalpy of boiling water at one atmosphere is 2.257 MJ/kg, which is significantly greater than the density at which energy can be stored in most chemical batteries, which for lithium-ion batteries is about 0.5 MJ/kg. However, when expressed per unit volumes, the numbers are not that impressive, since the mass density of vapor at the atmospheric pressure is quite low.

Energy Stored in Molecules

Binding atoms in molecules is another place where internal energy can be stored. The physical process to retrieve the energy from molecules is to decompose them into atoms. An example of such process is **dissociation of water**,

$$H_2O \rightarrow 2H + O. \tag{1.20}$$

This reaction can be described as two chemical reactions: dissociation of hydrogen $H_2O \rightarrow H + OH$, requiring 493 kJ/mol, and dissociation of OH into O and H, requiring 424 kJ/mol. Temperatures in excess of several thousand degrees kelvin are needed to dissociate water molecules into hydrogen and oxygen atoms.

Energy Stored in Atoms

Binding forces of electrons in atoms are much stronger than binding forces of atoms in molecules, so the ionization energies are usually larger than the energy required to break molecular bonds. For example, the temperature at which hydrogen is ionized is about 160 000 K, exceeding significantly any temperatures that occur in earth-bound energy transformation processes.

Energy Stored in Nuclei

Protons and neutrons are bound by about 8 MeV per particle in a typical nucleus with many protons and neutrons. This huge energy can be retrieved in fission and fusion reactors. Nuclear energy is discussed separately in more detail in §6.

1.3.3 ENERGY IN CHEMICAL REACTIONS

The **enthalpy of reaction** (called also **heat of reaction**) ΔI_r is defined to be the amount of energy that is absorbed ($\Delta I_r > 0$) or released ($\Delta I_r < 0$) as heat in the course of a reaction that takes place at constant pressure. This quantity is provided in thermodynamic tables for various reactions. For example, consider the following reaction[4],

$$CaCO_3 \to CaO + CO_2. \tag{1.21}$$

From the thermodynamic tables we find the reaction enthalpy as $\Delta I_r = +178.3$ kJ/mol. Since $\Delta I_r > 0$, heat must be supplied to make the reaction go. Such reactions are called **endothermic reactions**.

Chemists do not tabulate ΔI_r for every possible chemical reaction, but rather, for all compounds of interest, they provide the **enthalpy of formation** denoted ΔI^f, which is equivalent to the system enthalpy change when building up a chemical compound out of its constituents. Once the enthalpies of formation for products and reactants are known, the reaction enthalpy can be found from the **Hess law**, which states that the reaction enthalpy for any reaction is equal to the sum of the enthalpies of formation of the products minus the sum of the enthalpies of formation of the reactants,

$$\Delta I_r = \sum_{products} \Delta I^f - \sum_{reactants} \Delta I^f. \tag{1.22}$$

Example 1.3: The Enthalpy of the Calcination Reaction

Calculate the reaction enthalpy for calcination knowing that the enthalpies of formation for $CaCO_3$, CaO and CO_2 are equal to –1207 kJ/mol, –635 kJ/mol and –394 kJ/mol, respectively.

Solution

Using Hess's law we have $\Delta I_r = \Delta I^f(CO_2) + \Delta I^f(CaO) - \Delta I^f(CaCO_3) = -635 + (-394) - (-1207)$ = +178 kJ/mol.

Combustion reactions are particularly important since they are involved once retrieving the energy from various chemical fuels. For example, the combustion reaction of methane is,

$$CH_4 + 2O_2 \to CO_2(g) + 2H_2O(l). \tag{1.23}$$

Here (g) and (l) symbols indicate that carbon dioxide is gas, whereas water is liquid after the reaction. This information is essential since the phase state of the combustion products determines the value of the reaction enthalpy.

The negative of the reaction enthalpy at 25 °C and 1 atm is defined as the **higher heating value** (HHV) of a fuel. This definition indicates that the latent heat of condensation of the water is included into HHV. Correspondingly, the lower heating value (LHV) is a measure of the heat released by combustion when the water vapor is not condensed.

Example 1.4: HHV and LHV for Combustion of Methane

Calculate the HHV and LHV for combustion of methane (CH_4).

Solution

HHV can be found from reaction $CH_4 + 2O_2 \to CO_2(g) + 2H_2O(l)$, where the enthalpies of formation for CH_4, CO_2 and $H_2O(l)$ are –74.8 kJ/mol, –393.5 kJ/mol and –285.8 kJ/mol, respectively. From Hess's law we get $\Delta I_r = 2\Delta I^f(H_2O(l)) + \Delta I^f(CO_2) - \Delta I^f(CH_4) = 2(-285.8) + (-393.5)$

---

[4]This reaction, in which carbon dioxide is driven from limestone, is known as calcination.

– (−74.8) = −890.3 kJ/mol. Thus HHV $\cong$ 890.3 kJ/mol. LHV is found in a similar way, but the enthalpy of formation of water vapor (−241.8 kJ/mol) is taken instead of water liquid. Thus $\Delta I_r = 2\Delta I^f(H_2O(g)) + \Delta I^f(CO_2) - \Delta I^f(CH_4)$ = 2(−241.8) + (−393.5) − (−74.8) = −802.3 kJ/mol. The difference between the HHV and LHV is equal to latent heat of condensation of two moles of water at 25 °C, which is exactly 88 kJ/mol. We note that the difference between the LHV and HHV is approximately equal to 10% of the higher heating value.

Combustion of various fuels is discussed in a more detail in §5.5.

1.3.4   ENERGY IN NUCLEAR REACTIONS

A **nuclear reaction** is a transformation of a nucleus caused by interactions with elementary particles, photons, or other nuclei, due to which from the initial nucleus a new nucleus is created. The reaction is usually accompanied with a release of energy, which manifests itself as a kinetic energy of emitted particles or photons. A particularly important role is played by reactions involving neutrons, since they are closely related to fission of nuclei and thus to a release of the fission energy. In nuclear reactor applications a relatively large group of reactions belong to **binary reactions**, in which two reactants are involved and the reaction yields two reaction products. In particular, a neutron elastic scattering with a nucleus belongs to this category. Nuclear reactions have to obey several conservation principles. During certain reactions the rest mass of the product nucleus is changed.

In any nuclear reaction reactants interact with each other and yield reaction products. These types of reactions can be schematically represented as

$$A + B + \cdots \rightarrow C + D + \cdots \tag{1.24}$$

where A, B, ... are reactants and C, D, ... are reaction products. In all such reactions several quantities are conserved such as the total energy, the linear momentum, the charge, the number of protons, the number of neutrons, and others.

A very common type of nuclear reaction is such in which two source nuclei react to form two product nuclei. This type of reaction can be written as

$$^{A_1}_{Z_1}X_1 + ^{A_2}_{Z_2}X_2 \rightarrow ^{A_3}_{Z_3}X_3 + ^{A_4}_{Z_4}X_4. \tag{1.25}$$

This reaction has to satisfy the conservation of the number of protons and neutrons as follows

$$A_1 + A_2 = A_3 + A_4, \tag{1.26}$$

$$Z_1 + Z_2 = Z_3 + Z_4. \tag{1.27}$$

Particularly important are binary nuclear reactions in which two nuclear particles (nucleons, nuclei or photons) interact to form different nuclear particles. Typically some light particle x (a nucleon or a nucleus) moves with some kinetic energy and interacts with a heavy nucleus to form a pair of product particles. Such reaction can be represented as,

$$x + X \rightarrow y + Y, \tag{1.28}$$

or in a shorthand notation as $X(x,y)Y$. For example, $^{14}_{7}N(\alpha, p)^{17}_{8}O$ represents a reaction in which a nitrogen nucleus, bombarded with alpha particles, converts into oxygen and yields a proton. The number of protons and neutrons is balanced on both sides of the reaction equation. This reaction was first reported by Rutherford and was the first nuclear reaction ever discovered.

In any nuclear reaction the total energy (that is the kinetic energy plus the rest-mass energy) must be conserved, i.e.,

$$\sum_{i \in \text{reactants}} \left(E_i + m_i c^2 \right) = \sum_{i \in \text{products}} \left(E_i' + m_i' c^2 \right), \tag{1.29}$$

where E_i and E_i' is the kinetic energy of ith reactant and product, respectively. Equation (1.29) infers that any change in the total kinetic energy of reactants (particles before the reaction) and products (particles after the reaction) must be accompanied by an equivalent change in the total rest mass of reactants and products. This change is represented by a so-called Q value of the reaction defined as,

$$Q = \sum_{i \in \text{products}} E_i' - \sum_{i \in \text{reactants}} E_i. \tag{1.30}$$

Combining Eq. (1.29) and Eq. (1.30) yields,

$$Q = \left(\sum_{i \in \text{reactants}} m_i - \sum_{i \in \text{products}} m_i' \right) c^2. \tag{1.31}$$

When $Q > 0$, the reaction is exothermic, and conversely, if $Q < 0$ the reaction is endothermic.

For binary reactions the Q value is given by

$$Q = [(m_x + m_X) - (m_y + m_Y)] c^2, \tag{1.32}$$

where m_x and m_X are the rest masses of reactants and m_y and m_Y are the rest masses of products. For most binary nuclear reactions the number of protons is conserved. By adding the same number of electron masses to both sides of Eq. (1.32), and neglecting differences in electron binding energy, the Q value can be written in terms of atomic masses as

$$Q = [(M_x + M_X) - (M_y + M_Y)] c^2, \tag{1.33}$$

where we follow a convention that $M(^A_Z X)$ denotes the rest mass of an atom, whereas $m(^A_Z X)$ denotes the rest mass of its nucleus. With this notation the rest mass of a proton can be written as $m_p \equiv m(^1_1 H)$.

Example 1.5: Q Value of a Nuclear Reaction

Calculate the Q value for the reaction $^9_4 Be(\alpha, n)^{12}_6 C$ knowing the following rest masses: $M(^9_4 Be)$ = 9.012182 u, $M(^4_2 He)$ = 4.002603 u, $M(^{12}_6 C)$ = 12.000000 u and $m_n \equiv m(^1_0 n)$ = 1.008664 u. Is the reaction endothermic or exothermic?

Solution

We find the difference in rest masses between reactants and products as $M(^9_4 Be) + M(^4_2 He)$ $-M(^{12}_6 C) - m(^1_0 n)$ = 0.006121 u. Thus Q = 931.5 MeV/u × 0.006121 u = 5.702 MeV > 0. The reaction is exothermic.

Nuclear Fission

Fission nuclear reactions are special type of reactions in which a very heavy nucleus (e.g. ^{235}U) splits into two lighter nuclei and some other particles such as neutrons and photons. Initially the heavy nucleus absorbs an incident particle (neutron) and stays in an excited state for about 10^{-14} s. The excited heavy nucleus is very unstable and scissions into two nuclear pieces (fragments) within 10^{-20} s. These fragments are repelling each other with a tremendous Coulombic force, leaving behind many orbital electrons. For example, at this stage, the neutron-induced fission of ^{235}U can have the following reaction sequence,

$$^1_0 n + ^{235}_{92} U \rightarrow ^{236}_{92} U^* \rightarrow Y_H^{+n} + Y_L^{+m} + (n+m)e^-, \tag{1.34}$$

where $^{236}_{92} U^*$ is the excited compound nucleus, Y_H—the heavy primary fission product and Y_L—the light primary fission product. The ionic charge n and m on the fission fragments is about 20.

After about 10^{-12} s following the scission, the fission fragments slow down, transferring their kinetic energy to the ambient medium and acquiring electrons to become neutral atoms. At this stage the fission reaction can be written as,

$$_0^1 n + _{92}^{235} U \rightarrow _{92}^{236} U^* \rightarrow Y_H + Y_L + v_p \left(_0^1 n\right) + \gamma_p. \tag{1.35}$$

Here v_p is the number of neutrons emitted from the primary fission fragments within about 10^{-17} s of the splitting apart of the compound nucleus. This neutrons are called prompt neutrons and v_p is any number between 0 and 8. Following the prompt neutron emission, the fission fragments decay to lower energy levels by emitting prompt gamma rays γ_p within about 2×10^{-14} s after the emission of prompt neutrons. The prompt energy released E_p is obtained from the mass deficit of reaction (1.35) as

$$E_p = \left[M\left(_{92}^{235} U\right) + m_n - M(Y_H) - M(Y_L) - v_p m_n\right] c^2. \tag{1.36}$$

Example 1.6: Prompt Energy of Fission of ^{235}U

Calculate the prompt energy release for the following fission reaction

$$_{92}^{235} U + _0^1 n \rightarrow _{54}^{139} Xe + _{38}^{95} Sr + 2 \left(_0^1 n\right) + 7 \left(\gamma_p\right).$$

Assuming that two prompt fission neutrons have a total kinetic energy of 5.2 MeV and the prompt gamma rays have a total energy of 6.7 MeV, find the total kinetic energy of initial fission fragments.

Solution

The mass deficit of the reaction is $\Delta M = M(_{92}^{235}U) + m_n - M(_{54}^{139}Xe) - M(_{38}^{95}Sr) - 2m_n = (235.043923 + 1.008665 - 138.918787 - 94.919358 - 2 \times 1.008665)u = 0.197113$ u. Thus, $E_p = 931.5$ MeV/u $\times 0.197113$ u $= 183.6$ MeV. The total kinetic energy of initial fission fragments can be found as $E_{Kff} = 183.6 - 5.2 - 6.7 = 171.7$ MeV.

Example 1.6 shows how the "prompt" energy released in a fission reaction can be calculated. This is the majority of the fission energy, which is released within the first 10^{-12} s. However, fission products are not stable and decay after some time to their final stable end-chain nuclei. For example, $_{54}^{139}Xe$ reaches the stable nuclide $_{57}^{139}La$ after three β^- decays, whereas $_{38}^{95}Sr$ reaches stable $_{42}^{95}Mo$ after four β^- decays. Thus the decay reactions for fission products can be written as,

$$_{54}^{139}Xe + _{38}^{95}Sr + 7\left(_{-1}^0 e\right) \rightarrow _{57}^{139}La + _{42}^{95}Mo + 7\left(_{-1}^0 \beta\right) + 7\left(\overline{v}\right). \tag{1.37}$$

Here $7\left(_{-1}^0 e\right)$ on the left-hand-side represents seven electrons acquired by decaying fission products to make each member of the decay chains electrically neutral. Equation (1.37) can be used to calculate the "delayed" fission energy (see Problem 1.9).

Nuclear Fusion

In addition to fission of heavy nuclei, fusion of light nuclei can be used for extracting nuclear energy. Some possible fusion reactions, along with their Q values, are as follows,

$$_1^2 H + _1^2 H \rightarrow _2^3 He + _0^1 n, \ Q = 3.27 \text{ MeV}, \tag{1.38}$$

$$_1^2 H + _1^2 H \rightarrow _1^3 H + _1^1 H, \ Q = 4.03 \text{ MeV}, \tag{1.39}$$

$$_1^2 H + _1^3 H \rightarrow _2^4 He + _0^1 n, \ Q = 17.59 \text{ MeV}, \tag{1.40}$$

$$\text{}^2_1\text{H} + {}^3_2\text{He} \rightarrow {}^4_2\text{He} + {}^1_1\text{H}, \ Q = 18.35 \text{ MeV}, \tag{1.41}$$

$$\text{}^3_1\text{H} + {}^3_1\text{H} \rightarrow {}^4_2\text{He} + 2{}^1_0\text{n}, \ Q = 11.33 \text{ MeV}, \tag{1.42}$$

$$\text{}^1_1\text{H} + {}^6_3\text{Li} \rightarrow {}^4_2\text{He} + {}^3_2\text{He}, \ Q = 4.02 \text{ MeV}, \tag{1.43}$$

$$\text{}^1_1\text{H} + {}^{11}_5\text{B} \rightarrow 3({}^4_2\text{He}), \ Q = 8.08 \text{ MeV}, \tag{1.44}$$

Even though all above-mentioned fusion reactions are exoergic (i.e. $Q > 0$), they occur with varying difficulty due to so-called threshold energies that are required to overcome the repulsive Coulomb forces between the two reactants.

All current experimental research into fusion power is based on the D-T fusion given by Eq. (1.40), which requires the lowest plasma temperature to overcome the Coulomb barrier problem. This reaction requires a radioactive tritium ($T_{1/2} = 12.3$ y) that can be obtained from lithium ${}^6_3\text{Li}$ interacting with neutrons.

PROBLEMS

PROBLEM 1.1

Express in SI units:
(a) area $A = 314$ sq in (square inch) $= x$ m^2
(b) volume $V = 18.6$ cu ft (cubic foot) $= x$ m^3
(c) velocity $v = 65$ mi/h (miles per hour) $= x$ km/h
(d) heat $Q = 1580$ BTU $= x$ kJ
(e) mass $m = 48$ lb (pound) $= x$ kg

PROBLEM 1.2

Express in SI units:
(a) density $\rho = 12.5$ lb/cu ft $= x$ kg/m^3
(b) pressure $p = 1100$ psi $= x$ MPa
(c) work $L = 2250$ Lb ft $= x$ kJ
(d) power $N = 210$ hp $= x$ kW
(e) heat transfer coefficient $h = 250$ BTU/(ft^2 h deg F) $= x$ W/(m^2 K)

PROBLEM 1.3

Calculate the number of ^{235}U atoms in 1 kg of natural uranium.

PROBLEM 1.4

Calculate the number of deuterium atoms in 1 kg of water.

PROBLEM 1.5

Estimate the mass of ^{235}U atom.

PROBLEM 1.6

What is the atomic weight of natural boron, when it is known that it consists of two stable isotopes ^{10}B and ^{11}B with isotopic abundances of 19.9 and 80.1 atom-percent, respectively?

PROBLEM 1.7

Calculate the energy per molecule required to dissociate H_2O into H and OH. Find needed data in the text.

PROBLEM 1.8

Calculate the Q value for the reaction $^{16}_{8}O(n,\alpha)^{13}_{6}C$ knowing the following rest masses: $M(^{16}_{8}O)$ = 15.994915 u, $M(^{4}_{2}He)$ = 4.002603 u, $M(^{13}_{6}C)$ = 13.003354 u and $m_n \equiv m(^{1}_{0}n)$ = 1.008664 u. Is the reaction endothermic or exothermic?

PROBLEM 1.9

Calculate the "delayed" fission energy for Eq. (1.37). Use the following data: $M(^{139}_{54}Xe)$ = 138.918787 u, $M(^{95}_{38}Sr)$ = 94.919358 u, $M(^{139}_{57}La)$ = 138.906348 u and $M(^{95}_{42}Mo)$ = 94.905842 u.

PROBLEM 1.10

Calculate the Q value of the following fusion reaction $^{2}_{1}D + ^{2}_{1}D \rightarrow ^{4}_{2}He$. Note that deuterons do not fuse into $^{4}_{2}He$ directly. How this reaction could be realized? Hint: consider a chain of reactions from those given by Eq. (1.38) through Eq. (1.44).

2 Energy Forms, Reserves, Supply, and Consumption

In the previous chapter we discussed various energy forms that are present in matter. In particular, we learned that energy can have many different forms and that it is located in various bindings present in matter, including nuclei, atoms, and molecules. Natural questions are how much of this energy is available, how this energy can be used by humans, and whether the energy supply and consumption are sustainable? To answer these difficult questions, we need to have a closer look at the past and present energy reserves, consumption, and production. Comparing the energy consumption rate with known energy reserves will give us a hint about the time horizon, when the reserves will be depleted. But even with an unlimited amount of energy available, the required energy transformations may incur negative environmental effects, such as air pollution and greenhouse gas emissions. These aspects are equally important as the sufficiency of energy reserves.

No doubt, one of the greatest current challenges of humankind is to assure access to clean, affordable, and useful energy in a sustainable manner. To address these challenges, one has to have a good understanding of past, current, and future trends in the energy sector. The purpose of this chapter is to contribute to this understanding. As the rest of the book is dealing with various aspects of energy transformation technologies, this chapter is devoted to the past and present status of the global energy system.

More specifically, in this chapter we discuss the main aspects of energy transformations that are present in energy supply and consumption paths. Firstly, we discuss the various energy forms as supplied and consumed. Next, we make an overview of the major energy supply paths. Finally, important from the sustainability point of view, we track the consumption processes and discuss their influence on the environment.

2.1 ENERGY FORMS

Energy transformation is a process in which one energy form changes, either naturally or using special devices called converters, into another one. Since the law of energy conservation states that energy is always conserved, these energy transformations are not changing the total amount of energy in the universe; however, energy of a certain form can be reduced, while the energy of another form is increased. For example, in a gasoline-driven car, chemical energy stored in fuel is transformed into an equal amount of mechanical and thermal energy. However, only the mechanical energy is treated as a desired form of energy in this transformation, whereas the thermal energy is treated as an energy loss.

Energy forms presented in §1.3 are not always adequate to be consumed. There are also other forms of energy that are more useful in everyday life. The most important energy forms from supply, storage, and consumption point of view are introduced in this section.

2.1.1 PRIMARY AND SECONDARY ENERGY

Primary energy (also called **natural energy**) is an energy form as present in the natural environment. Even though, on the earth-science scale, all primary energy resources can be considered as always changing, some of them are consumed at much higher rates than they can be geophysically renewed. Such primary energy resources are commonly termed as non-renewable resources. To this category belong oil, coal, natural gas, uranium, and thorium. Solar, wind, hydro (both falling and

DOI: 10.1201/9781003036982-2

flowing water), biomass, and geothermal energy are termed as renewable energy resources. Thus, both renewable and non-renewable primary energy resources exist.

Since not all forms of primary energy are suitable for practical use, they are often transformed or converted into other energy forms, which are termed as **secondary energy**. For example, coal or natural gas are used as fuel in power plants and are converted into heat and then partially converted into electrical energy. However, electricity is not directly a useful form of energy but is rather an energy carrier that can be converted into a useful form such as comfort heating or cooling, motion, sound, or light.

According to the first law of thermodynamics, the primary energy resources cannot be destroyed. However, they can be transformed into useless forms such as heat at low temperature. This process is taking place when fuel is burned in power plants and the energy is partly converted into electricity, whereas the rest is dumped to the environment, as the second law of thermodynamics dictates.

2.1.2 ENERGY CARRIER

Energy carrier is a substance, a phenomenon, or a system that contains energy in any form, which can be transformed into a useful or final energy form. Examples of energy carriers include electricity, electric batteries, capacitors, springs, pressurized air, dammed water, hydrogen, petroleum, coal, wood, and natural gas.

2.1.3 FINAL ENERGY

The energy form that is delivered to end users is called a **final energy**. Usually this form of energy is the primary energy after some required transformations, but this is not always the case. For example, natural gas is at the same time the primary and the final energy form, since it is delivered to end users without any transformation. Coal can be delivered to the end user in its natural form, and burned in domestic burners, but it can also be transformed and its energy can be delivered to end users as electricity. For centuries the wind and hydro energy was at the same time the primary and final energy. However, recently the mechanical energy of air and water is transformed into electricity using wind and hydro turbines, and this form of energy is delivered to end users as the final energy.

2.1.4 USEFUL ENERGY

The main useful energy forms are mechanical energy, thermal energy, light, and sound. These energy forms can be obtained from energy carriers using various devices, such as engines, electric motors, heaters, coolers, lamps, and loudspeakers.

2.1.5 ELECTRICITY

Electricity is an energy form that can be used in a very wide range of applications in industry, transportation, agriculture, and households. It can be used for running machines, lighting, heating, and cooling. Electricity is produced as a primary as well as secondary energy. **Primary electricity** is obtained from natural resources such as hydro, wind, solar, wave, and tidal power. These natural energy forms can be directly converted into electricity using generators. **Secondary electricity** is produced indirectly from such primary energy resources as coal, natural gas, and uranium, using combustion and fission, respectively, to produce heat. The heat is used in heat engines to generate mechanical (rotational) energy, which is next converted into electricity.

2.1.6 HEAT

Heat is an energy form that is primarily used for warming spaces and in industrial processes. Similarly to electricity, heat is produced as primary as well as secondary energy. **Primary heat**

is obtained directly from natural resources such as geothermal and solar thermal power. **Secondary heat** is obtained from burning primary combustible fuels, such as coal, oil and wastes, or from fission of uranium. Secondary heat is also produced by transforming electricity to heat in electric heaters, boilers, and heat pumps.

From the thermodynamic point of view, heat is thermal energy that is transferred from one system to another system. In other words, heat is thermal energy in transition between different systems and there is no meaning to the "heat content" of a single system. Moreover, our common experience shows that heat always flows from regions at higher temperature to regions at lower temperature. Thus the temperature difference is a "driving force" that generates heat. As a result a proper characteristic of heat should include not only the amount of heat expressed in energy units, but also a temperature at which the heat is delivered.

2.2 RESERVES OF ENERGY-CONTAINING MINERALS

The reserves of energy-containing minerals, such as oil, coal, natural gas, and uranium, are not known exactly. The estimates of these reserves are constantly changing, as new geological discoveries are taking place. Since the reserves are varying in a type of the mineral, its availability, and certainty of estimates, various definitions of mineral reserves are introduced.

2.2.1 FOSSIL FUELS

Within the broad concept of "oil reserves" there are several distinctions, as shortly described below.

Ultimate recoverable resources (URR) correspond to the total amount of oil that will ever be recovered and produced. This estimate is quite uncertain and frequently based on geology and physical laws. The ultimate recoverable resources are further broken down into three main categories: **cumulative production**, **discovered reserves**, and **undiscovered resources**. Cumulative production is an estimate of all of the oil produced to a given date, whereas discovered reserves are an estimate of future cumulative production from known fields. Discovered reserves are typically defined in terms of probability distribution and are broken down into proved, probable, and possible reserves.

There is no single, commonly accepted definition of **proved reserves**, but frequently this category corresponds to the estimated quantities of oil, which geological and engineering data demonstrate with reasonable certainty to be recoverable in future years from known reservoirs under current economic and operating conditions.

Probable reserves are often termed as **P50 reserves**, since these reserves are estimated to have better than 50% chance of being technically and economically producible. Similarly **possible reserves** (or "inferred" reserves) are sometimes referred to as **P10 reserves** or **P20 reserves**, indicating reserves which are estimated to have a significant, but less than 50% chance of being technically and economically producible. Similar classification is employed for the other fossil fuels, such as coal and natural gas. The proved world reserves of fossil fuels are given in Table 2.1. The reserves of oil and natural gas are shown to increase with over 50% during years 1998–2018. In 2018 their reserve-to-production ratio (calculated as a ratio of the reserves remaining at the end of any year divided by the production in that year) was around 50, and for coal 132.

2.2.2 URANIUM

Another system is used for **uranium resources**, for which a classification was introduced by the International Atomic Energy Agency [45]. Two main considerations are behind the classification: the confidence level of the estimates and the market-based cost of uranium recovering.

According to the current system, uranium resources are classified as either conventional or unconventional. **Conventional resources** are those that provide uranium production as a primary

Table 2.1
World Total Proved Fossil Fuel Reserves

Fuel	1998	2008	2017	2018	R/P Ratio in 2018[a]
Oil (10^9 barrels[b])	1 141.2	1 493.8	1 727.5	1 729.7	50
Natural gas (10^{12} m^3)	130.8	170.2	196.9	196.9	50.9
Coal (10^6 tonnes)					
Anthracite and bituminous				734 903	
Sub-bituminous and lignite				319 879	
Total coal				1 054 782	132

[a] Reserve-to-production (R/P) ratio is calculated as a ratio of the reserves remaining at the end of any year divided by the production in that year.
[b] 1 barrel = 0.1364 tonne (metric)
Source: [11]

product, co-product, or an important by-product (e.g. from the mining of copper and gold). **Unconventional resources** are those that contain uranium below a very low level and that do not have an established history of uranium production.

Conventional resources are further divided into four categories, with decreasing confidence level of occurrence: **reasonably assured resources** (RAR), **inferred resources** (IR), **prognosticated resource** (PR), and **speculative resources** (SP). In addition, all resource categories are defined in terms of costs of uranium recovered at the ore processing plants. This uranium resource classification system is the standard system for making official statistics presented in the Red Book [67]. The most recent data on world total uranium resources are shown in Table 2.2. Similarly as for fossil fuel, the reserve-to-production ratio is calculated as a ratio of uranium reserves remaining at the end of any year divided by the uranium production in that year. Including all identified reserves (both RAR and inferred), this ratio increased from 267.6 in 2015 to 291.0 in 2017.

2.2.3   OTHER MINERALS

There are many minerals that are important in energy transformations, even though they are not used as fuels. Arsenic, gallium, germanium, indium, and tellurium are examples of elements needed for production of photovoltaic solar cells. *Aluminium* and *rare-earth elements* (e.g., neodymium, dysprosium, and praseodymium) are used to build critical parts of wind turbines, such as *nacelle* and wind turbine generators, which need to be small, light, efficient, and powerful. Batteries play an important supporting role for intermittent renewable energy resources like wind and solar, allowing excess power to be stored for usage when needed. Modern batteries rely on several mineral commodities, particularly *cobalt*, *graphite*, *lithium*, and *manganese*. Current known reserves of selected minerals are presented in Table 2.3.

2.3   ENERGY SUPPLY

The current (year 2017) world total primary energy supply (TPES) by fuel is shown in Table 2.4. Crude oil, followed by coal and natural gas are three dominant fuel types, contributing with over 80% of the total production. The fraction of low-emission fuels in the total energy supply mix is still low and does not reach 20%.

Table 2.2
World Total Uranium Resources (in 1000 tU^a)

Category	2015	2017
RAR		
<USD 260/kgU	4 386.4	4 815
<USD 130/kgU	3 458.4	3 865
<USD 80/kgU	1 223.6	1 279.9
<USD 40/kgU	478.5	713.4
Inferred		
<USD 260/kgU	3 255.1	3 173
<USD 130/kgU	2 260.1	2 277
<USD 80/kgU	901.1	1 799.9
<USD 40/kgU	168.4	344.4
Identified (total)		
<USD 260/kgU	7 641.5	7 988
<USD 130/kgU	5 718.5	6 142
<USD 80/kgU	2 124.7	2 079.8
<USD 40/kgU	646.9	1 057.8
R/P Ratio^b	267.6	291.0^c

^{a} Metric units are used of tonnes (t) containing uranium (U), rather than uranium oxide (U_3O_8), where 1% U_3O_8 = 0.848% U.
^{b} Reserve-to-production (R/P) ratio is calculated as a ratio of the reserves remaining at the end of any year divided by the production in that year.
^{c} Based on estimated production in 2017.
Source: [67]

2.3.1 CRUDE OIL

Oil includes both primary (unrefined) and secondary (refined) products. Crude oil is the most important oil from which petroleum products, such as gasoline, lubricants, gas oil, or fuel oil, are manufactured. Oil remains the largest traded commodity worldwide with continuously growing supplies in absolute terms, even though its share in global total energy supply has been decreasing during the recent half-century, from over 45% in 1973 to about 35% in recent years. During years 1990–2017 the world oil production increased from 2 986.5 Mtoe to 3 893.6 Mtoe, where 1 Mtoe = 10^6 toe, and toe is a **tonne of oil equivalent** defined as

$$1 \text{ tonne of oil equivalent} = 41.868 \text{ GJ.} \qquad (2.1)$$

2.3.2 COAL

Coal continues to play a major role in the total energy production, with the second highest share (after oil) of 26.6% in 2017. Its importance for electricity production is even greater, contributing with 44.6% of the total. It also plays a crucial role in industries such as iron and steel. Even though the future of coal is uncertain due to growing concerns about air pollution and greenhouse gas emissions resulting from burning coal, its production is still growing, as shown in Table 2.5. Especially between 2000 and 2010, the production increased by more than 50%.

Coal is generally measured in tonnes (1000 kg) or short tons (907.2 kg), and its energy content is expressed in **tonnes of coal equivalent** (tce), defined as

$$1 \text{ tonne of coal equivalent} = 29.3076 \text{ GJ.} \qquad (2.2)$$

Table 2.3

World Total Selected Mineral Production and Reserves (thousands tonnes)

Mineral	2008 Production	2018 Production	2018 Reserves	2018 R/P Ratio[a]
Cobalt	83.6	158.1	6 569	41.5
Lithium	27.7	61.8	13 919	225.2
Natural graphite	965.2	895.6	306 700	342.5
Rare-earth metals	128.1	166.7	116 749	700.4

[a] Reserve-to-production (R/P) ratio is calculated as a ratio of the reserves remaining at the end of any year divided by the production in that year.
Source: [11]

Table 2.4

World Total Energy Production by Fuel in 2017

Fuel	Mtoe	Percent
Crude oil	4 477.212	31.9
Coal	3 773.421	26.9
Natural gas	3 162.893	22.5
Biofuels and waste	1 342.112	9.6
Nuclear	687.481	4.9
Hydro	351.029	2.5
Wind, solar, etc.	256.830	1.8
Heat	1 918.0	13.7
Total	14 034.897	100

Source: www.iea.org, accessed 2020-02-25.

However, when comparing many different fuels in statistical data, the energy content is primarily expressed in tonnes of oil equivalent (toe).

2.3.3   NATURAL GAS

Natural gas is taken from natural underground reserves. It contains mainly (85% or more) methane (CH_4), but it is not a chemically unique product, and comprises several other gases. After extraction from the gas field, or in association with crude oil as so called **associated gas**, it is processed to be ready for delivery. Natural gas which is released from underground coal mines is called **colliery gas** or **colliery methane**. This gas must be removed for safety reasons, and when used as a fuel, it is treated as natural gas production.

Natural gas is measured either by its energy content or volume. The energy content is expressed as **calorific value** (CV), also referred to as **heat value** (HV). It corresponds to the amount of heat released when the gas is burned. There are two sets of conditions under which gas can be measured by volume:

- Normal conditions measured at a temperature of 273.15 K and a pressure of 101 325 Pa.
- Standard conditions measured at a temperature of 288.15 K and a pressure of 101 325 Pa.

Table 2.5

World Total Coal Production by Type in Years 1990–2017 (ktoe)

Type	1990	2000	2010	2017
Anthracite	21 048	14 624	49 776	49 616
Coking coal	396 579	314 323	605 488	655 480
Lignite	262 908	205 028	201 896	196 431
Sub-bituminous coal	130 599	229 828	332 638	371 974
Other bituminous coal	1 399 506	1 508 555	2 465 214	2 493 328
Total (ktoe)	2 210 640	2 272 358	3 655 012	3 766 829
Total (EJ)	92.56	95.14	153.0	157.7

Source: www.iea.org, accessed 2020-03-25.

Example 2.1: Average Calorific Value of Natural Gas Import

Country A is importing 2 500 Mm^3 natural gas from country B and 4 200 Mm^3 natural gas from country C, with the respective calorific values of 36.1 MJ/m^3 and 41.2 MJ/m^3. Calculate the average calorific value of the natural gas imported by country A.

Solution

The total imported amount of gas is (2 500 + 4 200) Mm^3 = 6 700 Mm^3, and its combined energy content is (2 500 × 36.1 + 4 200 × 41.2) Mm^3 × MJ/m^3 = 263 290 × 10^{12} J. Thus, the average calorific value of the imported natural gas is 263 290/6 700 MJ/m^3 ≅ 39.30 MJ/m^3.

Estimated quantities of gas not yet produced, but which are recoverable with reasonable certainty in the future are referred to as gas reserves. Total proved reserves of natural gas are generally increasing from year to year due to new discoveries (see Table 2.1). At the end of 2018, the total world proved reserves were 196.9 trillion standard cubic meters (measured at 288.15 K and 101325 Pa), with standardized gross calorific value (GCV) of 40 MJ/m^3. This is a substantial increase from 130.8 trillion cubic meters reported at the end of 1998. If the reserves remaining at the end of any year are divided by the production in that year, the reserves-to-production ratio is obtained. This ratio is numerically equal to the number of years that those remaining reserves would last if production were to continue at that rate. The reserve-to-production ratio at the end of 2018 for the whole world was 50.9 years, and for the European Union 10.3 years [11].

2.3.4 BIOFUELS AND WASTE

In recent years, modern bioenergy accounts for roughly 10% of the world total energy production (see Table 2.4). Modern bioenergy does not include the traditional use of biomass for cooking and heating in inefficient open fires and simple cook stoves. It is rather used for electricity production and as biofuels in the transport sector, but also as fuel in heating sector, which remains the largest fraction of bioenergy.

Recent IEA's assessment[1] of 45 critical energy technologies indicates that bioenergy is on track to meet the climate goal as far as the power generation sector is concerned. In 2018, bioenergy electricity generation increased by over 8%, maintaining average growth rates since 2011 and exceeding the 6% annual growth rate needed through 2030 to reach the sustainable development scenario level

---

[1] www.iea.org/tcep/, accessed 2019-07-05.

set as 1168.2 TWh in 2030. Bioenergy power reached 595.2 TWh in 2018 and will increase to 764.1 TWh in 2023 according to IEA's forecast. However, development of transport biofuels is not on track, since its production expanded 7% year-on-year in 2018 (reaching 88.01 Mtoe), and 3% annual production growth is expected over the next five years. This falls short of the sustained 10% output growth per year needed until 2030 to align with the sustainable development scenario, which is estimated as 252 Mtoe[2].

2.3.5 NUCLEAR

Currently there are 441 nuclear power reactors in operation worldwide, with a total net installed capacity of 389 994 MWe. The leading countries are USA (95), France (57), and China (48). Additional 54 reactors, with a capacity of 57 441 MWe, are under construction. The leading countries constructing new reactors are China (11) and India (7), followed by Republic of Korea, Russia, and United Arab Emirates (constructing 4 new reactors each). Accumulated operational experience reached 18 487 reactor-years of operation[3].

During the recent five decades, the nuclear power has been significantly contributing to carbon-free electricity production. In 2017 nuclear reactors accounted for more than 10% of the global electricity production. In spite of positive contribution of nuclear power to combat air pollution and greenhouse gas emissions, its future is uncertain. In many countries nuclear power has trouble competing against other, more economic alternatives, such as natural gas and even modern renewables. Only successful development and implementation of new technologies, such as small modular reactors and fast breeder reactors, along with further improvements in safety, non-proliferation, and treatment of highly-active wastes, has potential to change these negative trends.

2.3.6 HYDRO

Hydropower is the world's largest source of carbon-free electricity generation and, at the same time, it plays an important role in improving power system flexibility. In 2019, the installed global hydropower capacity was 1 307 GWe[4]. In recent years, the electricity production share stays slightly above 16%. Even though the hydropower is a mature technology, it keeps evolving and growing. In the nearest future, the major part of global growth is expected to come from three large projects: two in China (16-GWe Wudongde and 10-GWe Baihetan projects) and one in Ethiopia (the 6.2-GWe Grand Renaissance project).

Hydropower theoretical reserve, estimated at about 39.1 PWh/y, includes all water resources carrying net kinetic or potential energy, which allow electricity generation under the technical, infrastructural, and ecological constraints. About 20 to 35% of the theoretical hydropower reserves have a technical potential, which is defined as the total energy that can be generated under the above-mentioned constraints. Asia has by far the largest hydropower technical potential, followed by Latin America, and North America. China has the highest existing energy generation and uses 24% of its potential. **Hydropower economic potential**, estimated at 8.7 PWh/y, includes energy capacity that is economically exploitable relative to alternative energy forms. Since the market of various available energy technologies is changing, the hydropower economic potential changes as well [54].

The power system flexibility is provided mainly by reservoir hydropower plants, since they rely on stored water in reservoirs. Very large reservoirs can provide stable inflow for months or even years, at the same time protecting against floods and providing irrigation services. Additional flexibility can be attained from pumped storage plants. Such plants pump water from a lower reservoir

---

[2] www.iea.org/tcep/transport/biofuels/, accessed 2019-07-10

[3] www.iaea.org, Power Reactor Information System, accessed 2020-05-25.

[4] www.iea.org, accessed 2020-05-25.

into an upper reservoir when electricity supply exceeds demand. When electricity is needed due to high demand, water is released to flow back from the upper reservoir through turbines to generate electricity.

2.3.7 WIND

The global wind power capacity is the fastest growing of all renewable electricity-generating technologies in absolute terms and it reached 650.8 GWe in 2019[5]. The onshore wind turbines, operating at sites with lower wind speeds, have become bigger with taller hub heights, and larger rotor diameters. The offshore wind turbines take advantage of better wind resources than land-based units. As a result, the modern wind turbines are able to achieve significantly more full-load hours compared to previous designs (see Problem 2.1).

2.3.8 SOLAR

Solar power is using sunlight to generate usable energy forms, such as electricity and heat. Solar photovoltaic (PV) is the fastest-growing technology (31% increase in 2018), representing the largest generation growth rate of all renewable technologies, including wind and hydropower[6], whereas solar thermal power, used for both electricity and heat generation, exhibits a rather flat development.

2.4 POWER SECTOR

In modern economies the reliable delivery of electricity and heat is essential. This vital sector of energy supply is supported by the power sector, which contains fleets of power plants, generating electricity and heat using various fuels. At the same time, the power sector remains to have tremendous influence on the environment, partly through a strong dependence on mining of fuels and raw materials, and partly because of significant greenhouse gas emissions. In 2017, the power sector was based on combustible fuels at about 67%. Three fossil fuels: coal, natural gas, and oil, are the main energy resources used by the sector.

Facilities that are used to generate electricity and heat, both combined or separately, are called **power plants**, or **power stations**. There are three types of power plants:

- Electricity-only plants which produce electricity only.
- Combined heat and power plants (CHP) which generate heat and electricity simultaneously.
- Heat-only plants which generate heat only.

Power plants can have different functions, such as production of electricity or heat (or both) and supplying it to the public grid, or production of electricity or heat for own use. In the latter case the enterprise produces electricity or heat in support of its own business different from energy production. The structure of global energy used in power plants in 2017 is presented in Table 2.6. The fossil fuels (oil and oil products, coal, and natural gas) contribute with 72.6% of the total 5 313 Mtoe energy used. About 43% of the total input energy was converted into electricity, whereas the rest (2 772 Mtoe) were power losses. The world total electricity and heat generation using different fuels during years 1990 to 2017 is shown in Table 2.7 and Table 2.8, respectively.

---

[5]World Wind Energy Association, www.wwindea.org, accessed 2020-05-25.

[6]www.iea.org/fuels-and-technologies/solar, accessed 2020-05-25.

Table 2.6
World Total Power Stations in 2017

Fuel	Mtoe	Percent
Oil and oil products	226	4.3
Coal	2 374	44.6
Natural gas	1 264	23.7
Biofuels and waste	200	3.8
Nuclear	687	13.0
Hydro	351	6.6
Wind, solar, etc.	139	2.6
Geothermal	72	1.4
Total	5 313	100

Source: www.iea.org, accessed 2020-05-19.

2.5 ENERGY CONSUMPTION

For statistical purposes, the energy consumption is presented by different sectors. Usually four main sectors are distinguished: industry, transport, agriculture and households, and non-energy use of fuels.

The industry sector is divided into several branches, such as, e.g., aluminium production, cement production, iron and steel manufacture, and chemical industry. However, only this fuel which is used for heating belongs to the industry sector, whereas coal and hydrocarbons used in the chemical industry and steel and iron production are reported as the non-energy use of fuels.

The transport sector includes various modes of transport such as rail, road, air, pipelines[7], and navigation. The energy consumed in this sector is used for transport activities, such as passenger and goods transport.

The energy consumption in agriculture typically includes forestry and fishing, whereas statistics of energy consumption in households varies from country to country, but most often it includes consumption of electricity, gas, and district heat.

The non-energy uses of fuels represent this fraction of the total fuel consumption which is used as raw materials for the manufacture of non-fuel products. The use of the hydrocarbon content of fuels as raw materials is an activity which is entirely confined to the refining and petrochemical industries.

The structure of current (year 2017) consumption of various types of fuels is shown in Table 2.9.

There are many technologies that are vital for global development and that are consuming significant amounts of energy. Production of such goods as aluminium, cement, iron and steel, pulp and paper, and various chemicals belongs to this category.

2.5.1 ALUMINIUM PRODUCTION

Aluminium production is highly energy intensive, with demand of about 14.2 MWh of electric energy per tonne of aluminium[8]. Primary aluminium[9] production involves two steps: (1) refining the *bauxite ore* into aluminium oxide (process known as the *Bayer process* or *alumina refining*), (2)

---

[7]According to IEA; Eurostat treats this consumption as part of the energy sector's own use.

[8]Source: www.iea.org

[9]Primary aluminium is aluminium tapped from electrolytic cells or pots during the electrolytic reduction of metallurgical alumina (aluminium oxide).

Table 2.7
World Total Electricity Generation by Fuel in Years 1990–2017 (GWh)

Fuel	1990	2000	2010	2017
Coal	4 429 511	5 994 499	8 666 205	9 863 339
Oil	1 324 817	1 206 768	977 326	841 876
Natural gas	1 750 266	2 759 718	4 838 767	5 882 825
Biofuel	105 479	114 407	280 559	481 529
Waste	24 142	49 543	89 029	114 043
Nuclear	2 012 902	2 590 624	2 756 288	2 636 030
Hydro	2 191 675	2 695 854	3 530 272	4 197 299
Geothermal	36 426	51 989	68 106	85 384
Solar PV	91	994	32 222	443 554
Solar thermal	663	526	1 645	10 848
Wind	3 880	31 348	341 384	1 127 319
Tide	536	546	513	1 044
Other	19 939	22 049	33 939	36 022
Total (GWh)	11 902 317	15 520 865	21 618 265	25 723 095
Total (EJ)	42.85	55.88	77.83	92.60

Source: www.iea.org, accessed 2020-05-21.

converting aluminium oxide to pure aluminium (known as the *aluminium smelting process* or the *Hall-Heroult process*).

Aluminium can be recycled endlessly without loss of its properties and this makes it a key contributor to a more resource-efficient economy. The current end-of-life recycling rates can be as high as 90% in transport and construction sectors and more than 60% in the packing industry.

2.5.2 CEMENT PRODUCTION

Cement is a binding, powdery substance used for civil engineering construction. It is made of limestone, sand or clay, bauxite and iron ore, and may include shells, chalk, marl, etc. When cement is mixed with water, it can bind sand and gravel into a hard, solid mass called **concrete**. Limestone, sand, and clay contain the essential elements (calcium, silicon, aluminium, and iron) required to make cement. The raw materials are crushed and blended in the right proportion, and subsequently they are ground to a powder. The next step is called **sintering**, and is essentially heating of the material without melting in a rotating furnace at high temperature reaching 1480 °C. This causes material changes into glassy, red-hot cinders called clinker. The clinker is cooled and ground into a fine grey powder together with a small amount of gypsum, resulting in the final product—cement. Fossil fuels continue to provide the majority of energy in the cement sector, with alternative fuels such as biomass and waste accounting for only 6% of thermal energy used in 2017[10]. Apart from considerable energy demand, cement production is causing significant CO_2 emissions due to both the combustion of fuels and the decomposition of limestone in the clinker production process. Reduction of CO_2 emissions could be achieved by electrifying production and by facilitating the capture of process CO_2 emissions (that is emissions from limestone decomposition during the clinker production).

---

[10]Source: www.iea.org

Table 2.8
World Total Heat Generation by Fuel in Years 1990–2017 (TJ)

Fuel	1990	2000	2010	2017
Coal	4 833 069	4 331 117	5 295 352	5 978 057
Oil	2550935	1159776	838851	526230
Natural gas	8 102 240	6 236 076	6 772 792	6 175 340
Biofuel	173 815	213 965	448 739	627 335
Waste	85 809	200 129	334 205	454 656
Nuclear	43 545	19 106	27 352	26 771
Geothermal	15 403	18 314	26 112	39910
Solar thermal	6	24	192	1 815
Other	93 228	65 198	410 535	537 483
Total (TJ)	15 898 052	12 243 705	14 154 130	14 367 597
Total (EJ)	15.90	12.24	14.15	14.37

Source: www.iea.org, accessed 2020-05-21.

2.5.3 IRON AND STEEL

Global crude steel production in 2017 was 1690 million tonnes. Driven by population and industrial growth, global demand for steel has been growing strongly in recent years. The steel production is currently highly reliant on coal, which supplies 75% of energy demand, with energy intensity 19.8 GJ/t in 2017[11].

Making Steel

Currently there are two major commercial processes for **making steel**:

- Primary steelmaking that involves converting liquid iron from a blast furnace and steel scrap into steel, or melting scrap steel or direct reduced iron in an electric arc furnace.
- Secondary steelmaking, most commonly performed in ladles, which involves refining of the crude steel before casting.

The blast furnace, which is the first step in producing steel from iron oxides, uses coke, iron ore, and limestone to produce pig iron. Blowing oxygen through molten pig iron lowers the carbon content of the alloy and changes it into steel. The electric arc furnace (EAF) is different from the blast furnace as it produces steel by using an electrical current to melt scrap steel and (or exclusively) direct reduced iron. Gas burners may be used to assist with the melt down of the scrap pipe in the furnace.

New smelt reduction technologies based on coal or hydrogen plasma can cut emissions from coke production. Direct reduction technologies based on natural gas, hydrogen or electricity could reduce emissions considerably compared with the conventional blast furnace-coke oven route.

---

[11]Source: www.iea.org

Table 2.9
World Total Final Consumption by Fuel in 2017 (Mtoe)

Fuel	Industry	Transport	Other	Non-Energy	Total
Oil products	315	2 589	433	634	3 971
Oil	6	0	0	9	15
Coal	818	0	152	51	1 021
Natural gas	568	105	644	186	1 503
Biofuels and waste	207	84	747	-	1 038
Geothermal	1	0	13	-	14
Solar/tide/wind	0	0	31	-	31
Electricity	769	31	1 037	-	1837
Heat	138	0	152	-	290
Total	2 822	2 809	3 209	880	9 720

Source: www.iea.org, accessed 2020-05-19.

2.5.4   PULP AND PAPER

Electricity demand in pulp and paper production in 2017 was 2.08 EJ. Biomass materials are the main feedstock to the pulp and paper industry (1.99 EJ). Other energy resources include coal (0.89 EJ), oil (0.31 EJ), and gas (1.16 EJ)[12].

2.5.5   CHEMICALS

In 2018, chemical sector's share of total primary demand for oil and gas was 14% and 8%, respectively. This makes the chemical industry the single largest industrial consumer of both oil and gas. In terms of CO_2 emissions, chemicals production is on the third place after iron and steel, and cement. This is largely because around half of the sectors energy input is consumed as feedstock, the emissions of which are calculated downstream in other sectors (e.g. waste and agriculture). The main chemicals include plastics, fertilizers such as ammonia, with demand about 170 million tonnes per year, and methanol, with main end use for formaldehyde, employed to produce several specialist plastics, but also for additive applications[13].

2.5.6   ENERGY SERVICES

Modern civilization that we live in today is more and more relying on functions that require access to reliable energy resources. This includes transportation, heating, cooling, data centers, and other applications. All these functions are called **energy services**, since access to energy resources is critical for these functions to be available. In this section we shortly present and discuss the most important energy services.

Aviation

Aviation provides transportation services and in 2018 it contributed with about 8200 billion passenger-km. In the same year, CO_2 emissions from aviation accounted for around 2.5% of global energy-related emissions, corresponding to 918 Mt of CO_2[14]. Currently efforts are undertaken to use

[12]Source: www.iea.org

[13]Source: www.iea.org

[14]Source: www.iea.org, accessed 2020-05-16.

sustainable aviation fuel, and biofuel blends are already successfully used on commercial flights. There is also a growing interest to derive liquid fuel for aviation by using non-fossil electricity. However, this type of fuel is still more expensive than the fossil-based jet kerosene, and an adequate policy support would be required for a complete transition.

International Shipping

World seaborne trade volumes rose to 11 billion tonnes in 2018, of which 7.8 billion tonnes were classified as dry cargo, such as coal, ores, grains, pallets, bags, crates, and tankers. Crude oil, the most transported cargo in the 1970s, accounted for less than 20% of the goods delivered by sea[15]. In 97%, the fuel used in international shipping originated from fossil fuels, and this incurred 695 Mt of CO_2 emissions. The major sustainable development challenge is to reduce the emissions by replacing the traditional fuel with a low-carbon fuel and by improving the energy efficiency. Interest exists in using alternative fuels such as ammonia, hydrogen or advanced biodiesel. However, the main obstacle in transition to such fuel is their high cost and lack of proper infrastructure.

Rail Transportation

Rail transportation contributed with 0.3% of global direct emissions in 2018. At the same time, it is one of the most energy-efficient transport modes, corresponding to 8% in passenger transport and 7% in freight, but using only 2% of energy used globally in the transportation sector[16].

The average, minimum, and maximum energy intensity of various passenger transport modes in 2018 is shown in Table 2.10.

Table 2.10

Energy Intensity of Passenger Transport in 2018, MJ/passenger-km

Transportation Mode	Average	Minimum	Maximum
Rail	0.2	0.1	0.8
Two/three-wheelers	0.5	0.4	0.8
Buses and minibuses	0.7	0.4	1.1
Cars	1.8	0.8	2.9
Aviation	1.8	1.0	3.1
Large cars	2.7	1.0	3.7

Source: www.iea.org/reports/tracking-transport-2019, accessed 2020-05-17.

The rail sector has great potential to provide benefits for the energy sector and for the environment. It can lower transport energy use and reduce carbon dioxide and other pollution emissions. Today 75% of passenger rail transport uses electric trains, and it is the only mode of transport that is widely electrified. It provided 2.82 trillion passenger-km using electric trains and 1.23 trillion passenger-km using diesel trains.

Trucks and Buses

CO_2 emissions from heavy-duty vehicles have increased on average 2.2% annually since 2000 and reached 1856 Mt in 2018. The largest contribution to these emissions comes from trucks that account

---

[15] Source: stats.unctad.org, accessed 2020-05-16.

[16] Source: www.iea.org, accessed 2020-05-17.

for more than 80% of this growth. This is much worse development scenario as observed for light-duty vehicles. The main reason for this difference is that light-duty vehicles are covered in about 85% by strong policy as far as the fuel economy standards are concerned. For heavy-duty vehicles only 50% are covered by such policy. In general, transportation is responsible for 24% of direct CO_2 emissions from fuel combustion, where road vehicles (cars, trucks, buses, and two- and three-wheelers) account for nearly 75% of transport CO_2 emissions.

Electric Vehicles

The total global stock of electric cars reached 5.12 millions in 2018, and increased by 1.98 millions only in 2018. Most of the new electric cars are sold in China (over 1 million in 2018), followed by Europe (385 000) and the United States (361 000). The highest market shares have three Nordic countries: Norway (46%), followed by Iceland (17%) and Sweden (8%). However, the global share of electric cars is still very low and in 2018 it was 0.6%[17].

The most expensive component in electric vehicles remains a battery, even though the average lithium battery price fell 18% from 2017 to 2018 and reached USD 176 per kWh. Majority of a battery production is still provided by rather small plants, with capacities of 3 GWh to 8 GWh per year. However, the capacities of new plants are growing and the three largest battery factories in operation in 2020 have capacity of 20 GWh/year and account for roughly 21% of the total installed capacity. The growing size of automotive battery market should bring the benefits of economies of scale and the price of batteries should continue decreasing.

Electric cars are still more expensive than similar-sized conventional cars. They cost more to purchase, and only in quite limited range of cases this can be compensated by an economic advantage by reduced operational costs such as fuel. All these cost challenges are primarily linked to the battery. Technological improvements such as more compact batteries with longer ranges, extra durability, and the capacity to charge at very high power (such as fast charging from 100 kW to 1 MW) will also influence the level of acceptance of electric cars.

Data Centers and Networks

The global electric demand of data centers is rapidly increasing and reached 198 TWh in 2018. As the world becomes increasingly digitalized, data centers and data transmission networks reached about 1% of the global electricity demand each. The increase in energy demand is, however, not as rapid as growth of these services thanks to significant improvements in energy efficiency.

Energy Storage

The purpose of energy storage is to reduce imbalances between energy demand and energy production. The needed for storage capacity has significantly increased with growing intermittent electricity from solar power and wind power. The total energy storage capacity was 8 GWh in 2018, and it was doubled from 2017.

In general there are electrochemical, electrical, mechanical, and thermal energy storage techniques. Examples of electrochemical energy storage include batteries and hydrogen. Batteries can provide energy storage in a range of small-scale to large-scale applications. The applications span from portable electronics such as mobile phones and computers to large grid-based applications for stabilization and load balancing of the intermittent power supply. The lithium-ion chemistry can be used for low- and moderate-energy-density applications, which are typical for portable applications. For sustainable energy infrastructure of the future, the greatest prospect for electrochemical

---

[17]Source: www.iea.org, accessed 2020-05-16.

storage is mostly on the stabilization of frequency and voltage in dealing with hourly and daily fluctuations [71].

The three main types of mechanical energy storage systems are pumped hydro, flywheel, and compressed air. The pumped hydro storage is characterized by its flexibility, low maintenance cost, and long lifetime. The main components of the pumped hydro storage are hydro turbine, pumping system, and upper reservoir. Water is pumped from the lower reservoir to the upper reservoir when there is high supply of electricity exceeding the demand. When the electricity demand exceeds the supply, water from the upper reservoir is used to turn hydro turbines to produce electricity in the same way as a regular hydro power plant.

The flywheel energy storage system is storing kinetic energy by using a rotating mass. A most studied application of the flywheel energy storage is to store breaking power in locomotives, trains, and cars.

The compressed air energy storage system contains electric motor, compressor, air storage tank, heat exchanger, expander, and generator. When the electricity supply is higher than the demand, ambient air is compressed in the compressor and it is stored in the storage tank. During high electricity demand, the compressed air is expanded in the expander to generate electricity in the generator.

Thermal energy storage allows the storage of heat and cold to be used later. It can help to optimize use of the thermal energy when there is a mismatch between heat generation and use. Different methods of storage are possible, based on sensible energy storage (in, e.g., air or water), latent heat storage (using phase change materials), and thermochemical energy storage (based on chemical reactions and sorption systems).

Hydrogen

Hydrogen is considered to be an energy carrier with great potential for future. On the one hand, it can be produced from almost all energy resources. On the other hand, it is a versatile energy carrier that can be used in many sectors, in particular, in transportation. One such application is in fuel cell electric vehicles. Such vehicles are still in an early phase of development and the global stock of fuel-cell driven electric vehicles is very low.

At present 70 million tonnes of hydrogen is used, mostly for oil refining and chemical production. This hydrogen is produced from fossil fuels, with significant associated CO_2 emissions.

Production of Hydrogen

There are many different sources of hydrogen and ways for producing it for use as a fuel. The most common methods for producing hydrogen are steam reforming of natural gas, partial oxidation of methane, and coal gasification. Other methods of hydrogen production include biomass gasification and electrolysis of water.

Water electrolysis is the process of using electricity to split water into hydrogen and oxygen in a device called an **electrolyzer**. Electrolyzers consist of an anode and a cathode separated by an electrolyte. There are different ways the electrolyzers operate depending on the type of electrolyte material involved.

In a **polymer electrolyte membrane** (PEM) electrolyzer, water reacts at the anode to form oxygen and positively charged hydrogen ions ($2H_2O \longrightarrow O_2 + 4H^+ + 4e^-$), whereas electrons flow through an external circuit. Hydrogen ions selectively move across the PEM to the cathode, where they combine with electrons from the external circuit to form hydrogen gas ($4H^+ + 4e^- \longrightarrow 2H_2$).

Alkaline electrolyzers operate via transport of hydroxide ions (OH^+) through the electrolyte from the cathode to the anode with hydrogen being generated on the cathode side. Various electrolytes have been used such as a liquid alkaline solution of sodium or potassium hydroxide or a solid alkaline exchange membrane.

Hydrogen produced via electrolysis can result in zero greenhouse gas emissions, if the electricity used is generated from low carbon energy options, such as wind power or nuclear power.

Electrolyzers are cross-cutting technology that enable the production of clean-burning and storable hydrogen fuel from electricity and water. These technologies are well-known and long used in variety of industry sectors. New projects increasingly opt for polymer electrolyte membrane (PEM) technology, which is at an earlier stage of development than the alkaline electrolyzers.

The existing natural gas infrastructure can be used to deliver hydrogen to end users, mostly for heating purposes. Hydrogen can be blended with natural gas with an increasing fraction starting from a few percent to about 20 percent in the nearest future. There are also plans to include low-carbon hydrogen in industrial applications.

Heating

Heat is the largest energy end-use. Providing heat for homes, industry, and other applications accounts for around half of total energy. More than half of heat produced is consumed in industry as process heat, for drying, and for industrial hot water uses. Renewables provide only about 10% of the heat.

Cooling

2075 TWh energy was used for cooling in 2018. This is the fastest-growing end use of energy in buildings. Modern air conditioners are highly efficient, but the air conditioners currently mostly in use, and still purchased in many regions, are two- or three times less efficient than the most modern ones. Thus the improved performance of practically used air conditioners is the main challenge for the nearest future.

Lighting

Lighting efficiency improved considerably since 2000. The phase-down of incandescent lamps is prompting global technology shift toward more efficient technologies such as fluorescent lamps. In 2018, LED achieved the same share of global residential sales as less-efficient fluorescent lamps: 40%. As LED costs continue to fall, this technology should dominate in future sales.

Luminous Efficacy

Luminous efficacy is a measure of how well a light source produces visible light. It is the ratio of luminous flux to power, measured in lm/W (lumen per watt) in the SI units.

For example, incandescent 100 W/230 V lamp has overall luminous efficacy of 13.8 lm/W whereas light-emitting diode (LED) of 5–16 W/230 V has the efficacy in a range 75–120 lm/W.

Sales of incandescent lamps, with efficacies of around 13 lm/W, have dropped to less than 5% of the market. Halogen and compact fluorescent lamp sales peaked around 2015 and have since declined to about 50% of the residential market. New minimum performance standards imposed by governments for lighting can be expected. For example, in 2018, EU Member States voted to phase out inefficient halogen lamps and compact fluorescent lamps in 2021, while introducing minimum performance and quality standards for LED lamps and luminaires.

2.5.7   ENERGY EFFICIENCY AND ENVIRONMENT PROTECTION

Energy efficiency is a measure of an amount of energy that is needed to provide products and services. High energy efficiency is achieved when the required services and products can be delivered with lowest possible energy consumption. Energy efficiency can be achieved in many different ways. Combined heat and power systems use the heat from power plants (which, otherwise, would be treated as waste) to provide heating and hot water to nearby buildings and facilities. Such solutions can significantly increase the overall energy efficiency to 70–80%. Smart grids, efficient buildings (so called zero-energy buildings), efficient use of fuels in transportation are other examples of energy efficiency improvements.

Efficient use of energy has also a beneficiary influence on the environment, since with increasing energy efficiency, less energy needs to be produced, and the resulting emissions are reduced. However, additional active environmental-protection measures are still needed to reduce emissions of greenhouse gases.

Carbon Capture, Utilization and Storage

Carbon capture, utilization and storage (CCUS) is an emission reduction technology. It captures CO_2 from fuel combustion or industrial processes, transports this CO_2 via ship or pipeline, and either use as a resource to create products or services, or storage deep underground in geological formations. At the end of 2018 only two large-scale CCUS power projects, with combined capacity of 2.4 million tonnes of CO_2 per year, were operational. This technology is still far off-track of the sustainable development goal level of 350 $MtCO_2$/year in 2030. The key innovation goal for CCUS technology is the reduction of energy penalty (and hence the cost of CO_2 capture).

Smart Grids

Smart grids comprise a broad mix of technologies to modernize electricity networks, extending from the end user to distribution and transmission. They have potential to:

- provide technologies for better monitoring, control and automation,
- unlock system-wide benefits,
- reduce outages,
- shorten response time.

On the user side, smart grids can enable demand flexibility and consumer participation in the energy system, through demand response (e.g., by using smart meters), electric vehicle charging, and self-produced distributed generation and storage. The monitoring of power usage at the level of individual consumers in real time would allow for adjusting the pricing and usage patterns to optimize network performance.

Demand Responses

Demand response polices are considered to be a key pillar to expand renewable energy production. Increase of the system flexibility through numerous small-scale devices leads to more system capacity to host variable renewables while accelerating the electrification of heating, cooling, transport, and industry at lower cost.

An example of such devices includes **smart meters**, which are next-generation, digital, and wireless-enabled gas and electricity meters. They track the energy consumption like old meters, but display it digitally on a screen in kilowatt hours and in currency units. This information is automatically transmitted to the energy supplier each day. It opens for potential savings and reduction of environmental impact. Behind-the-meter energy storage paired with power generation can act as both generation and load. This solution can bring additional important benefits to customers, utilities, and the broader power system.

Another example of a flexibility-enabling system is a **virtual power plant**, which is a system that integrates several types of power resources to give a reliable overall power supply. The resources can include micro combined heat and power (co-generation of up to 50 kW), natural gas-fired reciprocating engines, small-scale wind power plants, photovoltaics, run-of-river hydroelectric plants, small hydro, biomass, backup generators, and energy storage systems. The system has benefits such as the ability to deliver peak load electricity or load-following power generation on a short notice.

PROBLEMS

PROBLEM 2.1

The total installed capacity of wind power in 2017 was 540 840 MWe[18]. Using Table 2.7 with the statistics on electricity generation, calculate the global capacity factor of wind power during that year. Hint: the global capacity factor is defined here as a ratio of the actual electricity generation to a theoretical maximum electricity generation if the installed capacity was used in 100% during the whole year.

PROBLEM 2.2

The total installed capacity of nuclear power in 2017 was 391 720 MWe[19]. Using Table 2.7 with the statistics on electricity generation, calculate the global capacity factor of nuclear power during that year. See hint provided in Problem 2.1.

PROBLEM 2.3

The total installed capacity of hydropower in 2017 was 1 267 GWe[20]. Using Table 2.7 with the statistics on electricity generation, calculate the global capacity factor of hydropower during that year. See hint provided in Problem 2.1.

PROBLEM 2.4

The total installed capacity of solar PV power in 2017 was 403 GWe[21]. Using Table 2.7 with the statistics on electricity generation, calculate the global capacity factor of solar PV power during that year. See hint provided in Problem 2.1.

PROBLEM 2.5

Assume that all electricity generated in 2017 by burning coal is to be replaced by wind turbines. Calculate what would be the required total installed capacity of the wind power, taking the effective wind power capacity factor found in Problem 2.1. Estimate the number of wind turbines with capacity 10 MWe each that would be required.

PROBLEM 2.6

The uranium required to run nuclear reactors, as of 1 January 2017, was 62 825 tU [67]. Using the electricity production data from Table 2.7, estimate the required uranium if all electricity generated in 2017 by burning coal was instead generated in nuclear reactors.

---

[18] wwindea.org

[19] iaea.org

[20] www.hydropower.org/publications/2018-hydropower-status-report, accessed 2020-05-25.

[21] iea-pvps.org/wp-content/uploads/2020/01/2018_iea-pvps_report_2018.pdf, accessed 2020-05-25.

3 Elements of Sustainability

The concept of sustainability is not new, but its meaning has evolved with the development of human civilization. In early human history, the development and decline of particular societies was determined by ability to provide food, shelter, and safe living conditions. However, already during the industrial revolution of the 18th and 19th centuries it became clear that, for further development of the human civilization, access to affordable energy resources is necessary. Initially the vastly growing energy needs were met with fossil fuels only, and in particular, with coal. As a result of this growing usage of fossil fuels two facts became evident. Firstly, as clearly demonstrated by the energy crises in 1973 and 1979, the amount of non-renewable energy resources is limited. Secondly, as already noted by many in mid-20th century, the environmental costs associated with the usage of fossil fuels become greater than the social benefits resulting from the increased energy usage. Presently, in the beginning of 21st century, there is increasing global awareness of the threat posed by burning of fossil fuels, and resulting from it, anthropogenic pollution of the environment and greenhouse effect.

3.1 SUSTAINABILITY GOALS

Even though the main focus of this book is on the sustainable energy transformation, it is instructive to pay some attention to the notion of sustainability in general. Sustainability issues are hugely complex and, so far, impossible to completely understand and resolve by humans. Learning to live sustainably on Earth is going to require enormous advances in our understanding of the natural world and our relation with it.

The first logical step to address the sustainability issues is the definition of the goals of the sustainable development. The 2030 Agenda for sustainable development, adopted by all United Nations Member States in 2015, provides a list of seventeen such goals[1].

1. No Poverty.
2. Zero Hunger.
3. Good Health and Well Being.
4. Quality Education.
5. Gender Equality.
6. Clean Water and Sanitation.
7. Affordable and Clean Energy.
8. Decent Work and Economic Growth.
9. Industry, Innovation and Infrastructure.
10. Reduced Inequalities.
11. Sustainable Cities and Communities.
12. Responsible Consumption and Production.
13. Climate Action.
14. Life Below Water.
15. Life on Land.
16. Peace, Justice and Strong Institutions.
17. Partnership.

The goals represent the economic, environmental and social aspects of the desired development of our civilization. Each of the goals creates a tremendous challenge itself, since it involves interactions

---

[1] https://sustainabledevelopment.un.org, retrieved 2020-02-20.

DOI: 10.1201/9781003036982-3

between humans and the environment. Human-environment systems are known to be very complex, with constantly changing conditions, and with interacting feedbacks between different parts of the system.

One method to deal with complex systems is by breaking them down into simple components that can be modeled and understood separately. For example, the Newton laws of dynamics could be applied for isolated simple systems in laboratories, and when well-understood, could be applied to explain motions of planets in our solar system. However, it is not possible to apply a similar approach to develop sustainability models. The main reason for this limitation is that the sustainability problems, which always include human-environment systems, involve complex non-linear interactions with feedbacks. Such systems cannot be broken into simple components and be tested in a laboratory environment. Due to the complex feedbacks, they have to be modeled as a whole.

A Coupled Human-Environment System

A **coupled human-environment system** (known also as a coupled human and natural system, CHANS) is a system with dynamic two-way coupling between human systems and natural systems, such as between social or economic systems and hydrological, biological or geological systems. This concept reflects the idea that there is a mutual relation between humans and the environment and they cannot be treated separately. The research related to CHANS is relatively new and no well-established mathematical frameworks have been developed yet. However, the aim is to obtain a more quantitative approaches, including analytical and numerical models to study and quantify the highly nonlinear dynamics often present in CHANS.

Sustainability models are currently under development, but to be precise and useful, they must inevitably be extraordinarily complex, involving many scientific disciplines. In particular, almost every sustainability challenge we face will require new mathematical tools [20]. For instance, the economic issues, which are substantial parts of sustainability, rise their own modeling challenges. However, the financial crises in 2008 dramatically demonstrated that our current models of economy are unable to forecast the nearest future. Furthermore, economics and the environment are interacting with each other in a complex, and very often unpredictable manner, as was the case in Japan after the 2011 earthquake and tsunami.

3.2 ENVIRONMENT

Many of the sustainability goals are related to the well being of the natural environment, including the atmosphere, biosphere, hydrosphere and lithosphere. In the following sections we present the main characteristics of these important parts of the environment.

3.2.1 ATMOSPHERE

Earth's atmosphere is a gaseous layer surrounding the planet Earth, consisting of a mixture of gases, commonly known as air, and containing primarily nitrogen (78.084%), oxygen (20.946%) and argon (0.934%).

Earth's atmosphere can be divided into five main layers, with varying temperature, pressure and density. The average temperature of the atmosphere at Earth's surface is between 14 and 15 °C. The average atmospheric pressure at sea level is defined by the International Standard Atmosphere as 101325 Pa. The density of air at sea level is about 1.2 kg/m^3 and can be calculated from the equation of state for air from measurements of temperature, pressure and humidity. The main characteristics of the atmosphere layers are provided in Table 3.1.

Table 3.1
Layers of Earth's Atmosphere

Layer	Height, km	Properties
Troposphere	0 to 12	Nearly all weather and most clouds appear in this layer. Air pressure and temperature drop when the height increases.
Stratosphere	12 to 50	The ozone layer, which contains ozone molecules that are absorbing high-energy ultraviolet light from the Sun, is located in this layer. The temperature is initially constant and then increases with the increasing height.
Mesosphere	50 to 80	Most meteors burn in this layer. The temperature decreases with increasing height and is about -90 °C near the top of the layer.
Thermosphere	80 to 700	High-energy X-rays and UV radiation from the Sun are absorbed, rising the temperature to above 2000 °C.
Exosphere	700 to 10000	This layer, containing lightest atmospheric gases (helium, carbon dioxide, and atomic oxygen) can be considered a part of outer space.

In addition to nitrogen, oxygen and argon, the Earth's atmosphere contains **noble gases** (neon, helium, krypton and xenon), **greenhouse gases** (water vapor, carbon dioxide, methane, nitrous oxide and ozone), many substances of natural origin (dust, pollen and spores, sea spray and volcanic ash) and various industrial pollutants (chlorine and fluorine compounds, mercury vapour and sulphur compounds).

Anthropogenic pollutants and greenhouse gases are of particular concern, since they have negative human health and Earth's climate effects. Air pollution is a major cause of premature death and disease, and is the single largest environmental health risk in many countries. Heart disease and stroke are the most common reasons for premature death attributable to air pollution, followed by lung diseases and lung cancer. The International Agency for Research on Cancer[2] has classified air pollution in general, and particulate matter (PM) in particular, as carcinogenic.

Greenhouse gases are absorbing and emitting radiant energy within the thermal infrared range and are causing the greenhouse effect on Earth. Without greenhouse gases the average temperature of Earth's surface would be about -18 °C (rather than the present average of about 15 °C)[3].

The concentration of carbon dioxide in the atmosphere has been growing in parallel with increasing fossil fuel consumption in the world. The record of carbon dioxide concentration from 1860 to 2020 is shown in Fig. 3.1. Interestingly, estimates performed before year 1977 by Bacastow and Keeling agree quite well with selected measured concentrations in 2012 (passing 400 ppm threshold) and in January 2020, when the concentration reached 413.4 ppm[4]. The mean surface temperature anomaly[5] recorded for the Northern Hemisphere from 1850 to present is shown in Fig. 3.2. The temperature anomaly curves shown in the figure were obtained in year 1977, using measured data before that year and forecasts beyond 1977 [86]. According to the estimates, the global mean temperature anomaly (using the mean from 1850 to 1977 as a reference) in 2019 should be about 1.77 K, whereas the actual measured temperature anomaly is 1.13 K. Nevertheless, the measured data beyond year 1977 are placed well within the uncertainties indicated in the figure. Forecasting the future temperature anomaly is very difficult and requires application of extremely complex models.

---

[2]https://www.iarc.fr

[3]www.giss.nasa.gov, retrieved 2020-02-18

[4]www.co2.earth, retrieved 2020-03-03.

[5]By anomaly we mean here the difference between the current year mean temperature and the average temperature over a prescribed period of time in the past.

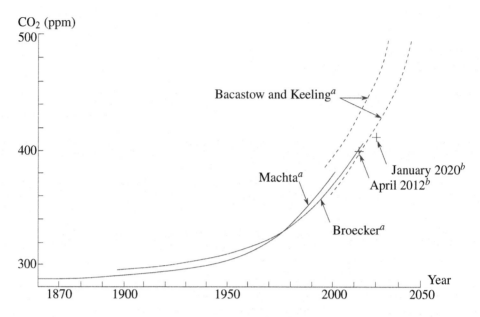

Figure 3.1 Measured and estimated carbon dioxide concentration in the atmosphere. Source: [a] W. W. Kellogg, *Effects of Human Activities on Global Climate* (Geneva, Switzerland: World Meteorological Organization (Tech. Note 156), 1977), created after [86]; [b] www.co2.earth, retrieved 2020-03-03.

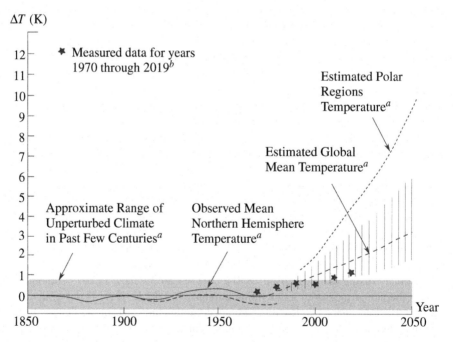

Figure 3.2 The mean surface temperature recorded for Nothern Hemisphere from 1850 to 1977 (solid line), and one estimate of the course that it might have taken without the addition of anthropogenic carbon dioxide. Source: [a] W. W. Kellogg, *Effects of Human Activities on Global Climate* (Geneva, Switzerland: World Meteorological Organization (Tech. Note 156), 1977), created after [86]; [b] data.giss.nasa.gov retrieved 2020-03-03 using as a reference the mean temperature between 1880 and 1975.

3.2.2 BIOSPHERE

The **biosphere** is a relatively thin life-supporting stratum that includes all ecosystems on Earth. It extends a few kilometers into the atmosphere to the deep ocean trenches. Life forms are also found inside soil and rocks. Humans are affecting the biosphere in various ways, such as by farming lands and raising plant-eating animals.

Multiple processes are taking place in the biosphere. Some of them are related to human presence and activities, and some have a natural character, such as the biological processes including consumption, death, decomposition, and food supply. The main processes related to the human behavior include:

- Anthropogenic release of substances, water, or energy.
- Introduction or extermination of species (e.g. introduction of crayfish in lakes or extermination of the European bison).
- Use of water for other purposes than drinking, e.g. washing, irrigation, and energy transformation.

In an assessment of human-environment interactions it is necessary to understand how the ecosystems function to be able to predict the impact of human activities on the biosphere.

3.2.3 HYDROSPHERE

The **hydrosphere** includes water (liquid, vapor or ice) that is on Earth's surface, underground, and in the air. It is estimated that there is about 1.386×10^9 km^3 water on Earth, of which about 97.5% is saline and 2.5% is freshwater. Most of the freshwater (68.7%) is locked up in ice and glaciers, 30.1% is in the ground and 1.2% on Earth's surface. Most of the surface water (69%), in turn, is locked in ice and permafrost, 20.9% in lakes, 3% in the atmosphere, and 0.49% in rivers. The rest is found in the soil moisture, living things, and swamps and marshes [77].

3.2.4 LITHOSPHERE

The **lithosphere** on Earth is composed of the crust and the portion of the upper mantle that behaves elastically on time scales of thousands of years or greater. There are two types of lithosphere with varying thickness [68]:

- Oceanic lithosphere, associated with the oceanic crust existing in the ocean basins, with thickness in the range from 50 to 140 km.
- Continental lithosphere, associated with continental crust, with thickness in the range from 40 to about 280 km.

The Earth's crust is made up of about 80 elements, which occur in over 2000 different compounds and minerals. However, only 8 of these elements exist in the crust in significant amounts. These are oxygen (46.60%), silicon (27.72%), aluminium (8.13%), iron (5%), calcium (3.63%), sodium (2.83%), potassium (2.59%) and magnesium (2.09%) [41].

Minerals could be pure elements, but usually they are made of many different elements combined. For example, quartz is made up of silicon and oxygen. This mineral is used for glass, in electrical components, optical lenses, and in building stones.

3.3 ECONOMIC SUSTAINABILITY

Many sustainability goals have clear economic dimensions. For example, such goals as "No Poverty" and "Decent Work and Economic Growth" are closely related to topics covered by mainstream economics. In this book we focus on three main economic aspects. Firstly, we will discuss cost estimations for various energy transformation technologies. In particular, we will introduce principles to calculate a levelized cost of electricity. Such estimations are necessary to guide politicians and investors to chose most economical solutions. These aspects will be discussed in later chapters. In this section we will make a short overview of such economy-related topics as the role of economy in sustainability and the currently used economic instruments to promote environmental protection.

3.3.1   ROLE OF ECONOMY IN SUSTAINABILITY

One of the goals in the 2030 Agenda addresses directly the economic aspects of the sustainable development. This goal deals with the needs for decent work and economic growth. However, almost all goals have some economic dimensions, since they require investments of human and material resources.

In fact, one of the greatest challenges for sustainable development is to find a balance between all its goals, consistent with generally satisfactory economic performance. So far, the economic, environmental, social and political trade-offs between the sustainability goals are not well-understood. This is mainly due to the tremendous complexity of the underlying processes, but also because of lacking trust-worthy economical models. Such models should include the environmental and social aspects, and should be able to forecast the economical development in the nearest (or even long-term) future.

3.3.2   WAYS TO PROMOTE ENVIRONMENTAL PROTECTION

There are various tools that can be used by governments to promote a protection of the environment, such as regulations, taxes and trade with emission fees.

Regulations

The Clean Air Act introduced in the US in 1970s is an example of a regulation introduced by a government to cope with air pollution. The essence of the Clean Air Act was to set the ambient air quality standards for atmospheric concentrations of sulphur dioxide, nitrogen oxides (NO_x), and particulate matter (PM). Such regulations forced all coal-fired power plants to introduce means to reduce the emission of pollutants. Practically all coal-fired power plants in the world have today equipment to reduce the pollutant emissions below allowed levels.

In using a regulation, the government forces power plant owners to a certain action, providing a detailed description how to reduce emissions, which technology to use, and to which level the emissions should be reduced. This helps the companies to implement the regulations, since otherwise, some companies would have difficulties to develop own solutions.

The regulating body can also choose a combination of directives and regulations that will guarantee the economic efficient, but this is rather a rare case. In reality, most regulations concerned with the environmental protection are not perfect. Several analyses have shown that the environmental protection measures introduced by means of regulations are very expensive.

Thus, instead of regulations, market-based solutions are often preferred. Such solutions are based on various ways to introduce emission fees that will depend on the level of pollution. Two such solutions are taxes and cap-and-trade programs.

Tax

The basic idea of a tax-based regulation is that companies should pay tax proportional to pollutant emissions. In that way the external effects (pollution of the environment) are included in the internal costs of the company.

Cap-and-Trade Program

Cap and trade is a regulatory program to reduce emissions of pollutants, such as carbon dioxide, as a result of industrial activity. Basic mechanisms of the program are as follows. A government issues a limited number of annual permits that allow companies to emit a certain amount of carbon dioxide. The "cap" part of the program is the total amount, or limit, of emissions that is permitted. This limit is decided by the government and as a rule is getting stricter after time. On the one hand, if a company produces a higher level of emissions than its permits allow, a tax must be paid. On the other hand, a company that reduces its emission can sell, or "trade", unused permits to other companies.

Since the number of permits is reduced each year, they are becoming more expensive. The cap-and-trade system has as a goal to create incentives for companies to invest in clean technology as it becomes cheaper than buying permits.

Cap-and-trade systems have been already created in several places. In 2005, the European Union created the world's first international cap-and-trade program. It is estimated that the emissions from sectors covered by the system have been significantly reduced and were 29% lower in 2018 than when the program started. Similarly, the state of California in the US introduced its own cap-and-trade program in 2013 and connected its system with a similar system in Quebec in Canada in 2014. The program was initially limited to in total fewer than 400 power plants, large industrial plants, and fuel distributors. California's emissions from sources subject to the cap declined 10% between the program launch in 2013 and 2018.

3.3.3 CLIMATE CHANGE

Energy systems can influence the climate on a large scale by the production of heat and by emission of carbon dioxide. Other factors of concern include the discharge of particulate matter, water vapor, and the resulting changes in albedo. Further emissions of CO_2 and other greenhouse gases, resulting mainly from burning of fossil fuels, most probably will lead to global warming and other climate changes in the nearest century. Forecasts obtained from climate models show that the mean temperature of the Earth's surface can increase by 2 to 4 K during nearest hundred years. A more detailed discussion of the influence of energy systems on the environment in general, and on the global climate in particular, is provided in §16.

Climate change is a global problem that requires cooperation at an international level. In response to this need World Meteorological Organization and the United Nations Environment Programme created the Intergovernmental Panel on Climate Change (IPCC)[6] in 1988. The IPCC is currently the United Nations body for assessing the science related to climate change, its impacts and future risks, and options for adaptation and mitigation. The assessment results are published on a regular basis in assessment reports that provide overviews of the state of knowledge concerning the climate change.

PROBLEMS

PROBLEM 3.1

Which of the 17 goals of the sustainable development are related to energy transformation?

---

[6]www.ipcc.ch

PROBLEM 3.2

Sustainability is often represented diagrammatically as an intersection of three pillars: social equity, economic viability, and environmental protection. Describe the role and impact of energy transformation for each of the pillars.

PROBLEM 3.3

Give examples how new economics could include natural capital (ecological systems) and human capital (human health, education, and skills) in sustainability assessment models.

PROBLEM 3.4

What are the concentrations of hydrogen and helium in Earth's atmosphere? Explain why they are so rare, even though they are the most common elements in the universe.

PROBLEM 3.5

Estimate by how much the global sea level would rise if all the ice melted. Assume that the global sea area will not change and, before melting, all ice is over land only.

PROBLEM 3.6

Estimate the contribution to global sea level rise until 2100 if Greenland's glaciers continue to melt at a rate 195 cubic kilometers per year.

4 Mechanical and Electromagnetic Energy

Mechanical and electromagnetic energy are two energy forms that have the greatest influence on our everyday life. Mechanical energy is required for transport of people, as well as food, raw materials and other goods. In 2016, 31.6% of the Total Final Consumption (TFC) of energy in the world was used for transportation, including personal and commercial, land, water and air transport. All vehicles used in transportation need to carry their fuel with them, unless they are connected to any energy-supplying grid. This requirement favors the use of fossil fuel such as gasoline, which has high energy density and can be easily combusted. For this reason, 65.1% of the final consumption of oil products and 7.1% of the final consumption of natural gas are used for transportation. Thus, it is essential to study how energy is used to put a vehicle in motion and keep it in motion (for example, a car on a road or an airplane in the air), to possibly reduce the energy needs irrespective of fuel option used. In this chapter we will discuss how the creation and losses of mechanical energy take place and how the mechanical energy can be stored.

Electromagnetic energy plays an increasingly important role in our daily energy consumption. In 2016 the world electricity consumption reached 20 863 TWh (5.795 EJ), with 41.6% of industrial, 27.2% residential and 22% commercial use. The agriculture and transport use of electricity was 2.9% and 1.7%, respectively.

Since devices that use electromagnetic energy are clean and convenient, their use is widespread and it will increase in the future. In this chapter we discuss the fundamental aspects of the electromagnetic energy, which are important for its generation, transmission, storage, and utilization.

4.1 FORCES AND FIELDS

The classical law of dynamics as developed by Newton provides a relationship between a force $\mathbf{F}$ acting on a body with mass m and its acceleration $\mathbf{a} = d\mathbf{v}/dt$. The second law of dynamics formulated by Newton expresses this relationship as $\mathbf{F} = m\mathbf{a}$. We know today that this relationship is only valid when the mass of the body is constant. However, as shown by Einstein, the mass m varies with its speed. This effect is, however, incredibly small in our everyday world, and the Newton formulation can be applied for the most of the cases discussed in this book.

Thus an action of a force on a body with a certain mass is closely related to the changes of the velocity of the body and, as a consequence, with changes of its kinetic energy. Out of four forces existing in the surrounding us world, the gravitation and the electromagnetic forces are of greatest importance. These forces are closely related to the gravitational and electromagnetic fields. These fields, as well as the forces belonging to them, are shortly presented in the present section.

4.1.1 A FORCE

A **force** is an influence on an object from another object or field. **Newton's second law of motion** describes the action of force $\mathbf{F}$ on an object with mass m as

$$\mathbf{F} = \frac{d\mathbf{p}}{dt} = \frac{d(m\mathbf{v})}{dt}. \tag{4.1}$$

This equation, originally stated by Newton, says that the rate of change of body's *linear momentum* $\mathbf{p}$ equals the force $\mathbf{F}$ applied to it. Equation (4.1) is always valid and represents one of the fundamental laws of nature. Its correctness was confirmed by Einstein's special relativity theory, in which the

DOI: 10.1201/9781003036982-4

mass of a body is not constant, and varies by increasing with the body's speed v. However, for bodies moving with speeds much less than the speed of light, the change of the mass is negligibly small and can be neglected. In such situations, for one-dimensional motion, the second law of dynamics is written as,

$$F = m\frac{d^2x(t)}{dt^2} = m\ddot{x}, \tag{4.2}$$

where shorthand notation $\ddot{x} \equiv d^2x(t)/dt^2$ is used to denote the acceleration. Here $x(t)$ represents the location of mass m at time t. In a similar manner, the shorthand notation for the mass velocity is $\dot{x} \equiv dx(t)/dt$.

The two major forces that are resulting from a distant gravitational and electromagnetic interactions between bodies are given as follows,

$$\mathbf{F}_g = -\frac{G_N m_1 m_2}{r^2}\hat{\mathbf{r}}, \tag{4.3}$$

where $\mathbf{F}_g$ is a gravitational attraction force exerted by mass m_1 on mass m_2 pointing from mass m_2 to mass m_1, r is the distance between the centers of the masses, $\hat{\mathbf{r}}$ is a unit vector pointing from mass m_1 to mass m_2 and $G_N = 6.674\,30(15)\times10^{-11}$ $m^3 \cdot kg^{-1} \cdot s^{-2}$ is the **Newtonian constant of gravitation**.

$$\mathbf{F}_{em} = \frac{1}{4\pi\varepsilon_0}\frac{q_1 q_2}{r^2}\hat{\mathbf{r}}, \tag{4.4}$$

where $\mathbf{F}_{em}$ is an electromagnetic force exerted by a body with an electric charge q_1 on a body with an electric charge q_2, $\hat{\mathbf{r}}$ is a unit vector pointing from charge q_1 to charge q_2, and ε_0 is the **vacuum electric permittivity**. The vacuum electric permittivity is defined as $\varepsilon_0 = 1/\mu_0 c^2 = 8.854\,187\,8128(13)\times10^{-12}$ $F\,m^{-1}$, so that $1/(4\pi\varepsilon_0) \cong 8.988\times10^9$ $N\,m^2\,C^{-2}$. In SI units, charge is measured in coulombs (C) and can be either positive or negative. From Eq. (4.4) it follows that same-sign charges repel and opposite-sign charges attract.

From observations we know that one additional force appears when charges are in motion. This force is conveniently described in terms of the magnetic field and is discussed in §4.3.3.

Example 4.1: Gravitational Acceleration on Earth Surface

Calculate the gravitational acceleration on Earth surface, knowing the mass of Earth $M_\oplus \approx 5.97\times10^{24}$ kg and Earth's radius $R_\oplus \approx 6370$ km.

Solution

The gravitational force acting on mass m on the Earth's surface is equal to $F = mg$, where g is the gravitational acceleration. The magnitude of the same force can be found from Eq. (4.3), assuming $m_1 = m$, $m_2 = M_\oplus$ and $r = R_\oplus$. Thus $mg = G_N m M_\oplus / R_\oplus^2$, from which we obtain $g = G_N M_\oplus / R_\oplus^2 \approx 9.8193$ $m \cdot s^{-2}$. This can be compared with the standard gravitational acceleration $9.806\,65$ $m \cdot s^{-2}$.

4.1.2   A FIELD

A field is a local mechanism that mediates a force. The gravitational field is responsible for the gravitational force (4.3), whereas force (4.4) is mediated by the electric field. When charges are in motion, magnetic fields are involved as well, leading to the magnetic force (discussed in §4.3.3).

An **electric field** is describe by a vector $\mathbf{E}(\mathbf{x}, t)$ at every point in space $\mathbf{x}$ and time t. To start with, we consider a stationary electric field $\mathbf{E}(\mathbf{x})$. This field exerts a force $\mathbf{F}$ on any charge q given as,

$$\mathbf{F} = q\mathbf{E}. \tag{4.5}$$

Assuming that the electric field results from a presence of charge Q located at point $\mathbf{x}_0$ and comparing Eqs. (4.4) and (4.5) we get

$$\mathbf{E} = \frac{Q}{4\pi\varepsilon_0 r^2}\hat{\mathbf{r}}, \tag{4.6}$$

where $r = |\mathbf{x} - \mathbf{x}_0|$. The electric field at location $\mathbf{x}$ around a set of charges Q_i located at points $\mathbf{x_i}$ is a superposition of the individual fields and is thus

$$\mathbf{E} = \frac{1}{4\pi\varepsilon_0}\sum_i Q_i \frac{\mathbf{x} - \mathbf{x}_i}{|\mathbf{x} - \mathbf{x}_i|^3}. \tag{4.7}$$

The electric field can be represented graphically by field lines that are parallel to the direction of $\mathbf{E}$. As shown in Fig. 4.1, the field lines begin on positive charges and end at negative ones.

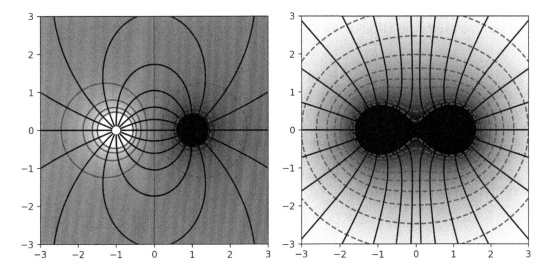

Figure 4.1 Electric field produced by a pair of two opposite charges (left) and a pair of two negative charges (right).

4.2 MECHANICAL ENERGY

A mechanical energy of a body results either from its motion (kinetic energy) or from its position in fields (potential energy). This type of energy is particularly relevant to transport, when it is required to put a vehicle in motion. In this section we discuss the two basic forms of the mechanical energy: the kinetic energy and the potential energy.

4.2.1 KINETIC ENERGY

The kinetic energy of an object of mass m moving with speed v is given by

$$E_K = \frac{1}{2}mv^2. \tag{4.8}$$

Example 4.2: Kinetic Energy of a Car

Calculate the kinetic energy of a car at total mass of 2000 kg, moving at a speed of 120 km/h.

Solution

The speed of the car is $v = 120$ km/h $= 120 \times 1000/3600$ m/s ≈ 33.3 m/s. The kinetic energy is found as $E_K = (2000$ kg$)(33.3$ m/s$)^2/2 \approx 1.11$ MJ.

4.2.2 POTENTIAL ENERGY

The potential energy is energy stored in a configuration of objects that interact through forces. For a particle moving a distance $\mathrm{d}\mathbf{x}$ in a field under action of force $\mathbf{F}$, the change in potential energy $\mathrm{d}E_P$ is given by

$$\mathrm{d}E_P = -\mathrm{d}E_K = -\mathbf{F} \cdot \mathrm{d}\mathbf{x}, \tag{4.9}$$

where E_K is the corresponding change of the kinetic energy. The signs in the equation suggest that the potential energy increases when the particle moves in a direction opposite to the force direction. If this is the case, then $\mathbf{F} \cdot \mathrm{d}\mathbf{x} < 0$ and thus $\mathrm{d}E_P > 0$. For a body with mass m moving in Earth's gravitational field close to the Earth's surface the gravity force is $F = -mg$. When the body moves from the Earth's surface to a location at a distance z above the surface, its potential energy increases as,

$$E_P(z) = \int_0^z \mathrm{d}E_P = \int_0^z mg\mathrm{d}z' = mgz. \tag{4.10}$$

We see that $E_P(0) = 0$ and the gravitational potential energy grows linearly with the height. The integral given by Eq. (4.10) does not depend on the integration path. Forces for which the integral (4.10) is independent of path are known as **conservative forces** since they conserve the sum of kinetic and potential energy. Examples of **non-conservative forces**, also known as **dissipative forces**, are friction and electromotive force induced by a changing magnetic field.

Combining Eqs. (4.10) and (4.3), the gravitational potential energy of a body with mass m at a center-to-center distance $r = |\mathbf{r}|$ from a body with mass M is

$$E_P(r) = -\frac{G_N M m}{r}. \tag{4.11}$$

The zero value of this potential energy is defined to be at an infinite distance $r \to \infty$. Once the objects approach each other their kinetic energy increases and the potential energy decreases by the same amount.

Example 4.3: Escape Velocity from Earth's Surface

Calculate the initial velocity of a body with mass m that is required for the body to reach an infinite distance from Earth with zero kinetic energy.

Solution

The total energy of the body moving with velocity v at distance z from Earth surface is

$$E = \frac{1}{2}mv^2 - \frac{G_N m M_\oplus}{R_\oplus + z}, \tag{4.12}$$

where G_N is the Newtonian constant of gravitation, $M_\oplus$ is the Earth's mass and $R_\oplus$ is the Earth's radius. For $z \to \infty$ both the kinetic and the potential energy become zero, thus the total energy is also zero. From the energy conservation principle, the total energy is zero as well when $z = 0$, thus the corresponding body velocity is obtained as,

$$v_\oplus = \sqrt{\frac{2G_N M_\oplus}{R_\oplus}}. \tag{4.13}$$

Substituting data yields $v_\oplus = 11.2$ km/s.

Example 4.4: Potential and Kinetic Energy of the Moon

Calculate the potential and kinetic energy of the Moon in its orbit around Earth. Given: the mass of the Moon $M_{\oslash} \approx 7.3 \times 10^{22}$ kg, the mass of the Earth $M_{\oplus} \approx 5.97 \times 10^{24}$ kg and the average radius of the Moon orbit $r_{\oslash} \approx 3.8 \times 10^8$ m.

Solution

The potential energy of the Moon is

$$E_P = -\frac{G_N M_{\oslash} M_{\oplus}}{r_{\oslash}} \approx -7.7 \times 10^{28} \, \text{J}. \tag{4.14}$$

The gravitational potential energy is negative and it would become zero if the Moon was moved to an infinite distance from the Earth. The Moon has also a kinetic energy as it revolves around the Earth. Taking into account that the centripetal acceleration acting on the Moon $-\omega^2 r_{\oslash} = -v_{\oslash}^2/r_{\oslash}$ must be equal to the gravitational acceleration at the Moon's orbit $-G_N M_{\oplus}/r_{\oslash}^2$, the kinetic energy of the Moon is

$$E_K = \frac{1}{2} M_{\oslash} v_{\oslash}^2 = \frac{1}{2} \frac{G_N M_{\oslash} M_{\oplus}}{r_{\oslash}} = -\frac{1}{2} E_P \approx 3.85 \times 10^{28} \, \text{J}. \tag{4.15}$$

4.2.3 WORK AND POWER

Energy transferred to an object by an action of a force is called **work**. An object gains energy when an external force does a positive work on it. An object loses energy when a negative work is performed, or equivalently, when the object does work on surroundings. When an object moves with a differential distance vector $d\mathbf{x}$ under the influence of a force $\mathbf{F}$, the differential work dL done by the force is

$$dL = \mathbf{F} \cdot d\mathbf{x} = F \, dx \cos(\mathbf{F}, d\mathbf{x}), \tag{4.16}$$

where $\cos(\mathbf{F}, d\mathbf{x})$ is a cosine of the angle between vectors $\mathbf{F}$ and $d\mathbf{x}$. Note that the work is zero when the vectors $\mathbf{F}$ and $d\mathbf{x}$ are mutually perpendicular.

The rate at which the work is done is termed a **power**. We have

$$N = \frac{dL}{dt} = \mathbf{F} \cdot \frac{d\mathbf{x}}{dt} = F v \cos(\mathbf{F}, \mathbf{v}), \tag{4.17}$$

where N is the power, $\mathbf{v} = d\mathbf{x}/dt$ is the velocity vector and $v = |\mathbf{v}|$.

4.2.4 LINEAR AND ANGULAR MOMENTUM

The **linear momentum** of an object of mass m moving with velocity $\mathbf{v}$ is defined as

$$\mathbf{p} \equiv m\mathbf{v}. \tag{4.18}$$

For a system containing several constituents with certain momenta $\mathbf{p}_i$, the total momentum of the system is $\mathbf{p}_{tot} = \sum_i \mathbf{p}_i$. For an isolated system the total momentum $\mathbf{p}_{tot}$ is conserved. The kinetic energy of an object with mass m and momentum $\mathbf{p}$ is $E_K = |\mathbf{p}|^2/(2m)$.

An **angular momentum** of a single particle moving in respect to an arbitrary fixed point (stationary in an inertial frame of reference) with a linear momentum $\mathbf{p}$ is defined as

$$\mathbf{L} \equiv \mathbf{r} \times \mathbf{p}, \tag{4.19}$$

where $\mathbf{r}$ is the distance vector from the fixed point to the particle location. The SI unit of the angular momentum is kg·m^2/s.

We also define a **torque T** in respect to the same point as

$$\mathbf{T} \equiv \mathbf{r} \times \mathbf{F}, \tag{4.20}$$

where **F** is the force acting on the particle. The SI unit of the torque is N·m. A time differential of the angular momentum given by Eq. (4.19) is as follows,

$$\frac{d\mathbf{L}}{dt} = \frac{d\mathbf{r}}{dt} \times \mathbf{p} + \mathbf{r} \times \frac{d\mathbf{p}}{dt} = \mathbf{r} \times \frac{d\mathbf{p}}{dt} = \mathbf{r} \times \mathbf{F} = \mathbf{T}, \tag{4.21}$$

since

$$\frac{d\mathbf{r}}{dt} \times \mathbf{p} = \mathbf{v} \times m\mathbf{v} = 0. \tag{4.22}$$

Thus we obtained an important result

$$\frac{d\mathbf{L}}{dt} = \mathbf{T}. \tag{4.23}$$

The above equation states that the rate of change of the angular momentum of a body is equal to the torque acting on the body. The equation is equivalent to the second law of dynamics given by Eq. (4.1) and is called the **law of conservation of the angular momentum**. It should be noted that this conservation law is valid not only for particles moving along closed orbits, but is more general.

When a particle at distance r from the origin revolves with constant angular velocity ω about the origin, its speed is $v = r\omega$ and its centripetal acceleration is $\mathbf{a} = -\omega^2 \mathbf{r}$. The magnitude of the linear momentum of the revolving particle is $p = mv = mr\omega$ and the magnitude of the angular momentum is $L = rp = mr^2\omega$.

4.2.5 MECHANICAL ENERGY LOSSES

In addition to the conservative forces discussed so far, there are *dissipative forces*, which transform kinetic energy into thermal energy, that is, random motion of molecules. To this category belong:

- Friction force appearing between two bodies, which are in a direct contact and which move with different velocities.
- Drag force arising on a solid surface in relative motion within a fluid (for example, a car moving in air).
- Rolling resistance when one solid is rolling on the surface of another one (for example, a car tires and the road).

For any rigid body (such as, e.g., a car) or non-rigid body (such as, e.g., a liquid droplet), moving in a surrounding it fluid, a force that acts in the direction opposite to the relative velocity vector between the body and the fluid is created. This force is called the **drag force**, and is given by the following equation,

$$\mathbf{F}_D = -\frac{1}{2} C_D A \rho v^2 \hat{\mathbf{v}}, \tag{4.24}$$

where $\hat{\mathbf{v}}$ is the unit relative velocity vector, A is the body's reference area (often defined as the cross-section area of the body in a plane perpendicular to $\hat{\mathbf{v}}$), ρ is the fluid density and C_D is a **drag coefficient**. The drag coefficient depends on the shape of the body and the relative velocity. This coefficient is most often determined experimentally and its value is found from Eq. (4.24), when all other parameters in the equation are measured. For example, the average modern passenger car achieves drag coefficient of 0.25 to 0.30, and a typical passenger aircraft has a drag coefficient around 0.03.

Once the drag force is known, the energy transferred to the fluid from the body moving by a distance dx is $dE = \mathbf{F}_D \cdot d\mathbf{x}$, and the rate at which the body loses the kinetic energy is $dE/dt =$

$\mathbf{F}_D \cdot d\mathbf{x}/dt = \mathbf{F}_D \cdot \mathbf{v}$. Integrating this expression over a distance D travelled by the body during time T yields

$$\Delta E = \int_0^T (dE/dt)dt = \int_0^D \mathbf{F}_D \cdot d\mathbf{x} = \frac{1}{2}C_D AD\rho v^2. \tag{4.25}$$

We note that the energy loss due to the drag force changes as the square of the relative velocity. In §8 the various aspects of the presence of fluid mechanics in energy transformation will be discussed more thoroughly.

Example 4.5: Energy Losses due to Air Resistance

Calculate the energy loss of a passenger car traveling 330 km with mean speed 90 km/h. The car cross-section area in the plane perpendicular to the driving direction is approximately 2.7 m^2 and its drag coefficient is C_D=0.33. Assume air density $\rho = 1.2$ kg/m^3.

Solution

The total energy loss due to air resistance is found from Eq. (4.25) as

$$\Delta E_{air} = \frac{1}{2}C_D AD\rho v^2 = \frac{1}{2} \times 0.33 \times 2.7 \text{ m}^2 \times 330\ 000 \text{ m} \times 1.2 \frac{\text{kg}}{\text{m}^3} \times \left(90\ 000 \frac{\text{m}}{3\ 600 \text{ s}}\right)^2.$$

After calculations, we get $\Delta E_{air} \cong 110.3$ MJ.

4.2.6 MECHANICAL ENERGY STORAGE

Attempts have been made to develop methods for a mechanical energy storage. The most widespread method is based on the *pumped-storage hydropower* (PSH), which is a configuration of two water reservoirs at different elevations. PSH can either consume energy by pumping water from the lower to the upper reservoir. The potential energy stored in the upper reservoir can be retrieved as water moves down through a turbine. The over-all efficiency of the PSH is 65–70%.

Pumped-storage method requires large land areas and a quite specific topography. For more compact applications a **flywheel storage** can be considered. Flywheels are devices specifically designed to efficiently store rotational kinetic energy. They resist changes in rotational speed by their moment of inertia. The stored rotational kinetic energy is given as

$$E_{rot} = \frac{1}{2}I\omega^2, \tag{4.26}$$

where I is the moment of inertia of a rotating body (expressed in kg·m^2) and ω is the angular velocity (expressed in rad/s).

Example 4.6: Energy Storage in a Flywheel

Calculate the amount of energy stored in a flywheel made from a uniform steel disc with mass $m = 5$ kg and radius $R = 0.3$ m, rotating at 4×10^4 rpm.

Solution

The moment of inertia of a solid disk of density ρ, radius R and height H is

$$I = \int_0^R 2\pi r H \rho r^2 dr = \frac{\pi}{2}\rho H R^4, \tag{4.27}$$

and the mass is

$$m = \int_0^R 2\pi r H \rho dr = \pi \rho H R^2. \tag{4.28}$$

Thus we have $I = mR^2/2$. At 4×10^4 rpm $\omega = 40\,000 \times 2\pi/60 \approx 4200$ s^{-1}. Substituting the given data yields

$$E_{rot} = \frac{1}{2} \frac{mR^2}{2} \omega^2 = \frac{1}{2} \times \frac{5 \text{ kg} \times (0.3 \text{ m})^2}{2} \times \left(4200 \text{ s}^{-1}\right)^2 \cong 1.985 \text{ MJ}.$$

4.3 ELECTROMAGNETIC ENERGY

An **electromagnetic energy** results from forces and fields existing between electrically charged objects. In general, human use of electromagnetic energy involves four stages: generation, transmission, storage and utilization.

Devices that use electromagnetic energy are compact, clean and convenient. Electromagnetic energy can be transported over great distances with relatively small losses.

Mechanical and electrical energy can be transformed into one another efficiently using devices called electric motors and generators. In contrast, transforming thermal or chemical energy into electromagnetic energy involves significant losses.

4.3.1 ELECTROSTATICS

A branch of physics that studies electric charges at rest is called **electrostatics**. More specific, electrostatics is concerned with electric fields, electric forces and electric potentials resulting from a presence of stationary electric charges.

The electric field, resulting from a static charge $\rho(\mathbf{x})$, with units C/m^3, that is continuously distributed throughout space, satisfies the following relationships:

$$\nabla \cdot \mathbf{E}(\mathbf{x}) = \frac{\rho}{\varepsilon_0}, \tag{4.29}$$

and

$$\nabla \times \mathbf{E}(\mathbf{x}) = 0. \tag{4.30}$$

Both these equations are sufficient to uniquely determine the electric field $\mathbf{E}(\mathbf{x})$ resulting from any static charge distribution. Once the electric field is known, the **electrostatic voltage difference** between two points $\mathbf{x}_1$ and $\mathbf{x}_2$ can be determined as

$$V(\mathbf{x}_2) - V(\mathbf{x}_1) = -\int_{\mathbf{x}_1}^{\mathbf{x}_2} \mathbf{E} \cdot d\mathbf{x}. \tag{4.31}$$

The force exerted by the electric field on a charge q can be found from Eq. (4.5). The potential in Eq. (4.31) is measured in units of energy per unit charge called volts, where 1 V = 1 J/C.

A **conductor** is a material containing charges that are free to move. Metals, such as iron, aluminium and copper are conductors. In contrast, materials in which charges are not free to move about, such as rubber, glass and wood, are called **insulators**.

A **capacitor** is an electrical circuit element containing two separate conductors. A capacitor can be charged by connecting the two conducting elements to a battery, which creates a potential difference. The charge on the capacitor is then given as

$$Q = CV, \tag{4.32}$$

where C is called the **capacitance**, with the SI unit **farad**, where 1 F = 1 C/V. Equation (4.32) is merely stating that the charge on the capacitor is proportional to the voltage. For the classic **parallel plate capacitor** of two conducting plates of area A, separated by a gap of width d, filled with a dielectric material, the capacitance is given as,

$$C = \kappa \varepsilon_0 A/d, \tag{4.33}$$

where ε_0 is the permittivity of the vacuum and κ is the material-dependent *dielectric constant* ($\kappa_{air} = 1.00059$ at 1 atm).

Example 4.7: Energy Storage in a Plate Capacitor

A plate capacitor made of two square metal plates with 1 cm side length, separated by 1 mm gap, suffers a dielectric breakdown and conducts electric current in the presence of electric field greater than $E_{max} \approx 3.3 \times 10^6$ V/m. Calculate the maximum energy that can be stored in the capacitor and the associated energy density.

Solution

The capacitor has capacitance $C = \kappa \varepsilon_0 A/d \approx (1.00059)(8.854 \times 10^{-12}$ F/m$)(0.01$ m$)^2/(0.001$ m$)$ $\approx 0.885 \times 10^{-12}$ F. The maximum possible voltage to which the capacitor can be charged is $V_{max} = E_{max}d \approx 3.3 \times 10^3$ V, with the maximum energy $CV^2/2 \approx 4.8 \times 10^{-6}$ J, and energy density $CV^2/(2Ad) \approx 50$ J/m^3. This value is very small compared, for example, to the energy density of gasoline, which is about 32 GJ/m^3.

4.3.2 ELECTRIC CURRENT

In electric circuits, **electric currents** measure the net rate at which charges pass a given point in a wire

$$I = \frac{dQ}{dt}.$$ (4.34)

In the SI units, the electric current is measured in *amperes*, which belongs to the fundamental SI units (see Table E.1).

The rate of current flow I through any conductor is proportional to the voltage drop V across the conductor with resistance R as the proportionality constant

$$V = IR.$$ (4.35)

This relationship is known as **Ohm's law**. The SI unit of of the resistance is **ohm**, where $1\ \Omega = $ V/A. In general, the voltage and the current in Eq. (4.35) can depend on time.

When an electric charge is driven through a resistor, the work that is done is dissipated in the collisions between the electrons and the molecules that make up the resistor's material. Thus the electromagnetic energy is transformed into thermal energy of the resistor and the dissipation of power in a resistor is called the **Joule heating**. The rate of power dissipation $N_{Joule}(t)$ is determined by **Joule's law** as follows

$$N_{Joule}(t) = V(t)I(t) = I^2(t)R = \frac{V^2(t)}{R},$$ (4.36)

where Ohm's law has been used. Resistive heating described by Joule's law is almost 100% efficient and is frequently used in household devices. However, in many applications the Joule heating is a source of energy losses.

Example 4.8: Energy Losses in Bulbs

Compare energy loses in incandescent, LED, and fluorescent bulbs, all with roughly the same output of energy as visible light, corresponding to 100 W incandescent bulb.

Solution

A 100 W incandescent bulb draws 100 J per second from the electric grid. The energy is converted into thermal energy by the electrical resistance of the filament of the bulb, and only 2.6

W of power is radiated as light. Thus the incandescent bulb dissipates $(100 - 2.6)/100 \times 100\%$ = 97.4% of energy. A compact fluorescent bulb can produce the same amount energy in visible light while drawing 20 to 30 W from the grid, thus it dissipates 87 to 91.3% of electromagnetic energy. A light emitting diode (LED) draws only 16 W to emit the same amount of visible light, thus it dissipates about 83.8% of energy.

Example 4.9: Flashlight Battery

An old-fashioned flashlight takes two AA batteries connected in series. Each battery has a voltage drop of $V = 1.5$ V between terminals and stores roughly 10 kJ of energy. The flashlight bulb has a resistance of about 5 Ω. Determine the rate of power use and the approximate time that a pair of batteries will last.

Solution

With the batteries connected in series, their voltage adds to 3 V. The rate of power dissipation is $N_{Joule} = V^2/R = (3 \text{ V})^2/(5 \text{ Ω}) = 1.8$ W. Thus the batteries should last for roughly 20 kJ/1.8 W ≈ 3.1 hours.

Moving charges are described locally by the current density $\mathbf{j}(\mathbf{x},t)$, which represents the moving charge per unit area and unit time in the direction of unit vector $\hat{\mathbf{j}}$. The local form of Ohm's law is then

$$\mathbf{j} = \sigma \mathbf{E} = \mathbf{E}/\rho, \tag{4.37}$$

where σ is the **electric conductivity** (or *specific conductance*, expressed in the SI unit system by siemens per meter, S/m, where **siemens** is redundantly defined as a reciprocal of ohm, $S \equiv \Omega^{-1}$), and $\rho = 1/\sigma$ is the **resistivity** (expressed in SI units by ohm-meter, Ω m). The following relationship is valid between a wire resistance R, its length l, cross-section area A, and material resistivity ρ:

$$R = \rho l/A. \tag{4.38}$$

Thus the wire resistance varies in direct proportion to its length and inversely with its cross-section area.

Resistive losses are particularly important during transmission of electricity from power plants to customers over high-voltage power lines. The technology of electric power transmission is complicated and its understanding requires such concepts as *impedance* and *reactance*. Even though some long-distance transmission lines use the direct current (DC), most electric transmission lines use alternating current (AC). In a simplified form the voltage for AC varies with time as

$$V(t) = V_0 \cos \omega t. \tag{4.39}$$

From Ohm's law the current varies with time as $I(t) = (V_0/R) \cos \omega t$. Thus, the time-averaged power dissipation to Joule's heating is

$$\langle N_{Joule} \rangle = \langle I(t)V(t) \rangle = \frac{1}{2} \frac{V_0^2}{R} = \frac{V_{RMS}^2}{R}, \tag{4.40}$$

where $V_{RMS} \equiv \sqrt{\langle V(t)^2 \rangle} = V_0/\sqrt{2}$ is the root-mean-square voltage.

A power transmission circuit consists of a power plant generating AC voltage $V(t) = V_0 \cos \omega t$, a load with resistance R_L, and the transmission lines with total resistance R_T. The transmission lines and load resistances are in series, thus the current in the circuit is

$$I(t) = \frac{V(t)}{R_T + R_L} = \frac{V_0}{R_T + R_L} \cos \omega t = I_0 \cos \omega t, \tag{4.41}$$

where $I_0 = V_0/(R_T + R_L)$. Since the voltage drop over the transmission lines is $V_T(t) = R_T I(t)$, the time-averaged power dissipation due to Joule's heating is

$$\langle N_{Joule} \rangle = \langle I(t) V_T(t) \rangle = \frac{1}{2} I_0^2 R_T. \tag{4.42}$$

Similarly, the load power is obtained as

$$\langle N_L \rangle = \langle I(t) V_L(t) \rangle = \frac{1}{2} I_0^2 R_L. \tag{4.43}$$

The ratio of the dissipated power in the transmission lines to the load power is now obtained as

$$\frac{\langle N_{Joule} \rangle}{\langle N_L \rangle} = \frac{R_T}{R_L}. \tag{4.44}$$

Since in general $R_T \ll R_L$ then $I_0 = V_0/(R_T + R_L) \simeq V_0/R_L$, and from Eq. (4.43) we get

$$R_L = \frac{1}{2} V_0^2 \langle N_L \rangle. \tag{4.45}$$

Combining Eqs. (4.44) and (4.45) yields

$$\frac{\langle N_{Joule} \rangle}{\langle N_L \rangle} = \frac{R_T}{R_L} \simeq \frac{2\langle N_L \rangle R_T}{V_0^2} = \frac{R_T \langle N_L \rangle}{V_{RMS}^2}. \tag{4.46}$$

This result indicates that the power loss fraction in transmission lines decreases with the square of the voltage V_{RMS}. It also explains why electric power is transmitted at as high voltage as possible.

Example 4.10: Energy Loss in a Power Transmission Cable

A power transmission cable of radius 1.75 cm consists of 45 strands of aluminium (reinforced by 7 strands of structural steel that do not carry current). The effective area of the aluminium conductors is 8 cm² and the resistivity of aluminium is 2.8×10^{-8} Ω m . This cable is used to carry power at V_{RMS} = 345 kV over a distance of 200 km. How much power can be transmitted by this cable if the maximum acceptable resistive loss in the cable is 2%?

Solution

First we compute the resistance of the transmission cable

$$R_T = \frac{\rho l}{A} = \frac{2.8 \times 10^{-8} \text{ Ω m} \times 200 \times 10^3 \text{ m}}{8 \times 10^{-4} \text{ m}^2} = 7.0 \text{ Ω}. \tag{4.47}$$

Using Eq. (4.46) we find $\langle N_L \rangle = 0.02 \times (3.45 \times 10^5 \text{ V})^2/7.0 \text{ Ω} \cong 340.1$ MW. Thus a single cable can deliver 340 MW with loss of 2% to Joule heating. Since power is usually carried in three cables, they can carry the full output of a 1 GW power plant.

4.3.3 MAGNETISM

Magnetic phenomena play a role in transformation of electrical energy into mechanical energy using motors and back to electrical energy using generators. Moving charges produce a magnetic field, which exerts a force on a moving charged particle. Like an electric field, the magnetic field $\mathbf{B}(\mathbf{x}, t)$ is described by a vector at point in space and time. The force acting on a charge q moving with velocity $\mathbf{v}$ in the magnetic field is

$$\mathbf{F} = q\mathbf{v} \times \mathbf{B}. \tag{4.48}$$

For stationary, space dependent electric and magnetic fields, the combined effects on a charge q are given by the **Lorentz force law** as follows,

$$\mathbf{F}(\mathbf{x}) = q\mathbf{E}(\mathbf{x}) + q\mathbf{v} \times \mathbf{B}(\mathbf{x}). \tag{4.49}$$

Since there are no elementary magnetic charges, the equivalent of Eq. (4.29) for the magnetic field is

$$\nabla \cdot \mathbf{B}(\mathbf{x}) = 0. \tag{4.50}$$

One consequence of the form of Eq. (4.50) is that the lines of magnetic field have no place to begin or end, and must therefore be closed curves.

For a stationary current and charge density, the **Ampere's law** in the following form is valid

$$\nabla \times \mathbf{B}(\mathbf{x}) = \mu_0 \mathbf{j}(\mathbf{x}), \tag{4.51}$$

where $\mathbf{j}(\mathbf{x})$ is the current density, determined by the local version of Ohm's law given by Eq. (4.37). The proportionality constant μ_0 appearing in Eq. (4.51) is known as the **magnetic permeability of the vacuum** and is equal to $1.256\ 637\ 062\ 12(19) \times 10^{-6}$ N A^{-2}.

Equations (4.50) and (4.51) uniquely determine the magnetic field associated with any time-independent current distribution. In particular, the magnetic field around a long straight wire carrying constant current I can be found as

$$\mathbf{B}_{wire} = \frac{\mu_0}{2\pi} I \frac{\hat{\mathbf{z}} \times \hat{\rho}}{\rho}, \tag{4.52}$$

where $\hat{\mathbf{z}}$ is a unit vector pointing along the wire and $\hat{\rho}$ is a unit vector pointing radially outward from the wire. Using the right-hand-rule for the cross product $\hat{\mathbf{z}} \times \hat{\rho}$, the fingers of the right hand wrap in the direction of the magnetic field if the thumb points along the direction of the current flow in the wire.

Another example of a simple magnetic field is the **solenoid**, which is a long wire wound in a helix around a cylinder of cross-sectional area A and length l. The cylinder can be empty or filled with a material of permeability μ, such as iron, for which $\mu \gg \mu_0$. The interior magnetic field in a solenoid points along the solenoid's axis and its strength is given by

$$\mathbf{B}_{sol} = \mu n I \hat{\mathbf{z}}, \tag{4.53}$$

where $n = N/l$ is the number of turns of wire N per unit length along the solenoid, I is the current and $\hat{\mathbf{z}}$ is a unit vector along the solenoid axis pointing in the positive current direction.

4.3.4 INDUCTION

Electric fields are produced when magnetic fields change and vice versa. Two different ways to produce electric field are as follows:

- When a wire loop moves into an increasing magnetic field.
- When a wire loop is fixed in a magnetic field that is constant in space but grows with time.

The strength of the produced electrical field is governed by **Faraday's law of induction**, which is closely related to the mechanisms behind the electric motors and generators that are discussed in the next subsection. Faraday's law of induction in a differential form is as follows

$$\nabla \times \mathbf{E} = -\frac{\partial \mathbf{B}}{\partial t}, \tag{4.54}$$

or in the integral form it is given as

$$\oint_C dx \cdot \mathbf{E} = -\frac{d}{dt} \int_S d\mathbf{S} \cdot \mathbf{B} = -\frac{d\Phi}{dt}. \tag{4.55}$$

Here C is the wire closed loop, S is the planar surface bounded by the wire loop and Φ is the flux of the magnetic field through the planar surface S given as $\Phi = \int_S d\mathbf{S} \cdot \mathbf{B}$. The left-hand-side of Eq. (4.55), $\oint_C dx \cdot \mathbf{E} = \mathscr{E}(C)$, is known as the **electromotive force**, or EMF. The minus sign in Eq. (4.55) indicates that the current generated by the EMF would act to reduce the magnetic flux through the C loop. This sign and the fact that an induced current acts to reduce any imposed magnetic field is known as **Lenz's law**.

An important feature of Faraday's law is that a wire wound N times around the loop C captures N times the flux

$$\mathscr{E} = -N\frac{d\Phi}{dt}. \tag{4.56}$$

Thus the EMF is enhanced N times.

An **inductor** is a circuit element that produces a voltage proportional to the time rate of change of the current passing through it,

$$V = -L\frac{dI}{dt}, \tag{4.57}$$

where L is the inductance measured in henrys (1 H = 1 V s/A). The inductance of a solenoid with n turns per unit length, volume $\mathscr{V}$, and core permeability μ is approximately

$$L_{sol} \cong n^2 \mathscr{V} \mu. \tag{4.58}$$

The energy stored in an inductor is $E_{EM} = LI^2/2$. The general form of this formula is

$$E_{EM} = \frac{1}{2\pi} \int |\mathbf{B}|^2 dV, \tag{4.59}$$

where the integration should be performed over the volume of the inductor.

4.3.5 ELECTRICAL DEVICES

Electrical devices are used to transform the electric energy into some other forms of energy, such as light, heat, or motion. For example, pumps and blowers are electrical devices which transform the electrical current in the fluid motion. The heater transforms the current into heat, whereas the electric bulb transforms the current into light. Electric motors and electric generators belong to the most important electric devices widely used in energy systems.

An electric motor is a machine that transforms electrical energy into mechanical energy. Electric motors are used in a wide range of devices, including cell phones, computers, blowers, pumps, elevators, and for propulsion of vehicles. Motors contain generally two magnets, the rotating rotor, and the stationary stator.

In the simplest motors the rotor is an electromagnet powered by direct current with a commutator and brushes switching the direction of the current flow depending on the angle of the rotor, while the stator uses a permanent magnet. For such simple motor the theoretical average power output when the motor is rotating at frequency v is $\langle N \rangle = 4nIABv$, where n is the number of windings of the wire loop in the rotor, I is the current in the wire, A is the area of the current loop, and B is the constant magnetic field of the stator. In real devises there will be losses due to friction and Joule's heating.

Example 4.11: Power Output of a Direct Current Motor

A simple small motor has a stator as a permanent magnet with field 0.025 T. The rotor has an electromagnet formed by wrapping a wire 750 times a single cylinder of radius 2.4 cm. The motor uses 350 mA current and runs at 3000 rpm. Neglecting frictional losses, calculate the average power output.

Solution

The average power output is found as $\langle N \rangle = 4nIABv = 4 \times 750 \times (0.35 \text{ A}) \times \pi \times (0.024 \text{ m})^2 \times (0.025 \text{ T}) \times (50 \text{ s}^{-1}) \approx 2.38 \text{ W}$.

4.3.6 ELECTROMAGNETIC ENERGY LOSSES

Electromagnetic energy is transformed into thermal energy due to Joule heating, as described by Eqs. (4.36) and (4.40). Additional losses are caused by magnetic hysteresis and eddy currents created in electric conductors, such as steel, when time varying magnetic fields pass through them. For almost all applications involving "soft" iron, eddy currents are the dominant source of loss. To reduce the eddy current loss, magnetic circuits of transformers and electric machines are almost always laminated, or made up of relatively thin sheets of steel. To further reduce losses the steel is alloyed with elements (often silicon) which deteriorate the electrical conductivity.

These losses occur in all types of electric devices such as motors, generators, transformers and similar. In general, electromagnetic energy losses are much less than thermodynamic losses, resulting from the second law of thermodynamics. Due to this sometime electric devices are treated as "ideal" lossless machines, where the output power is equal to the provided input power. When the loss of power is not negligible, it can be found as

$$N_{loss} = N_{in} - N_{out}, \tag{4.60}$$

where N_{in} is the provided input power and N_{out} is the obtained output power. For example, for an electric motor at steady-state the input power is

$$N_e = \sum_i V_i I_i, \tag{4.61}$$

which is just the sum of electric power inputs to the different phase terminals. The mechanical output power is just torque T times angular velocity ω,

$$N_m = T\omega. \tag{4.62}$$

The electrical motor loss is then found as

$$N_{loss} = N_e - N_m = \sum_i V_i I_i - T\omega. \tag{4.63}$$

The motor efficiency is defined as

$$\eta = \frac{N_m}{N_e} = 1 - \frac{N_{loss}}{N_e}. \tag{4.64}$$

The efficiency of electric motors varies with their design and operation conditions. Most electric motors operate at efficiencies between 70 and 90% [24].

4.3.7 ELECTROMAGNETIC ENERGY STORAGE

Large-scale storage of electromagnetic energy is an important and challenging problem. As already shown, the capacitive energy storage is quite limited and provides low energy densities. The main technology currently in practical use, but still under development, is the storage in re-chargeable batteries.

Batteries store energy in chemical form and use electrochemical reactions to convert energy to and from electrical form. Most home energy storage batteries, also used in consumer electronics, are lithium-ion. These batteries have a high energy density, relatively low self-discharge, and low maintenance. Their main disadvantages include ageing (even if not used), requirement of a protection circuit to maintain voltage and current within safe limits, and high manufacturing costs.

A lithium-ion battery uses cathode (positive electrode), an anode (negative electrode) and electrolyte as conductor. The cathode is metal oxide and the anode consists of porous carbon. During discharge, the ions flow from the anode to the cathode through the electrolyte and separator. As a result, the anode undergoes oxidation. The ion flow is reversed during charging and the cathode gains electrons.

Lithium-ion batteries come in many varieties and vary in performance and the choice of active materials. Originally the battery used coke as the anode, but since 1990s, most battery manufacturers shifted to graphite to attain a flatter voltage discharging curve. A future material that promises to enhance the performance of lithium-ion batteries is graphene.

PROBLEMS

PROBLEM 4.1

Calculate the potential energy of a body with mass 1 kg at 1 km distance from Earth's surface. Neglecting friction, calculate the body's velocity at the Earth's surface when freely falling from 1 km height.

PROBLEM 4.2

Calculate the potential energy of a body with mass 1 kg located at the Earth's surface if its potential energy at infinity is zero.

PROBLEM 4.3

Calculate the electrostatic potential energy of an electron and a proton which are located at a distance 1 Å apart. Assume that the electrostatic potential energy of the system is zero when the distance between the particles is infinity.

PROBLEM 4.4

A flashlight uses two AA batteries connected in series. Each battery has a voltage drop of $V = 1.5$ V, internal resistance of 0.5 Ω, and stores roughly 10 kJ of energy. The flashlight bulb has a resistance of roughly 5 Ω. Calculate (a) the rate of power use, (b) the approximate time that the batteries will last, (c) the fraction of power that is lost to Joule heating within the batteries.

PROBLEM 4.5

How much energy can one store on a parallel plate capacitor with $d = 1\mu$m, $A = 10$ cm^2, and $\varepsilon = 100\varepsilon_0$, assuming that the breakdown field of the dielectric is the same as for air?

PROBLEM 4.6

An electric motor operates at 1 200 rpm with an average torque of 0.01 Nm. Calculate: (a) the motor's power output, (b) the back-EMF from the rotor, if the motor is running on 1.2 A current.

5 Biological and Chemical Energy

Living organisms and all forms of activity of living cells require energy. Life on Earth depends on the availability of solar energy. Plants use photosynthesis to convert it into chemical energy, which is used to support own living processes and to grow. The energy accumulated by plants is directly or indirectly used by all other living organisms.

A human with no physical activity needs 1700 kcal during 24 hours just to support all essential life activities. This amount of energy can be provided by consumption and digestion of 200 g (800 kcal) of carbohydrates, 70 g (400 kcal) of protein and 55 g (500 kcal) of fat. In addition 16 moles of oxygen will be used and 13.7 moles of carbon dioxide will be produced due to cellular respiration. Humans with active lifestyle need 3000–3500 kcal, whereas physically hard-working individuals can need up to 5000 kcal.

Direct use of biological energy sources constitutes a significant fraction of human energy consumption. In addition to use for food, heating, and cooking, biomass derived from variety of sources is used for power generation. For example, sugarcane and maize are converted into ethanol and other biofuels.

Fossil fuels, such as coal, petroleum and natural gas, are also based on biological energy sources. However, unlike biological energy, their organic material was processed through natural systems over relatively long time periods.

In this chapter we discuss energy resources that have been transformed from solar energy by photosynthesis. We focus first our attention on the mechanisms of photosynthesis. Then we make a survey of biofuels and fossil fuels.

5.1 PHOTOSYNTHESIS

Photosynthesis is the biochemical mechanism by which most plants, most algae and cyanobacteria[1] capture and store solar energy as chemical energy. This chemical energy is stored in carbohydrate molecules, such as sugars, which are synthesized from carbon dioxide and water.

A chlorophyll molecule is the key player in the photosynthesis process in most plants. Chlorophyll is responsible for the green color of many plants and algae. It is concentrated within organisms in structures called chloroplasts. Plants are perceived as green since chlorophyll absorbs mainly the blue and red wavelength and reflects the green. There are several types of chlorophyll, but all of them share the chlorin-magnesium ligand[2]. Chlorophyll serves three functions: (i) to absorb light, (ii) to transfer the light energy by resonance energy transfer to a specific chlorophyll pair in the reaction center of the photosystems, (iii) to transfer absorbed energy to an electron in a process called charge separation, leading to biosynthesis.

The physics of photosynthesis process has many similarities with the energy collection process in a photovoltaic (PV) cell. Absorbed photons excite electrons similarly to the excitation of electrons across the band gap of silicon in PV solar cell. The excited electrons provide the energy that powers photosynthesis. However, in biological systems the electrochemical energy gathered from photosynthesis is stored in organic compounds composed primarily of carbohydrates, such as sugars and starches.

---

[1] Such organisms are called photoautotrophs.

[2] An ion or molecule that binds to a central metal atom to form a coordination complex.

Carbohydrates are the most important organic molecules from the energetic point of view. They belong to a relatively simple class of molecules formed from carbon, oxygen, and hydrogen that store and transport energy in biological systems. Most carbohydrates have the chemical formula $C_n(H_2O)_m$, which explains the name, since it can be seen as bond structures of carbon and water: "carbo-hydrate". They are divided into four chemical groups: monosaccharides, disaccharides, oligosaccharides, and polysaccharides. The first two ones are commonly referred to as sugars and include fructose (fruit sugar), glucose (starch sugar), sucrose (cane or beet sugar) and lactose (milk sugar). More complex carbohydrates include cellulose, hemicellulose and lignin, which comprise material used by plants for structure.

5.1.1   MECHANISMS OF PHOTOSYNTHESIS

There are two types of photosynthetic mechanisms: oxygenic photosynthesis and anoxygenic photosynthesis. Both these processes are quite similar, but oxygenic photosynthesis is the most common one and is present in plants, algae and cyanobacteria.

During **oxygenic photosynthesis**, light energy transfers electrons from water to carbon dioxide to produce carbohydrates. In this transfer the carbon dioxide is reduced (receives electrons) and the water becomes oxidized (loses electrons). In that way oxygenic photosynthesis functions as a counterbalance to respiration by taking in the carbon dioxide produced by all breathing organisms and reintroducing oxygen to the atmosphere.

Processes involved in photosynthesis are very complex and their full explanation is beyond the scope of this book. Thus we focus our attention only on a selected process and we neglect more complex mechanisms that are involved. A simplified diagram of oxygenic photosynthesis is shown in Fig. 5.1, where the sequence of reactions in which carbon dioxide is fixed by some plants is shown. The process is known as the Calvin-Benson-Bassham (CBB) cycle or just **Calvin cycle**. This cycle includes 13 biochemical reactions catalysed by 11 distinct molecules. The net result is the conversion of three CO_2 molecules into a three-carbon sugar triose-P using energy stored in the energy-transporting molecules adenosine triphosphate (ATP) and nicotinamide adenine dinucleotide phosphate (NADPH). This energy originates from the photosynthesis light reactions. Triose-P is then used to produce six-carbon sugars and sugar phosphates that store energy in stable forms such as glucose or sucrose. An enzyme known by an acronym RuBisCO catalyzes a critical step, the carboxylation of ribulose-1,5-P_2 (RuBP), in which CO_2 is captured from the atmosphere.

Unlike the oxygenic photosynthesis, the **anoxygenic photosynthesis** uses electron donors other than water. The process typically occurs in bacteria such as purple bacteria and green sulphur bacteria, which are primarily found in various aquatic habitats.

Photosynthesis involves a complicated set of biochemical reactions with the following net conversion

$$n(8\gamma + CO_2 + H_2O) \rightarrow C_nH_{2n}O_n + nO_2. \tag{5.1}$$

The reaction indicates that, under action of $8n$ photons, n-water and carbon dioxide molecules produce a molecule of sugar and n molecules of oxygen. The sugars produced from photosynthesis can be combined to form starches and cellulose for energy storage and plant structure formation.

5.1.2   PHOTOSYNTHESIS EFFICIENCY

The efficiency with which solar energy is converted into biomass for most plants is a fraction of 1%. Only 40% of the energy in sunlight lies in the photosynthetically active frequency range. Even when photons are captured, energy is lost in transporting the electrons to the reaction, and in each stage of chemical reaction, so that the fraction of energy stored in a single CH_2O sugar component from eight incident photons at 680 nm is roughly 30%. This gives an upper theoretical limit of 12% on photosynthetic efficiency. However, under realistic conditions for standard photosynthesis,

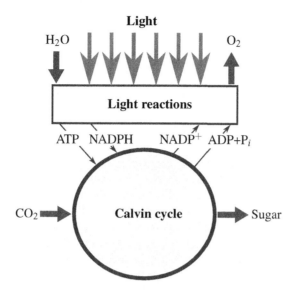

Figure 5.1 A simplified diagram of photosynthesis type C-3 binding CO_2 in Calvin's cycle, occurring in two stages: (1)—light reactions to capture the energy of light and make ATP and NADPH molecules, (2)-light-independent reactions using these products to capture and reduce carbon dioxide; ATP—adenosine triphosphate, NADPH—reduced $NADP^+$, $NADP^+$—nicotinamide adenine dinucleotide phosphate, ADP—adenosine diphosphate, P_i—inorganic phosphate.

the maximum efficiency realizable is closer to 5%, but typical efficiencies are much lower. This is because much of the energy transformed to sugars is used quickly for plant metabolism and other purposes. The energy that is stored in structural material within the plant, and which could be extracted by combustion or other means is generally substantially less than 1% of the incident sunlight, and typically around 0.25%. This low efficiency of photosynthesis can be demonstrated by the global rate of biomass production on Earth: 1660 EJ/y, in relation to the rate of energy use by humankind: 550 EJ/y.

5.2 FOOD ENERGY

Rate of energy production as food for human consumption is roughly 45 EJ/y, of which approximately one third goes to waste.

5.2.1 FOOD PRODUCTION

About 50×10^6 km^2 (roughly one third of the world's total land area) is used for agriculture, with 15×10^6 km^2 used for crops and the rest used as pastures and grassland for grazing. The three most important crops (providing about 60% of the world's food supply) are maize, rice and wheat. Taking into account the land used to grow these three crops and their corresponding yields, the efficiency of solar energy capture is on the order of 0.1%.

Example 5.1: Efficiency of Growing Maize

Estimate the efficiency of growing maize assuming that it is produced with harvest yield of 10 t/ha y, it has energy density of 15 MJ/kg and an average insolation is 200 W/m^2.

Solution

First we find the energy capture rate as $10\,000$ kg/ha y$\times 15$ MJ/kg $= 150$ GJ/ha y. The insolation gives 200 W/m$^2 \times 365$ days/y$\times 24$ hours/day$\times 3600$ s/hour$\times 10^4$ m^2/ha ≈ 63 TJ/ha y, thus the energy efficiency is $\sim 0.24\%$.

5.2.2 FERTILIZERS

Plants require nitrogen to grow, since nitrogen is incorporated in the structure of some critical to plants complex molecules including DNA and amino acids. In most natural situations, the limited supply of nitrogen is the rate-limiting factor for plant growth. The atmosphere contains nitrogen in the form of N_2, which is difficult to break due to triple bond in this diatomic molecule. Only certain microbes (cyanobacteria), active in a soil and crucial in the nitrogen cycle, can fix nitrogen from the atmosphere by converting nitrogen to ammonia: $N_2 + 8H^+ + 8e^- \rightarrow 2NH_3 + H_2$. However, high-yield plant grow is requiring synthetic ammonia that can be produced in the energy-intensive (roughly 36 MJ/kg) *Haber-Bosch process*. It is estimated that up to one half of the food energy consumed is made possible by the fixed nitrogen in fertilizers. However, intensive agriculture and monoculture (growing a single crop over a wide area for several consecutive years) lead to decreased soil quality and increased populations of pests. Due to that increased use of pesticides is needed, adding to further environmental problems.

5.3 BIOENERGY

Bioenergy is energy made from organic matter. Any organic material which has absorbed sunlight and stored it in the form of chemical energy is considered as biomass. This can include wood, energy crops, and agricultural and food waste. The term bioenergy also covers transport fuels produced from organic matter. In this section we discuss biomass and such biofuels as biogas, ethanol, and biodiesel.

5.3.1 BIOMASS

Biomass refers to any type of organic material derived from living or recently living organisms that is burned for heat, light, and cooking. This direct use of energy from biomass has served humankind for thousands of years and still represents roughly 8% of the global energy use. Many types of biomass can be digested by bacteria, producing biogas, primarily methane, which can be used for heating and power production.

Biomass can be upgraded into a better and more practical fuel by various conversion processes, using thermal, chemical, biochemical, and electrochemical mechanisms.

Thermal conversion processes use heat as the dominant mechanism to upgrade biomass. The basic alternatives are torrefaction, pyrolysis, and gasification. Thermal gasification is carried out by heating biomass in the presence of oxygen, air, and/or water vapor. **Pyrolysis** is the thermal decomposition of materials at elevated temperatures in an inert atmosphere. It involves a change of chemical composition and is irreversible. A mild form of pyrolysis, with temperatures typically between 200 and 300 °C is called **torrefaction**.

Biomass can be converted into multiple commodity chemicals using chemical conversion processes, similar to those developed for fossil fuels, such as the *Fischer-Tropsch synthesis*.

In biochemical conversion, micro-organisms are used to perform the conversion process to break down the molecules of which biomass is composed, including such processes as anaerobic digestion, fermentation and composting.

Electrochemical oxidation can be used for direct conversion of biomass to electric energy. This can be performed in various types of *fuel cells*, such as direct carbon fuel cell, direct ethanol fuel cell, direct formic acid fuel cell, L-ascorbic acid fuel cell, and microbial fuel cell.

Example 5.2: Land Need Estimation for Directly-Burned Biomass

Estimate the land needed to supply sustainably 40 EJ/y (according to IEA, this corresponded to the global energy use that came from the combustion of biomass in 2010) by direct burning of biomass consisting of crops and trees that accumulated energy through photosynthesis at a rate of around 5 t/ha y, with typical energy density of 12 MJ/kg.

Solution

From given data we obtain the energy supply rate of 5 t/ha y $\times$ 12 MJ/kg = 60 GJ/ha y. Thus the land area needed to supply 40 EJ/y sustainably is: $(4 \times 10^{19}$ J/y)/$(6 \times 10^{10}$ J/ha y) $\approx 6.7 \times 10^8$ ha $\approx 6.7 \times 10^6$ km^2. This corresponds to roughly 16.5% of all forested land area on Earth.

5.3.2 BIOGAS

Thermal gasification is performed by heating biomass in the presence of oxygen, air, and/or water vapor. This leads to break down of the molecular structure of the biological material and production of a gas that contains varying fractions of methane, CO_2, CO, H_2, and water vapor.

Biological syngas is obtained by incomplete combustion, when the amount of oxygen present is insufficient for complete combustion. Such gas contains substantial amounts of H_2 and CO.

Pyrolysis, which is heating in the absence of oxygen or air, releases gas containing H_2, light hydrocarbons (e.g. methane or ethane) and liquid hydrocarbons.

Biogas, as distinct from biological syngas, is produced from organic material through anaerobic digestion, that is broken down by certain types of bacteria (e.g. mesophilic bacteria or thermophilic bacteria) without presence of oxygen. The main components of biogas are methane and CO_2.

The two principal methods for biogas production are the capture of biogas from landfills and the production of biogas in closed containers known as anaerobic digesters. Solid woody material is generally not suitable for biogas production as it contains lignin, which cannot be anaerobically digested by standard bacteria. Biogas has potential as an *energy carrier* with a range of applications from domestic usage (for cooking and heat) to grid-scale electricity production.

5.3.3 ETHANOL

Ethanol (C_2H_5OH) can be obtained from simple sugars through fermentation, that is, biological transformation of sugars to alcohols, acids, or gases.

For glucose, the chemical transformation is

$$C_6H_{12}O_6 \rightarrow 2C_2H_5OH + 2CO_2, \tag{5.2}$$

and for sucrose in presence of water we have

$$C_{12}H_{22}O_{11} + H_2O \rightarrow 4C_2H_5OH + 4CO_2. \tag{5.3}$$

The energy density of ethanol is approximately twice that of simple sugars due to release of CO_2 during fermentation. Since ethanol is a liquid at ambient temperature and pressure, it is a convenient fuel to transport and use. Pure ethanol has a mass density of 0.789 kg/l, and an energy density of 23.4 MJ/l. This is substantially less than energy density of gasoline (32 MJ/l), but ethanol is a more stable molecule with less tendency to knock and has octane number 108.6.

The largest current contribution to world ethanol production is maize-based ethanol produced in the United States (54×10^9 l/y) and the second largest is sugarcane-based ethanol produced in Brazil (24×10^9 l/y). Both maize and sugarcane can also be used as food or animal feed, thus research is performed to produce ethanol from other organic materials. One of such materials is cellulose, which is used for plant structure. An efficient mechanism for producing cellulosic ethanol from the cellulose materials in non-food plants would make possible the extraction of energy from much of the biomass available on Earth. However, cellulosic ethanol production is still at an early stage, and has only been carried out in small-scale test projects.

An important metric for biofuels as an energy resource is the ratio of the output energy to energy input; a parameter known as *Net Energy Balance* (NEB). For example, with modern agricultural methods, sugarcane ethanol can be produced at a yearly rate of 180 GJ/ha y. This production, however, requires substantial energy inputs for fertilizers, planting, harvesting, and the resulting ethanol transport and distribution. Analyses indicate that for typical Brazilian sugarcane plant, the net energy balance is NEB $\equiv$ (total energy output)/(total energy input) ~ 8. This can be compared to estimates of the net energy balance for maize-based ethanol from US production plants that lies in the range ~ 1.2–1.6. This much poorer performance of maize-based ethanol stems from the fact that the complex carbohydrates in maize (starches amylopectin and amylose that cannot be directly fermented) require more energy in processing as compared to sugarcane. For cellulosic ethanol production the value of NEB is still unclear, but optimistic estimates suggest that it may be even higher than for sugarcane ethanol.

5.3.4 BIODIESEL

Biodiesel is a liquid fuel produced from vegetable oil and animal fat using purely chemical processes. Plant and animal oils can be processed to yield fatty acid esters with properties very similar to the hydrocarbons in conventional diesel fuels. Since biodiesel can be produced from a wide range of different feedstocks, a substantial variation in the distribution of molecules in different batches of biodiesel exists. Typical energy density of biodiesel fuel is 37 MJ/kg, which is about 10% lower than for standard diesel fuel.

The United States ranks number one and Brazil ranks second as the largest producers of fuel ethanol and biodiesel. They accounted for 84% of global ethanol production in 2017, and 26% of biodiesel production. In both countries, fuel ethanol and biodiesel are blended with fossil transport fuels, and in Brazil unblended fuel ethanol also competes directly with gasoline at the pump. However, as production costs for biofuels, gasoline, and diesel vary by country, the consequent difference in break-even oil price for ethanol and biodiesel, as well as policy measures, affect biofuel competitiveness in both countries.

To assess the relative competitiveness of biofuels with petroleum products, gasoline and diesel production costs must be compared with those of ethanol and biodiesel. In the second half of 2017 the costs of production of gasoline and ethanol in Brazil were 0.61 and 0.60 USD/l, respectively. The corresponding costs in the United States were 0.46 and 0.61 USD/l. Similar comparisons for diesel and biodiesel give 0.58 and 0.80 USD/l in Brazil, and 0.47 and 0.83 USD/l in the United States[3].

5.4 FOSSIL FUELS

The main fossil fuels that have driven development of the modern world, and that still play a major role in human energy use are coal, petroleum, and natural gas. In §2 we discussed the reserves and energy supply from fossil fuels. In this section we present a general characteristic of fossil fuels.

5.4.1 COAL

The coal is primarily composed of carbon (C), with some content of hydrogen (H), sulphur (S), oxygen (O) and nitrogen. The major coal ranks, together with their composition (by mass-percent of the indicated element) and heat content, are given in Table 5.1.

---

[3]www.iea.org/newsroom/news/2019/march/how-competitive-is-biofuel-production-in-brazil-and-the-united-states.html, accessed 2019-07-05.

Table 5.1
Coal Ranks

Coal Rank	Volatile, %	C, %	H, %	O, %	S, %	Heat Content, MJ/kg
Lignite	45–65	60–75	5.8–6.0	17–34	0.5–3	<28.47
Flame coal	40–45	75–82	5.8–6.0	>9.8	~1	<32.87
Gas flame coal	35–40	82–85	5.6–5.8	7.3–9.8	~1	<33.91
Gas coal	28–35	85–88	5.0–5.6	4.5–7.3	~1	<34.96
Fat coal	19–28	88–90	4.5–5.0	3.2–4.5	~1	<35.38
Forge coal	14–19	90–91	4.0–4.5	2.8–3.2	~1	<35.38
Nonbaking coal	10–14	91–92	3.8–4.0	2.8–3.5	~1	35.38
Anthracite	7–12	>92	<3.75	<2.5	~1	<35.30

In addition to the elements indicated in Table 5.1, coal contains some traces of other elements as well, such as mercury (Hg)—0.10 ± 0.01 ppm, arsenic (As)—1.4–71 ppm and selenium (Se)—3 ppm.

5.4.2 PETROLEUM

The dominant role of oil in the energy consumption results from its attractive features such as the high energy density (42–44 MJ/kg), and easy transport and storage.

Crude oil contains a wide mix of different hydrocarbons such as paraffins, cycloparaffins, olefins, and aromatic hydrocarbons, as they are called in the petroleum industry. Scientific names of ingredient molecules are alkanes, cycloalkanes, alkenes, and arenes.

Crude oil is a basic raw material to obtain various products such as naphtha, gasoline, kerosene, diesel, and heavier fuel oils (like heating oils for housing). To separate the various products, a distillation process is used in oil refineries, where crude oil is heated and the different chains are pulled out by their vaporization temperature.

The chains from C7H16 through C11H24 are blended together and used for gasoline. Next is kerosene, in the C12 to C15 range, followed by diesel fuel and heavier fuel oils.

The **gasoline** (or petrol), used as a fuel in motor vehicles, is a complex mixture of about 500 different molecules. It includes hydrocarbons and non-hydrocarbon additives, and is tuned to allow high engine compression ratios and optimize engine efficiency, while minimizing some negative impacts on the environment.

5.4.3 NATURAL GAS

Most **natural gas** resources contain primarily methane with smaller fractions of ethane, propane, and butane, and traces of higher hydrocarbons, nitrogen, CO_2, water vapor, oxygen, hydrogen sulfide, and even helium.

5.5 COMBUSTION

The energy from biofuels and fossil fuels is mainly harvested through combustion. In simple terms, the process of **combustion** is burning a substance in air. In more general terms, combustion is an exothermic reaction in which a compound reacts completely with an oxidant. Assuming an ideal combustion, every carbon atom in the compound turns into CO_2, and every hydrogen atom into H_2O. In practice, however, combustion is not complete. For example, some carbon can be left as ash, and some carbon may be partially oxidized to carbon monoxide (CO).

Stoichiometric combustion of hydrocarbon in oxygen is described with the following chemical equation:

$$C_xH_y + (x + \frac{y}{4})O_2 \rightarrow xCO_2 + \frac{y}{2}H_2O. \tag{5.4}$$

Practically all combustion takes place using air as the oxygen source. Treating all non-oxygen components in air as nitrogen, the combustion equation becomes,

$$C_xH_y + (x + \frac{y}{4})O_2 + 3.77(x + \frac{y}{4})N_2 \rightarrow xCO_2 + \frac{y}{2}H_2O + 3.77(x + \frac{y}{4})N_2. \tag{5.5}$$

Here the nitrogen to oxygen volume ratio is 3.77, since the volume fraction of oxygen in air is 20.95%. Even though the nitrogen does not participate in the combustion reaction, its inclusion shows the composition of the resultant flue gas.

Other substances begin to appear in significant amounts when the flame temperature is above 1600 K (then nitrogen may oxidize to NO or NO_2), or the combustion is incomplete (then CO and H_2 molecules may appear).

5.5.1   COMBUSTION OF GASOLINE

Detailed analysis of the **gasoline combustion** is very tedious due to its complex composition. For simplicity, we assume that the gasoline is 100% iso-octane C_8H_{18}. The complete combustion of two moles iso-octane with oxygen can be represented with the following reaction

$$2C_8H_{18} + 25O_2 \rightarrow 18H_2O(l) + 16CO_2. \tag{5.6}$$

Here we assume that H_2O is in the liquid phase after the combustion. Using Hess's law we obtain the enthalpy of combustion for one mole of the iso-octane as

$$\Delta I_c(C_8H_{18}) = 8\Delta I^f(CO_2(g)) + 9\Delta I^f(H_2O(l)) - \Delta I^f(C_8H_{18}). \tag{5.7}$$

Substituting the following enthalpies of formation: $\Delta I^f(CO_2(g)) = -394$ kJ/mol, $\Delta I^f(H_2O(l)) = -286$ kJ/mol and $\Delta I^f(C_8H_{18}) = -250$ kJ/mol, we obtain $\Delta I_c(C_8H_{18}) = -5.476$ MJ/mol. Since the molar mass of iso-octane is 114 g/mol, its specific energy density (positive by convention) is $E_c = 47.5$ MJ/kg.

For gasoline, which is a mixture of hydrocarbons, the heat of combustion (taken as negative of the higher heat value, HHV) is generally taken to be about 44 MJ/kg.

The proportion of fuel to air during combustion is very important. The mixture is stoichiometric when there is precisely enough oxygen to complete the combustion of all hydrocarbons. The mixture is lean when there is excess oxygen, and rich if there is insufficient oxygen for complete combustion. For gasoline the stoichiometric air/fuel ratio is around 14.7:1 by mass.

5.5.2   COMBUSTION OF ETHANOL

The chemical reaction describing **combustion of ethanol** in oxygen is as follows,

$$C_2H_5OH + 3O_2 \rightarrow 3H_2O(l) + 2CO_2. \tag{5.8}$$

Here we assume that H_2O is in the liquid phase after the combustion. Using Hess's law we obtain the enthalpy of combustion for one mole of the ethanol as,

$$\Delta I_c(C_2H_5OH) = 2\Delta I^f(CO_2(g)) + 3\Delta I^f(H_2O(l)) - \Delta I^f(C_2H_5OH). \tag{5.9}$$

Substituting the following enthalpies of formation: $\Delta I^f(CO_2(g)) = -394$ kJ/mol, $\Delta I^f(H_2O(l)) = -286$ kJ/mol and $\Delta I^f(C_2H_5OH) = -278$ kJ/mol, we obtain $\Delta I_c(C_2H_5OH) = -1.368$ MJ/mol. Since the molar mass of ethanol is 46 g/mol, its specific energy density (positive by convention) is $E_c = 29.7$ MJ/kg. As can be seen, ethanol has a lower energy density than both iso-octane and gasoline.

5.5.3 COMBUSTION OF COAL

Coal is burned in power stations for production of electricity and useful heat. Coal is prepared for burning by crushing the rough coal to pieces less than 5 cm in size. The coal is then transported from the storage yard to in-plant storage silos by conveyor belts. In plants that burn pulverized coal, silos feed coal to pulverizers (coal mills) that take the larger 5 cm pieces, grind them to the consistency of talcum powder, sort them, and mix them with primary combustion air which transports the coal to the boiler furnace and preheat the coal in order to drive off excess moisture content. A 500 MWe plant may have six pulverizers, five of which can supply coal to the furnace at 250 tonnes per hour under full load.

Chemical reaction describing **coal combustion** includes carbon (C), sulphur (S), hydrogen (H), oxygen (O), nitrogen (N), and moisture (H_2O). Neglecting all non-combustible components, the coal is represented as C-H and the combustion reactions are approximated with the following formulas:

$$2C - H + \frac{3}{2}O_2 \rightarrow H_2O + 2CO \text{ (incomplete)}, \tag{5.10}$$

$$2C - H + \frac{5}{2}O_2 \rightarrow H_2O + 2CO_2 \text{ (complete)}. \tag{5.11}$$

Energy content of coal is in a range E_c = 18–35 MJ/kg, though most coal used for the electricity production has energy content 20–25 MJ/kg. Table 5.1 contains the heat content for major coal ranks.

Fluidized Bed Combustion

To reduce plant emissions, **Fluidized Bed Combustion** (FBC) technology has been under development since 1980s. In this method, strong jets of air are passed upward through coal, keeping the fuel particles in a turbulent, fluid-like motion during the combustion process. Due to more efficient heat transfer, combustion occurs at lower temperature (typically 800–900 °C), which suppresses the formation of nitrogen oxides (NO_x).

Integrated Gasification Combined Cycle

In an **Integrated Gasification Combined Cycle** (IGCC) plant, a coal gasification system is integrated into a combined cycle plant. Coal gasification begins with **steam reforming**, in which high-temperature steam is passed over a bed of coke forming a mixture of hydrogen and carbon monoxide known as **water gas**:

$$C + H_2O(g) \rightarrow H_2(g) + CO. \tag{5.12}$$

The above reaction is endothermic (ΔI_r = 131 kJ/mol) and requires a temperature above 1000 K to proceed. If additional steam is available, the water gas will react exothermically to produce hydrogen and carbon dioxide, in the following water-gas shift reaction,

$$CO(g) + H_2O(g) \rightarrow H_2(g) + CO_2(g). \tag{5.13}$$

The resulting gas is thus a mixture of H_2, CO, and CO_2. The whole process is complicated and expensive, but some improvement of the cycle efficiency is obtained.

5.5.4 COMBUSTION OF HYDROGEN

The **combustion of hydrogen** with oxygen produces water as its only product,

$$2H_2 + O_2 \rightarrow 2H_2O. \tag{5.14}$$

This reaction can be realized in direct combustion process or in a fuel cell.

The importance of reaction (5.14) is that it has zero emissions, and would be a desirable replacement for combustion of fossil fuels. However, hydrogen is not as energy-dense as other fuels. In addition, while burning in air, there will be some emissions of NO_x particles. NO_x emission is not present in fuel cells, however.

PROBLEMS

PROBLEM 5.1

Estimate the total food energy needed for human population assuming a diet of 2400 kcal/person/day. Compare to the global food production rate stated in the text.

PROBLEM 5.2

The US has roughly 2.7×10^6 km^2 of arable land. In 2014, a fraction of the land was used to grow maize for ethanol at energy rate 100 GJ/ha y. This corresponded to 9% of the total gasoline energy consumption in 2014, which was 16.6 EJ/y. Find the fraction of arable land used in 2014 to grow maize for ethanol. Calculate the fraction of arable land that would be needed if the gasoline was completely replaced by ethanol.

PROBLEM 5.3

Taking rate of energy in ethanol produced from sugarcane as 180 GJ/ha y, and assuming average insolation for growing sugarcane of 250 W/m^2, calculate the corresponding efficiency of the solar energy conversion.

PROBLEM 5.4

TNT, or trinitrotoluene, is an explosive with the chemical formula $C_6H_2(NO_2)_3CH_3$. Calculate its specific energy density E_c (MJ/kg) assuming the following combustion reaction: $C_6H_2(NO_2)_3CH_3 + \frac{21}{4}O_2 \rightarrow 7CO_2 + \frac{5}{2}H_2O(g) + \frac{3}{2}N_2$. Given: the enthalpy of formation of TNT -63 kJ/mol and its molecular mass 0.277 kg/mol. Find other needed data in the text.

PROBLEM 5.5

TNT, or trinitrotoluene, is an explosive with the chemical formula $C_6H_2(NO_2)_3CH_3$. Calculate its specific energy density E_r (MJ/kg) assuming the following explosion (disintegration) reaction: $C_6H_2(NO_2)_3CH_3 \rightarrow \frac{5}{2}H_2O + \frac{7}{2}CO + \frac{7}{2}C + \frac{3}{2}N_2$. Given: the enthalpy of formation of TNT -63 kJ/mol and its molecular mass 0.277 kg/mol. Find other needed data in the text.

PROBLEM 5.6

Calculate the energy released per one carbon atom in the carbon combustion reaction $C + O_2$, if $\Delta I^f(CO_2) = -394$ kJ/mol. $Ans.E_r \approx 4.08$ eV.

PROBLEM 5.7

Calculate the energy released per one hydrogen molecule in the hydrogen combustion reaction $2H_2 + O_2 \rightarrow 2H_2O(l)$, if $\Delta I^f(H_2O(l)) = $ -286 kJ/mol. $Ans.E_r \approx 2.96$ eV.

6 Nuclear Energy

When discussing energy in nuclear reactions in §1, we already pointed out that this type of reactions provide directly or indirectly almost all energy used on Earth. The solar energy, originating from nuclear fusion reactions, can be harvested directly from thermal solar plants or photovoltaic cells. Indirectly, wind-, hydro-, fossil- and bioenergy result from solar insolation on Earth. Nuclear decay processes are responsible for the existence of the geothermal energy. Finally, nuclear fission and fusion can be created artificially and used for production of electricity.

In this chapter we discuss in more detail the various ways the nuclear energy manifests itself. We start with a description of the energy production in stars. Next we analyze the methods to produce useful energy from fission and fusion. Finally we discuss the origin of the geothermal energy due to radioactive decay.

6.1 BINDING ENERGY OF A NUCLEUS

In general terms, **binding energy** is the energy required to disassemble a whole into its separate parts. To investigate a binding energy (BE) of a nucleus, we need to compare the nucleus with its constituents, that is, Z protons and $A - Z$ neutrons. The formation of such a nucleus is described by the reaction,

$$Z \text{ protons} + (A - Z) \text{ neutrons} \rightarrow \text{nucleus} \left({}^A_Z X \right) + \text{BE}. \tag{6.1}$$

The binding energy is determined from the change of mass between the left-hand side and the right-hand side of the reaction equation in terms of the mass defect as,

$$\text{Mass Defect} = \text{BE}/c^2 = Z m_p + (A - Z) m_n - m \left({}^A_Z X \right). \tag{6.2}$$

Here $m \left({}^A_Z X \right)$ is the nuclear mass, m_p is the mass of proton and m_n is the mass of neutron. The nuclear mass in Eq. (6.2) can be replaced with the atomic mass (which is usually provided with great accuracy in tables—see Appendix C) by adding and subtracting Z electron masses from the right-hand side of the equation. Then, neglecting the binding energy of electrons, we get,

$$\text{BE} \left({}^A_Z X \right) = \left[Z M \left({}^1_1 H \right) + (A - Z) m_n - M \left({}^A_Z X \right) \right] c^2. \tag{6.3}$$

Here $M \left({}^A_Z X \right)$ is the atomic mass of element X and $M \left({}^1_1 H \right)$ is the mass of the hydrogen atom.

Example 6.1: Binding Energy of a Nucleus

Calculate the binding energy of the nucleus in ${}^4_2 \text{He}$.

Solution

Taking the atomic data provided in Appendix C we have mass defect = BE/c^2 = $2M({}^1_1\text{H})$ + $2m_n - M({}^4_2\text{He})$ = $2 \times 1.0078250 + 2 \times 1.0086649 - 4.0026032 = 0.0303766$ u. Thus $\text{BE}({}^4_2\text{He})$ = Mass Defect (u) $\times$ 931.5 (MeV/u) = 28.296 MeV.

The binding energy can be obtained from a semi-empirical mass formula (SEMF) as follows [61],

$$\text{BE} \left({}^A_Z X \right) = 15.56A - 17.23A^{2/3} - 0.7 \frac{Z^2}{A^{1/3}} - 23.28 \frac{(A - 2Z)^2}{A} + \frac{12}{A^{1/2}} \delta. \tag{6.4}$$

Here BE is expressed in MeV, A and Z are the mass number and the atomic number, respectively, and δ is equal to +1 for even $N = A - Z$ and Z, -1 for odd N and odd Z, and 0 for odd $A = N + Z$.

DOI: 10.1201/9781003036982-6

Equation (6.4) can be particularly useful when the table with atomic masses is not available and Eq. (6.3) can not be used. Figure 6.1 shows a BE/A ratio (binding energy per nucleon) versus mass number curve.

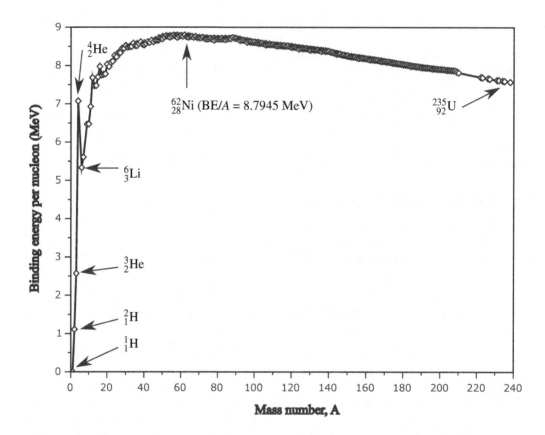

Figure 6.1 The average binding energy per nucleon BE/A of the nucleons in a nucleus versus the atomic mass number A for the naturally occurring nuclides.

The curve suggests that energy can be extracted from nuclei in two ways: if two light nuclei are joined (or fused) and by splitting a very heavy nucleus into two or more light nuclei.

As an example, consider the following reaction,

$$^2_1\text{H} + ^2_1\text{H} \rightarrow ^4_2\text{He}. \tag{6.5}$$

From Eq. (6.3) we find that BE($^2_1$H) = 2.225 MeV and BE($^4_2$He) = 28.296 MeV (see Example 6.1). Thus, the increase in binding energy of 4 nucleons involved is $28.296 - 2 \times 2.225 \cong 23.85$ MeV. This is the energy emitted when the fusion reaction takes place.

Splitting (or fission) of a heavy nucleus is another way of extracting nuclear energy. To obtain an approximate idea of how much energy is released in a fission event, we assume that $^{235}_{92}$U fissions into two nuclei with $A \simeq 117$. From Fig. 6.1 we get approximately that the binding energy per nucleon increases from 7.7 to 8.5 MeV/nucleon. Thus, the total fission energy released in the fission reaction is about $235 \times (8.5 - 7.7) \simeq 210$ MeV.

6.2 ENERGY TRANSFORMATION IN STARS

Our Sun has produced radiant energy at a rate 4×10^{26} W for the last few billion years. All this power arises from fusion reactions that occur in the core of the Sun. 173 PJ (173×10^{15} J) of energy from the Sun enters Earth's atmosphere every second. Solar radiation provides almost all energy used on Earth, except for nuclear, geothermal and tidal energy. Even fission energy requires the stars to produce heavy nuclides like uranium.

Stars are composed mostly of primordial matter (consisting of 90% hydrogen, a few percent helium and less than one percent heavier elements) created after first few minutes after the "big bang". Stars are "burning" hydrogen into helium, which later is converted into heavier elements, up to $A \simeq 56$, after which fusion reactions are not exothermic.

During early stage of a life of a star, hydrogen is fused to produce helium,

$$^1_1 H + ^1_1 H \rightarrow ^2_1 D + \beta^+ + \nu, \ Q = 0.42 \text{ MeV}. \tag{6.6}$$

This proton-proton reaction involves the weak force instead of the nuclear force and, consequently, it occurs very slowly, as measured in the nuclear time-scale. Actually reaction (6.6) has so low probability that it has never been observed on Earth. The reaction rate is thus about $R_{pp} \simeq 10^8 \text{ cm}^{-3}\text{s}^{-1}$ and the average lifetime of the proton in the core of the Sun is $\Delta t_p = N_p / R_{pp} \simeq 3 \times 10^{10}$ y. Here $N_p \simeq 10^{26} \text{ cm}^{-3}$ is the average hydrogen nuclide density in the core of the Sun. This slow p-p reaction controls the rate at which a new star consumes its hydrogen fuel and is responsible for long life of stars.

The deuterium product obtained in reaction (6.6) very rapidly absorbs a proton through the reaction

$$^2_1 D + ^1_1 H \rightarrow ^3_2 He + \gamma, \ Q = 5.49 \text{ MeV}, \tag{6.7}$$

which is followed by

$$^3_2 He + ^3_2 He \rightarrow ^4_2 He + 2^1_1 H + \gamma, \ Q = 12.86 \text{ MeV}. \tag{6.8}$$

The net result of reactions (6.6) through (6.8) is the transformation of four protons in $^4_2$He, as illustrated in Fig. 6.2 and summarized by the following reaction,

$$4\left(^1_1 H\right) \rightarrow ^4_2 He + 2\gamma + 2\beta^+ + 2\nu, \ Q = 26.72 \text{ MeV}. \tag{6.9}$$

Once the hydrogen in a mature star's core is consumed, the burning of helium begins and the star becomes a so called red giant. In this phase of a star's life the core shrinks, the temperature increases and gravitational energy is converted into thermal energy. When the core temperature is high enough, the increased thermal motions of He overcome the repulsive Columbian forces and helium fusion begins. The effective path to heavier nuclei begins with the ternary reaction

$$3\left(^4_2 He\right) \rightarrow ^{12}_6 C, \ Q = 7.27 \text{ MeV}. \tag{6.10}$$

After formation of $^{12}_6$C, heavier elements can be produced through fusion reactions with $^4_2$He,

$$^4_2 He + ^{12}_6 C \rightarrow ^{16}_8 O + \gamma, \ Q = 7.16 \text{ MeV}, \tag{6.11}$$

$$^4_2 He + ^{16}_8 O \rightarrow ^{20}_{10} Ne + \gamma, \ Q = 4.73 \text{ MeV}, \tag{6.12}$$

$$^4_2 He + ^{20}_{10} Ne \rightarrow ^{24}_{12} Mg + \gamma, \ Q = 9.31 \text{ MeV}, \tag{6.13}$$

As a star ages it generates heavier nuclei, and it shrinks to create ever increasing core temperatures to sustain fusion of the heavier nuclei. With an iron core, the star reaches the end of fusion chain and becomes unstable. A less massive star, like our Sun, slowly shrinks into white dwarf. A more massive star becomes a supernova due to implosion. What often remains of a star after this collapse is a small ($\simeq 40$ km in diameter) neutron star. However, if mass of the star is sufficiently great, a black hole is formed.

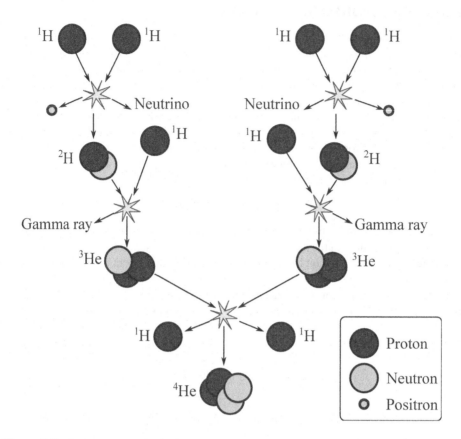

Figure 6.2 Fusion reactions in the Sun.

6.3 CHARACTERISTICS OF THE NUCLEAR FISSION

Some introductory information on nuclear fission reactions has been provided in §1.3.4. In this section we discuss the fission products, neutron emission in fission and energy release in fission.

6.3.1 FISSION PRODUCTS

Any very heavy nucleus can be made to fission if hit sufficiently hard by some incident particle. However, only a few nuclides can fission by the absorption of a neutron with a negligible kinetic energy. Nuclides such as ^{235}U, ^{233}U and ^{239}Pu belong to this category and are called **fissile** nuclei. Nuclides that fission only when struck with a high-energy neutron (with a few MeV kinetic energy) are called **fissionable** nuclei. Both ^{238}U and ^{240}Pu belong to this category.

A typical outcome of a neutron-induced fission reaction of ^{235}U is shown in Fig. 6.3. The reaction shown in the figure can be written as $^{1}_{0}\text{n} + ^{235}_{92}\text{U} \rightarrow ^{92}_{36}\text{Kr} + ^{141}_{56}\text{Ba} + 3(^{1}_{0}\text{n}) + 6\gamma$, where $^{92}_{36}\text{Kr}$ and $^{141}_{56}\text{Ba}$ are the initial fission fragments, $3(^{1}_{0}\text{n})$ are three neutrons and 6γ are six gamma photons. The equation represents the situation after about 10^{-12} s following the nucleus scission. When fission fragments are stopped and become electrically neutral, they are termed fission products. The neutrons and gamma photons appearing on the right-hand side of the equations are called **prompt neutrons** and **prompt gamma photons**, respectively. As we can see, the number of neutrons and protons is conserved in the reaction. However, at this stage fission products are radioactively unstable since they still have too many neutrons compared to protons. Thus they undergo β^- decay until a stable end-nuclide is reached. The half-lives of the fission product daughters range from fraction of a second to many thousands of years.

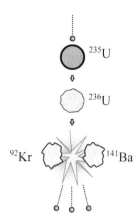

Figure 6.3 Fission reaction of $^{235}_{92}$U.

Fission products are decaying following many decay chains of variable lengths. The decay chain of historical importance is as follows,

$$^{140}_{54}\text{Xe} \xrightarrow[16\text{ s}]{\beta^-} {}^{140}_{55}\text{Cs} \xrightarrow[66\text{ s}]{\beta^-} {}^{140}_{56}\text{Ba} \xrightarrow[12.8\text{ d}]{\beta^-} {}^{140}_{57}\text{La} \xrightarrow[40\text{ h}]{\beta^-} {}^{140}_{58}\text{Ce} \quad \text{(stable).} \tag{6.14}$$

It is through this chain that the uranium fission was discovered by Hahn, Strassman, and Meitner, since they observed elements La and Ba in initially pure uranium sample bombarded by neutrons. Another example of an important decay chain for reactor operations is as follows,

$$^{135}_{52}\text{Te} \xrightarrow[19\text{ s}]{\beta^-} {}^{135}_{53}\text{I} \xrightarrow[6.57\text{ h}]{\beta^-} {}^{135}_{54}\text{Xe} \xrightarrow[9.10\text{ h}]{\beta^-} {}^{135}_{55}\text{Cs} \xrightarrow[2.3\times10^6\text{y}]{\beta^-} {}^{135}_{56}\text{Ba} \quad \text{(stable).} \tag{6.15}$$

This decay chain contains the nuclide $^{135}_{54}\text{Xe}$, which is a strong absorber of slow neutrons. A very little build-up of this nuclide in a reactor core can absorbed enough neutrons to stop the self-sustaining chain reaction.

In general, fission products can have atomic mass numbers between 70 and 170. However, not all fission products are equally likely. The probability that a fission fragment is a nuclide with mass number A is called the **fission chain yield** and is denoted by $y(A)$. Fission product yield for the thermal fission of $^{235}_{92}\text{U}$ is shown in Fig. 6.4. Similar curves are obtained for other fissile nuclides. We can notice that there is more likely to obtain asymmetric splitting, with mass numbers of 90–100 and 135–145 than a symmetric splitting with $A \simeq 118$.

6.3.2 NEUTRON EMISSION

The neutrons emitted by a fission reaction are of great importance for the practical generation of nuclear power. Neutrons are needed to cause other fission reactions, and if properly arranged, a sustained chain reaction can be established, as illustrated in Fig. 6.5.

Almost all the neutrons produced from a fission event are emitted within 10^{-14} s of the fission event. The number v_p of these prompt neutrons varies between 0 and 8, and its average $\overline{v_p}$ is typically about 2.5, but its precise value depends on the type of the fissile nuclide and the energy of the incident neutron.

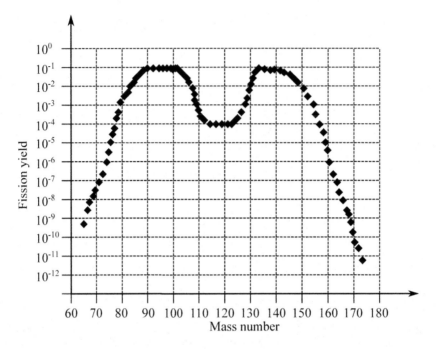

Figure 6.4 Fission product yield for the thermal fission of $^{235}_{92}$U.

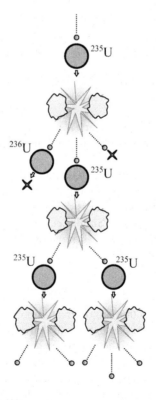

Figure 6.5 Fission chain reaction of ^{235}U nuclei.

Table 6.1

Total Fission Neutrons per Fission $\overline{v}$ and the Delayed Neutron Fraction $\beta \equiv \overline{v_d}/\overline{v}$

Nuclide	Fast Fission		Thermal Fission	
	$\overline{v}$	β	$\overline{v}$	β
^{233}U	2.62	0.0026	2.48	0.0026
^{235}U	2.57	0.0064	2.43	0.0067
^{239}Pu	3.09	0.0020	2.87	0.0022
^{241}Pu	-	-	3.14	0.0054
^{238}U	2.79	0.0164	-	-
^{232}Th	2.44	0.0203	-	-
^{240}Pu	3.3	0.0029	-	-

Source: [82]

A small fraction of fission neutrons (always less than 1% for a thermal fission) is emitted as delayed neutrons, and is produced as a neutron decay of a fission product daughter after β^- decay, any time from a few milliseconds to a few minutes after fission. The average number $\overline{v_d}$ of delayed neutrons emitted per fission depends on the fissioning nucleus and the energy of the neutron causing the fission.

The total average number of neutrons emitted after fission is thus $\overline{v} = \overline{v_p} + \overline{v_d}$, and the **delayed neutron fraction** is defined as $\beta \equiv \overline{v_d}/\overline{v}$.

The necessary condition to sustain a chain reaction is that, on average, at least one neutron, either prompt or delayed, will cause another fission in the chain reaction.

The average total number of neutrons (both prompt and delayed) released in a fission, $\overline{v}$, varies both with the energy of the neutron causing fission and with the type of the fissile nuclide. For $^{235}_{92}$U, this number is given as [26],

$$\overline{v}(E) = \begin{cases} 2.432 + 0.066E & 0 \leq E \leq 1 \\ 2.348 + 0.150E & E > 1 \end{cases}, \tag{6.16}$$

where E is the neutron energy in MeV. The values of $\overline{v}$ and β for several important nuclides are given in Table 6.1.

6.3.3 ENERGY RELEASED IN FISSION REACTIONS

The magnitude of energy ultimately released per fission is about 200 MeV. The amount of energy released per fission can be estimated by considering the binding energy per nucleon. However, since huge number of fissions occur in a nuclear reactor, we are interested in an average energy release per fission, averaged per all possible fission outcomes. The average energy release in the thermal fission of ^{235}U is shown in Table 6.2.

Table 6.2
The Average Energy Release in the Thermal Fission of ^{235}U

From	Energy from Fission (MeV)	Recoverable in Core (MeV)
Prompt:		
kinetic energy of the fission fragments	168	168
kinetic energy of prompt fission neutrons	5	5
fission γ-rays	7	7
γ-rays from neutron capture	-	3–9
Delayed:		
fission product β-decay energy	8	8
fission product γ-decay energy	7	7
neutrino kinetic energy	12	0
Total energy	207	198–204

Source: [26]

6.4 NUCLEAR FUSION

Basic information about possible fusion reactions was provided in §1.3. As already mentioned, all current experimental research into fusion power is based on the following D-T fusion reaction,

$$\ _{1}^{2}\mathrm{H} + _{1}^{3}\mathrm{H} \rightarrow _{2}^{4}\mathrm{He}(3.54\ \mathrm{MeV}) + _{0}^{1}\mathrm{n}(14.05\ \mathrm{MeV}). \tag{6.17}$$

The reaction is illustrated in Fig. 6.6. Tritium ($_{1}^{3}$H, T), needed for this reaction, is an unstable

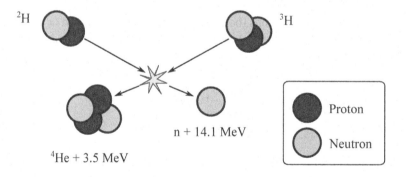

Figure 6.6 Deuterium-tritium fusion reaction.

hydrogen isotope. This radioactive nuclide ($T_{1/2} = 12.3$ y) can be generated by having the fusion neutron interact with lithium through the following reactions,

$$\ _{0}^{1}\mathrm{n} + _{3}^{6}\mathrm{Li} \rightarrow _{2}^{4}\mathrm{He} + _{1}^{3}\mathrm{H},\ Q = 4.78\ \mathrm{MeV}, \tag{6.18}$$

$$\ _{0}^{1}\mathrm{n} + _{3}^{7}\mathrm{Li} \rightarrow _{2}^{4}\mathrm{He} + _{1}^{3}\mathrm{H} + _{0}^{1}\mathrm{n}. \tag{6.19}$$

The first of the two reactions is exothermic and occurs for neutrons of any energy. However, the second reaction occurs only for high-energy neutrons and it is endothermic, consuming 2.466 MeV.

Deuterium ($_1^2$H, D), also known as heavy hydrogen, is one of two stable hydrogen isotopes. Deuterium has a natural abundance in Earth's oceans of about one atom in 6420 hydrogen atoms. Thus deuterium accounts for approximately 0.02% (or, on a mass basis, 0.03%) of all the naturally occurring hydrogen in the oceans.

6.5 RADIOACTIVE DECAY

A radioactive decay is a nuclear interaction that occurs spontaneously, and in which the nucleus of the parent atom $_Z^A$P is altered in some manner and one or more particles are emitted. Table 6.3 shows a summary of various types of radioactive decay.

Table 6.3
Summary of Various Types of Radioactive Decay

Decay Type	Reaction	Description
gamma (γ)	$_Z^A\text{P}^* \rightarrow {_Z^A}\text{P} + \gamma$	An excited nucleus decays to its ground state by the emission of a gamma photon.
alpha (α)	$_Z^A\text{P} \rightarrow {_{Z-2}^{A-4}}\text{D} + \alpha$	An α particle is emitted leaving the daughter with 2 fewer neutron and 2 fewer protons.
negatron (β^-)	$_Z^A\text{P} \rightarrow {_{Z+1}^A}\text{D} + \beta^- + \overline{\nu}$	A neutron in the nucleus changes into a proton. An electron (β^-) and an anti-neutrino ($\overline{\nu}$) are emitted.
positron (β^+)	$_Z^A\text{P} \rightarrow {_{Z-1}^A}\text{D} + \beta^+ + \nu$	A proton in the nucleus changes into a neutron. A positron (β^+) and a neutrino (ν) are emitted.

In all nuclear interactions, including radioactive decay, the following conservation laws are followed:

- Conservation of charge, in which the number of elementary positive and negative charges must be the same for reactants and reaction products.
- Conservation of the number of nucleons, i.e., A is always constant.
- Conservation of mass/energy (total energy).
- Conservation of linear momentum.
- Conservation of angular momentum.

Of the many radionuclide species present when the solar system was formed almost 5 billion years ago, still 17 very-long lived radionuclides exist on Earth. One of the most important is $_{19}^{40}$K, which has an isotopic abundance of 0.0118% (an element abundance of potassium is 0.0117%) and a half-life of 1.28×10^9 y.

Each naturally occurring radioactive nuclide with $Z > 83$ is a member of one of three long decay chains stretching through the upper part of the chart of nuclides. These radionuclides decay by α or β^- emission and they have the property that the number of nucleons A for each member of a given decay series can be expressed as $4n + i$, where n is an integer and i is a constant (0,2, or 3) for each series. The three series are named the thorium ($4n$), uranium ($4n + 2$), and actinium ($4n + 3$) series. There is no naturally occurring series represented by $4n + 1$, but it was recreated after ^{241}Pu was made in nuclear reactors.

The naturally occurring radioactive nuclides are primary sources of the geothermal energy. Figure 6.7 shows the $_{92}^{238}$U ($4n + 2$) natural decay series, containing 8 α decays (depicted by arrows pointing downward and to the left) and 6 β^- decays (depicted by arrows pointing downward and to the right).

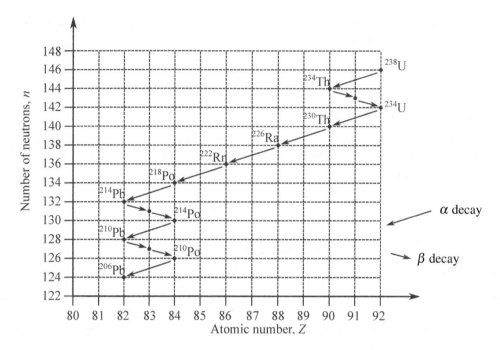

Figure 6.7   The $^{238}_{92}$U $(4n+2)$ natural decay series.

Example 6.2:

What is the energy released by the $^{238}_{92}$U $(4n+2)$ natural decay chain?

Solution

The chain contains 8 α decays and 6 β^- decays. Thus it can be represented in short as $^{238}_{92}$U$\rightarrow^{206}_{82}$Pb$+8(^4_2$He$)+6(^{\ 0}_{-1}$e$)+6(\overline{\nu})$. Although six β^- particles are emitted, six ambient electrons are absorbed by the decaying nuclides, which should appear on the left-hand side of the reaction equation. In addition we replaced α particles by He, adding in this way $8\times2 = 16$ electrons. These electrons should be subtracted from the right-hand side. However, in our estimation we neglect electron and neutrino masses. In this way we obtain the following released energy estimation $E_r = [M(^{238}_{92}$U$) - M(^{206}_{82}$Pb$) - 8\times M(^4_2$He$)]c^2 = [238.0507826 - 205.974449 - 8\times 4.002603]u\times$ 931.5 MeV/u $\simeq$ 51.7 MeV.

PROBLEMS

PROBLEM 6.1

Calculate the binding energy of the last neutron in $^{16}_8$O. Hint: consider the energy release in the following reaction $^{15}_8$O $+ ^1_0$n$\rightarrow^{16}_8$O.

PROBLEM 6.2

Determine the binding energy per nucleon (MeV/nucleon) for the nuclides: (a) $^{15}_8$O, (b) $^{16}_8$O, (c) $^{17}_8$O, (d) $^{56}_{26}$Fe, (e) $^{235}_{92}$U.

PROBLEM 6.3

Calculate the net energy release (in MeV) for each of the following fusion reactions: (a) $_1^2H + _1^2H \rightarrow _2^3He + _0^1n$, (b) $_1^2H + _1^3H \rightarrow _2^4He + _0^1n$.

PROBLEM 6.4

Estimate the available D-D fusion energy in a 2.5 dl glass of water. For how long would this energy cover the energy needs of a house with an average power consumption of 7.5 kW?

PROBLEM 6.5

What is the total energy emitted (both prompt and delayed) from the following fission reaction: $_0^1n + _{92}^{235}U \rightarrow _{36}^{90}Kr + _{56}^{142}Ba + 4(_0^1n) + 6\gamma$, where $_{36}^{90}Kr$ and $_{56}^{142}Ba$ are the initial fission fragments.

PROBLEM 6.6

Calculate the Q-values (in MeV) for each of the following beta-decay reactions: (a) $_{11}^{22}Na \rightarrow _{10}^{22}Ne + _{+1}^0e + \nu$, (b) $_{17}^{38}Cl \rightarrow _{18}^{38}Ar + _{+1}^0e + \bar{\nu}$.

PROBLEM 6.7

Estimate the energy released by: (a) the $_{92}^{235}U$ ($4n + 3$) natural decay chain, (b) the $_{90}^{232}Th$ ($4n$) natural decay chain.

PROBLEM 6.8

Estimate the energy released by the decay of the initial $_{54}^{139}Xe$ and $_{38}^{95}Sr$ fragments as they decay to their stable end points $_{57}^{139}La$ and $_{42}^{95}Sr$, respectively. Neglect anti-neutrino and electron masses, and the electron binding energy.

7 Thermal Energy

Thermal energy has been used through all of human history, and it still plays a central role in the global energy supply. Over 90% of the world's energy relies to a certain extend on thermal energy, including nuclear, bio- and fossil-fuelled power, and with such major exceptions as hydro-, wind- and solar photovoltaic power.

The main theme in this chapter is a transformation of the thermal energy into useful mechanical and electrical energy. This topic belongs to one of the most fundamental and subtle subjects of this book, namely *thermodynamics*. In particular, we learn from the thermodynamics that there is a physical limit on the efficiency with which thermal energy can be transformed into mechanical energy.

7.1 INTRODUCTORY DEFINITIONS

In this section we introduce the most basic concepts used in thermodynamics. Some important formulations and their practical implications are discussed in the following sections of the chapter.

A thermodynamic system is a content (both material and radiative) of a macroscopic volume that can be uniquely described by thermodynamic state variables, such as pressure, temperature, entropy and internal energy.

A typical **macroscopic volume** or **macroscopic system** contains a large number ($\mathcal{N} \sim 10^{23}$) of particles or molecules. **Thermal energy** of the system, $\mathcal{U}$, refers to collective energy contained in the relative motion of the particles against the center of mass of the system. It also contains energy needed for phase change of matter contained in the system from, e.g., solid to liquid or liquid to gas.

Pressure, p, is force per unit area directed perpendicular (or normal) to a surface. The momentum carried by moving molecules give rise to pressure on a macroscopic level once they collide with walls surrounding the system. Similarly we define *shear stress* as force per unit area directed parallel (or tangential) to the surface.

Temperature, T, is a relative measure of the amount of the thermal energy in a system. When temperature approaches *absolute zero*, the thermal energy of the system goes to zero.

Internal energy of a system, E_I, includes its thermal energy, together with binding energy of molecules, electrons in atoms and nucleons in atomic nuclei. Note that the internal energy and the thermal energy of a system only differ by an additive amount of energy, which is equal to the internal energy at absolute zero. In all thermodynamic considerations that we will study in this book, this additive amount of energy is considered to be constant. Thus, in thermodynamic processes, where only internal energy changes are of interest (rather than its absolute value), we will treat the internal energy and the thermal energy as interchangeable quantities.

Heat is thermal energy that is transferred from one system to another one. Experience shows that heat always flows from regions or systems at higher temperature to regions or systems at lower temperature.

7.1.1 THERMODYNAMIC CONTROL SYSTEMS

There are three useful types of thermodynamic control systems. The first one is an **isolated thermodynamic system**, which does not exchange any mass or energy with the surroundings. A perfectly insulated tank containing a fluid can be considered as an isolated thermodynamic system.

The second one, called a **control volume system**, has boundaries through which mass and energy can flow both into the system and out of the system. Thus the mass and energy in the control volume can change when the net mass and energy flow into the system is different from the net mass and

DOI: 10.1201/9781003036982-7

energy flow out of the system. Usually control volume has fixed shape and boundaries, but, in general, it can change the shape and its boundary can move. In thermodynamics, such system is also called **open thermodynamic system**. The surface enclosing the control volume is referred to as a **control surface**. A nuclear reactor core can be considered as a control volume system.

The third type of thermodynamic system is called a **control mass system**, in which energy freely flows through the boundaries, but mass does not cross the boundaries. Hence, the mass in the control mass system is constant, but not its energy. Similarly as for the control volume system, the shape and boundaries of the system can change. The volume of the control mass system is called a **material volume** and its boundary—a **material boundary**. In thermodynamics the control mass system is often referred to as a closed system. A closed, uninsulated tank with internal mixing blades can be considered as a control mass system.

7.1.2 STATE PARAMETERS

A state of any thermodynamic system can be uniquely determined by specific values of **state parameters** (also called *state functions* or *state variables*) of that system. State parameters are such macroscopic physical quantities of the system, which can be determined from measurements or analyses, and which do not depend on the history of the system. For example, work and heat depend on a system history and thus they are not state parameters. Temperature, pressure, volume, internal energy, enthalpy, and entropy are examples of state parameters.

We distinguish intensive and extensive state parameters. **Intensive parameters** do not depend on the size of the system and include such state parameters as temperature, pressure, and mass density. **Extensive parameters** (also called *additive parameters*) depend on the system size and include, e.g., energy, volume, and entropy.

Not all state parameters are independent of each other and there is always a limited number of state parameters that uniquely determine a system.

7.1.3 THERMODYNAMIC EQUILIBRIUM

Two systems are said to be in a **thermodynamic equilibrium** with each other when they are in mechanical, chemical, and thermal equilibrium with each other. At the thermodynamic equilibrium state, the properties of the system do not change with time and there is no flow of energy or material at macroscopic scales. Such condition is sometime referred to as the steady-state condition.

7.1.4 THERMODYNAMIC DIAGRAMS

Since in many situations the state of a thermodynamic system can be uniquely described by two state variables, this state can be represented as a single point on a plane, using the two state variables as coordinates. A continuous change of a system state from one point to another one on the plain will be represented by a curve connecting the two points.

Various possible pairs of state variables can be chosen as the coordinates. In power engineering applications where phase-change of the working fluid takes place, the most useful choices include the temperature-specific entropy (Ts) coordinates and the specific enthalpy-specific entropy (is) coordinates. For systems without phase change, pressure-volume (pV) system of coordinates can be used. We use thermodynamic diagrams in §7.4 and §7.5 to present thermodynamic processes and thermodynamic cycles.

7.1.5 THERMODYNAMIC PROCESSES

A **thermodynamic process** is a sequence of thermodynamic states that can be represented on a thermodynamic diagram with a curve connecting two points. The points represent the starting and the ending conditions of the thermodynamic process. We should note that any thermodynamic process is uniquely described only when the staring point, the ending point, and the path are known.

In general, for any pair of starting and ending points of a thermodynamic process, there exists an infinite number of possible paths connecting the two points. In practice it means that, to determine the process, it has to be additionally described by stating that, e.g., the process is isothermal, which means that the temperature in the system remains constant during the process.

The following processes are typically occurring in thermodynamic systems,

- isothermal, when the temperature is constant,
- isochoric, when the volume is constant,
- isobaric, when the pressure is constant,
- isentropic, when the entropy is constant.

In addition, we will also encounter adiabatic processes, when no heat transfer takes place.

All these processes are idealized and are difficult and even impossible to realize in practice. For example, isentropic process corresponds to a such ideal process during which no change of entropy takes place. Such process would be then *reversible*. However, all real process exhibit some increase of entropy due to, for example, presence of friction and thus are *irreversible*.

7.1.6 THERMODYNAMIC CYCLES

A *thermodynamic cycle* is such a thermodynamic process that starts and ends at the same point and that can be represented on a thermodynamic diagram with a closed curve. Thermodynamic cycles play important role in the theory of heat engines.

7.2 THE LAWS OF THERMODYNAMICS

There are four laws of thermodynamics. The numbering starts from zero, since the zero-th law was established long after the three others were in a wide use. The laws describe the way in which thermodynamic systems behave and they constitute fundamentals on which the thermodynamics is build as a field of science. The zero-th law formulates the conditions and consequences of a thermodynamic equilibrium. The first law of thermodynamics is concerned with the conservation of energy in an isolated system. The most challenging is the second law of thermodynamics stating that any isolated system will tend to a state in which the disorder will be maximized. Finally, the main consequence of the third law, known also as Nernst's theorem, is that the absolute zero temperature cannot be attained.

7.2.1 ZEROTH LAW OF THERMODYNAMICS

The **zero-th law of thermodynamics** states that if two thermodynamic systems are in thermal equilibrium with a third one, they are also in thermal equilibrium with each other. With this formulation the zero-th law can be viewed as an equivalence relation saying that systems in equilibrium have the same temperature. If two systems with different temperatures are put into a direct thermal contact, heat flows from the system with higher temperature to the system with lower temperature.

7.2.2 FIRST LAW OF THERMODYNAMICS

The **first law of thermodynamics** states that the increase in the energy of a system is equal to the amount of energy added by heating the system, minus the amount lost as a result of work performed by the system on its surroundings. We follow here a convention adopted by engineers that work performed by a system is positive, whereas work performed on a system is negative. This sign convention is opposite to that used in natural sciences.

The first law of thermodynamics is formulated as the energy conservation law. For systems that can only exchange heat and work with the surroundings (so called *closed systems*), it is expressed as,

$$dE_T = đQ - đL,\tag{7.1}$$

here $đQ$ is the differential heat added to the system, $đL$ is the differential work done by the system on surroundings and dE_T is the differential change of the total energy of the system. In particular, when there is no interaction of the system with the surroundings, the energy change dE_T is equal to zero. Thus, according to the first law of thermodynamics, the total energy of an isolated system is conserved. The symbol $đ$ in Eq. (7.1) represents a so-called **inexact differential**. Unlike the ordinary differential d, the inexact differential $đ$ cannot be integrated if the path of the process is unknown.

Energy balance is performed for a thermodynamic system within a certain control volume. The same control volume should be used for formulation of the mass balance, since the mass balance is a foundation of the energy balance. The control volume can have gates through which mass can be transferred from the surroundings to the system or from the system to the surroundings. Such system is called the open system. If the gates do not exist, the system is closed.

The total energy of the system can be defined as a sum of the kinetic energy E_K, potential energy E_P and internal energy E_I,

$$E_T = E_I + E_K + E_P.\tag{7.2}$$

Whereas the kinetic and the potential energies of the system are related to the macroscopic effects, the internal energy is entirely resulting from microscopic effects, such as,

- translational and rotational kinetic energy of molecules,
- oscillatory energy of atoms in molecules,
- potential energy in the field of attraction forces between molecules,
- chemical energy related to a possibility of reconstruction of molecules,
- nuclear energy.

In general, absolute value of the internal energy is very difficult to calculate. However, application of the first law of thermodynamics does not require such calculations. Instead, in thermodynamic analyses it is enough to find the internal energy difference between two states.

Sometimes a closed system is perfectly insulated and the total heat exchanged with the ambient is equal to zero. Such systems are called **adiabatic systems** and the first law of thermodynamics for such systems, on which work L has been done, has the following form,

$$\Delta E_T = -L,\tag{7.3}$$

where ΔE_T is the change of the system total energy.

The work L performed on the system (either adiabatic or diabatic one) can be divided into three parts, corresponding to the three components of the total energy of the system, as follows,

$$L = L_I + L_K + L_P,\tag{7.4}$$

where L_K and L_P represent the work that changes the system's kinetic and potential energy as a whole, respectively. In most cases of interest, however, the systems remain stationary and these work terms are equal to zero. Equations (7.2) through (7.4) suggest then that the work performed on a system influences its internal energy only.

Another classification of the system work is possible by considering the way in which it is performed. In turbines and pumps the work is transmitted to and extracted from the system by a shaft. We can call this type of work as the shaft work, L_{shaft}. Similarly we can define work transmitted through the control volume surface by normal and tangential (shear) forces and define L_{normal} and L_{shear}, respectively. Thus, the total work is,

$$L = L_{shaft} + L_{normal} + L_{shear}.\tag{7.5}$$

Based on the foregoing discussion, the first law of thermodynamics for a closed diabatic system under an action of work L can be written as,

$$\Delta E_T = Q - L_{shaft} - L_{normal} - L_{shear}. \tag{7.6}$$

It should be remembered, however, that the above equation can be expressed in terms of the internal energy only when applied to stationary systems, in which kinetic and potential energies do not change.

Dividing all terms in Eq. (7.2) by the total mass m contained in the system gives an expression for the specific total energy as a sum of the specific internal, kinetic, and potential energies,

$$e_T = e_I + e_K + e_P, \tag{7.7}$$

where $e_T = E_T/m$, $e_I = E_I/m$, $e_K = E_K/m$ and $e_P = E_P/m$. Similarly, Eq. (7.6) can be expressed per unit mass of the system as,

$$\Delta e_T = -l_{shaft} - l_{normal} - l_{shear}, \tag{7.8}$$

where an adiabatic system is assumed ($Q = 0$) and $l_{shaft} = L_{shaft}/m$, $l_{normal} = L_{normal}/m$ and $l_{shear} = L_{shear}/m$ represent shaft, surface normal and surface shear specific work, respectively.

A natural extension of the closed thermodynamic systems is an open system, which is exchanging mass with the surroundings. Let us assume that a substance with mass dm and total specific energy $e_T = e_I + e_K + e_P$ enters the system with pressure p. Since mass dm of a certain substance with density ρ has a volume $dV = dm/\rho$, this process requires an additional work to be performed against the system pressure equal to $dL_m = -pdV = -p/\rho dm$. Thus, the total energy change of the system is,

$$dE_T = e_T dm - (-p/\rho dm) = (e_T + p/\rho)dm = (e_I + p/\rho + e_K + e_P)dm. \tag{7.9}$$

The quantity

$$i = e_I + p/\rho \tag{7.10}$$

is called the **specific enthalpy** of the substance. Assuming several ports through which substance with mass m_j is flowing in and other ports, through which substance with mass m_k is flowing out, and assuming also exchange of heat and work with the surroundings, the change of the total system energy is obtained as,

$$\Delta E_T = Q - L_{shaft} - L_{normal} - L_{shear} +$$
$$\sum_{j \in in} (i + e_P + e_K)_j m_j - \sum_{k \in out} (i + e_P + e_K)_k m_k. \tag{7.11}$$

Many technical systems operate continuously and Eq. (7.11) should be expressed in the time rates rather than in the absolute values. Dividing the equation with Δt and taking $\Delta t \to 0$ yields,

$$\frac{dE_T}{dt} = q - N_{shaft} - N_{normal} - N_{shear} +$$
$$\sum_{j \in in} (i + e_P + e_K)_j W_j - \sum_{k \in out} (i + e_P + e_K)_k W_k, \tag{7.12}$$

where $q = \dot{Q}$ is the time rate of heat exchange (the thermal power exchange), $N = \dot{L}$—the work done per unit time (the mechanical power) and $W = \dot{m}$—the mass flow rate.

The corresponding mass conservation equation is as follows,

$$\frac{dm}{dt} = \sum_{j \in in} W_j - \sum_{k \in out} W_k, \tag{7.13}$$

where both conservation equations are defined for the same control volume.

Example 7.1:

A steam turbine operates under steady state conditions and delivers shaft power N = 423 MW with steam mass flow rate W = 500 kg/s. The inlet specific enthalpy and mean velocity of the steam are i_{in} = 2770 kJ/kg and U_{in} = 90 m/s, respectively. The corresponding steam parameters at the outlet from the turbine are i_{out} = 1900 kJ/kg and U_{out} = 230 m/s. Neglect potential energy of the steam streams. Calculate the turbine heat losses to the surroundings.

Solution

For steady-state system with only one inlet and one outlet, with no potential energy difference and producing only shaft work, Eq. (7.12) has the following form,

$$q = N_{shaft} - (i + e_K)_{in} W + (i + e_K)_{out} W.$$

Substituting the data, yields,

$$q = 4.23 \times 10^8 - 500 \times \left(2.77 \times 10^6 + \frac{90^2}{2} - 1.9 \times 10^6 - \frac{230^2}{2}\right) = -8 \times 10^5 \text{ W.}$$

Thus, the thermal loss from the turbine is 800 kW.

Since nuclear energy, which is a part of the internal energy, is the main source of the thermal energy generated in nuclear reactors, the internal energy in such systems can be partitioned into the nuclear, E_{Inuc}, and non-nuclear, E_{Innuc}, parts,

$$E_I = E_{Inuc} + E_{Innuc}. \tag{7.14}$$

The nuclear part of the internal energy is changing due to fission and its time rate of change is given as,

$$\frac{dE_{Inuc}}{dt} = \frac{dm_{fuel}}{dt}c^2 = -q_f, \tag{7.15}$$

where c is the speed of light and q_f is the fission power of the reactor. The minus sign results from the convention that the fission power is assumed to be positive when the internal energy decreases due to fission.

In a similar manner, the total mass in a reactor can be divided into fuel, m_{fuel}, construction material, m_{constr}, and coolant, $m_{coolant}$, parts,

$$m = m_{fuel} + m_{constr} + m_{coolant}. \tag{7.16}$$

The mass and energy conservation equation for nuclear reactors can now be written as,

$$\frac{dm_{coolant}}{dt} = \sum_{j \in in} W_j - \sum_{k \in out} W_k, \tag{7.17}$$

$$-q_f + \frac{dE_{Innuc}}{dt} + \frac{dE_K}{dt} + \frac{dE_P}{dt} = q - N_{shaft} - N_{normal} - N_{shear} + \\ \sum_{j \in in} (i + e_P + e_K)_j W_j - \sum_{k \in out} (i + e_P + e_K)_k W_k. \tag{7.18}$$

For a critical stationary reactor, when only the nuclear internal energy is changing with time and all other time derivatives are zero, and only the shaft mechanical power can be present, the equations become,

$$\sum_{j \in in} W_j - \sum_{k \in out} W_k = 0, \tag{7.19}$$

$$-q_f = q - N_{shaft} + \sum_{j\in in} (i + e_P + e_K)_j W_j - \sum_{k\in out} (i + e_P + e_K)_k W_k. \qquad (7.20)$$

Here q represents all energy losses due to heat convection and radiation through the boundaries of the reactor, N_{shaft} is the mechanical power provided to the main circulation pumps and q_f is the total fission power. An explanation of the sign convention in Eq. (7.20) is proper in this place. Since q represents energy losses only, its sign is always negative. If the shaft work is present, it represents the pumping power supplied to the reactor system and according to our convention, the sign of this power is also negative.

Example 7.2:

A BWR reactor operates at steady-state conditions and produces 850 kg/s saturated steam at pressure 7 MPa. Calculate the fission power of the reactor assuming that at the inlet to the reactor the feedwater is at 7.1 MPa pressure and has 10 K subcooling. Assume also that the total heat losses are equal to 5% of the fission power and that the pumping power is 5 MW. Neglect the potential and kinetic energy of the streams at the reactor inlets and outlets. What is the time rate of the fuel mass change due to fissions?

Solution

From water property tables we find the specific enthalpies at the inlet and outlet as i_{in} = 1219.7 kJ/kg and i_{out} = 2772.6 kJ/kg. Thus from Eq. (7.20) we have

$$-q_f = -0.05 q_f - (-5 \cdot 10^6) + 1219.7 \cdot 10^3 \times 850 - 2772.6 \cdot 10^3 \times 850.$$

Thus, the fission power is found as

$$q_f = -\left[5 \cdot 10^6 + (1219.7 - 2772.6) \cdot 10^3 \times 850\right]/0.95 = 1384.2 \text{ MW}.$$

Thus, the fission power of the reactor is found as 1384.2 MW. From Eq. (7.15) we can calculate the fuel mass time rate change as,

$$\frac{dm_{fuel}}{dt} = -q_f/c^2 = -1.54 \cdot 10^{-8} \text{ kg/s}.$$

This value of mass rate change corresponds to fuel mass reduction with 0.4857 kg/year.

7.2.3 SECOND LAW OF THERMODYNAMICS

The **second law of thermodynamics** states that there is no process that, operating in a cycle, produces no other effect than the subtraction of a positive amount of heat from one reservoir and a production of an equal amount of work. This formal statement (known as the Kelvin-Plank statement) says that it is impossible to construct a cyclic engine in which the only change after one cycle is the extraction of heat from one reservoir for the purpose of performing work. Such an engine is called a perpetual motion machine of the second kind. Thus according to the second law of thermodynamics a cycling engine can only deliver work by extracting heat from a high temperature heat reservoir if a second, colder reservoir is available, into which heat, and thus entropy, is transferred. The transfer rate of entropy is such that the total entropy increases. This limit was derived by Sadi Carnot long before the principles of thermodynamics were firmly established.

7.3 EQUATION OF STATE

Any state of a thermodynamic system can be described by a number of state (or thermodynamic) variables. As state variables we usually choose pressure p, specific volume (that is the volume per unit mass) υ and temperature T. A functional relationship between the state variables is called the **equation of state** of the thermodynamic system,

$$f(p, \upsilon, T) = 0, \tag{7.21}$$

where p is the pressure, υ is the specific volume and T is the temperature. Various formulations of the equation of state are discussed below.

7.3.1 THE IDEAL GAS LAW

An **ideal gas** is a theoretical gas that is composed of randomly moving and non-interacting with each other point particles. The ideal gas concept is very useful in thermodynamics since it allows to perform simplified calculations for gas processes and thermodynamic cycles. The following conditions are valid for the ideal gas:

1. it satisfies Clapeyron's equation of state,
2. it satisfies Avogadro's law,
3. it has a constant specific heat.

Clapeyron's equation of state is as follows

$$pV = mR_{sp}T, \tag{7.22}$$

where p is the system pressure, V—the system volume, m—the mass, T—the absolute temperature and R_{sp}—the specific gas constant. The specific gas constant is equal to the universal gas constant R divided by the molar mass of the gas M

$$R_{sp} = \frac{R}{M}. \tag{7.23}$$

In physics it is customary to express the ideal gas law in terms of the number of gas molecules $\mathcal{N}$ and the Boltzmann constant k_B as follows:

$$pV = \mathcal{N}k_BT = nRT, \tag{7.24}$$

where n is the number of moles. Here we note that

$$R = \frac{\mathcal{N}}{n}k_B = N_A k_B, \tag{7.25}$$

where $N_A \equiv \mathcal{N}/n$ is Avogadro's constant.

Avogadro's law states that equal volumes of all gases, at the same temperature and pressure, have the same number of molecules.

7.3.2 IDEAL GAS MIXTURES

According to the **Dalton law**, each component in an ideal gas mixture satisfies Clapeyron's equation as if it was occupying the whole system volume V and had the same temperature as the mixture T,

$$p_iV = m_iR_iT, \tag{7.26}$$

where p_i, m_i and R_i are the partial pressure, the mass and the specific gas constant of i-th gas component, respectively, and T is the mixture absolute temperature.

The total pressure of the gas mixture is then found as:

$$p = \sum_i p_i. \tag{7.27}$$

Substituting Eq. (7.26) into Eq. (7.27) yields,

$$p = \frac{T}{V} \sum_i m_i R_i = \frac{m}{V} R_m T. \tag{7.28}$$

Thus, the equation of state for the ideal gas mixture is as follows,

$$pV = mR_m T, \tag{7.29}$$

where we introduced here the mixture gas constant R_m defined as follows,

$$R_m = \frac{1}{m} \sum_i m_i R_i = \sum_i g_i R_i. \tag{7.30}$$

Here, $g_i = m_i/m$ is the mass fraction of i-th gas component in the mixture.

The mixture density is obtained as

$$\rho_m = \frac{m}{V} = \frac{1}{V} \sum_i \rho_i V_i = \sum_i r_i \rho_i, \tag{7.31}$$

where $r_i = V_i/V$ is the volume fraction of i-th gas component in the mixture.

7.3.3 VAN DER WAALS EQUATION OF STATE

The ideal gas law is not valid for real gases, since it treats gas molecules as point particles that do not interact except in elastic collisions. In other words, the repelling and attracting forces between molecules are neglected and the volume occupied by molecules is not taken into account. These limitations are removed in the van der Waals equation written as,

$$\left(p + \frac{a}{v_M^2} \right) (v_M - b) = RT, \tag{7.32}$$

where a and b are constants whose values depend on the gas, v_M is the molar volume and R is the universal gas constant.

Three main achievements of this relatively simple theory is that it predicts the experimentally observed transition between vapor and liquid, it offers correct prediction of critical behavior and it accommodates the metastable states. The latter refers to a possibility to compress gases beyond the limit where it would normally condense or to expand a liquid beyond a point where it would normally boil. However, the predictions of gas parameters using the van der Waals equation are usually in poor agreement with measured data and the equation is suitable to qualitative rather than quantitative analyses. For quantitative purposes, the principle of the corresponding states, discussed in the next section, offers a better alternative.

7.3.4 PRINCIPLE OF CORRESPONDING STATES

The principle of corresponding states is based on an observation that curves which describe relationships between various thermodynamic parameters for different substances are usually similar (as far as their shapes are concerned). This observation creates a possibility to replace different curves with a single universal curve. To this end it is necessary to represent all curves in a coordinate system using non-dimensional, or reduced, variables. The reduced parameters are defined as,

$$v_R = \frac{v}{v_{cr}}, \qquad p_R = \frac{p}{p_{cr}}, \qquad T_R = \frac{T}{T_{cr}} \quad \dots,$$

where v_{cr}, p_{cr}, and T_{cr} are the specific volume, pressure, and temperature at the critical point, respectively. In practical calculations one of the useful forms of the equation of state for real gases is the ideal gas equation modified by a so called compressibility factor Z,

$$pv = ZR_{sp}T \tag{7.33}$$

where $Z = f(p_R, T_R)$ is usually represented in diagrams or is given as a computational function.

Another popular form of the equation is a so-called third-order equation which is given as follows,

$$p = \frac{RT}{v_M - b} - \frac{a(T)}{(v_M + \lambda_1 b)(v_M + \lambda_2 b)} \tag{7.34}$$

where p is pressure, R is the universal gas constant, v_M is the molar volume, T is the absolute temperature and $a(T)$, b, λ_1 and λ_2 are gas-dependent parameters.

7.3.5 PHASE CHANGE

Phase change is such a process, during which matter undergoes transition from one state to another, for example, transition between solid, liquid, and gaseous states. During phase change of a given medium, certain properties of the medium change. For example, liquid may change into gas when heated and attaining a saturation temperature. Figure 7.1 shows various phase change paths between solid, liquid, gas, and plasma.

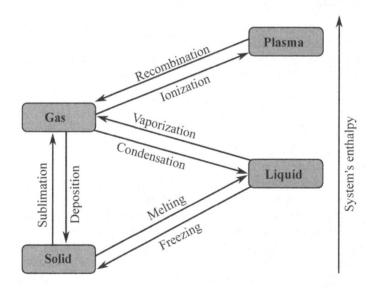

Figure 7.1   Phase change paths between solid, liquid, gas, and plasma.

A phase diagram represents various phases of a substance on a pressure-temperature plane. This is a very useful diagram, since it allows to check in what state matter exists under certain conditions. Figure 7.2 shows a phase diagram typical for CO_2. Similar diagrams can be created for other substances, such as water. The curves shown in the phase diagram divide the p-T plane into regions in which the substance exists in a specific phase. A phase change takes place when the point, representing the current conditions of the substance, crosses any of the lines. The line between the liquid and gas phases is particularly important, since it contains all pressure-temperature points, where evaporation and condensation occurs. The line starts at a triple point, and ends at a critical point.

The triple point is determined by the temperature and pressure, at which solid, liquid, and vapor phases coexist in equilibrium. The critical point is the highest temperature and pressure at which a pure material can exist in vapor-liquid equilibrium.

The **triple point of water** ($T = 273.16$ K, $p = 611.65$ Pa) is a point at which all three phases of water (solid, liquid, and vapor) coexist. The **critical point of water** indicates conditions, beyond which liquid-vapor phase transition does not exist.

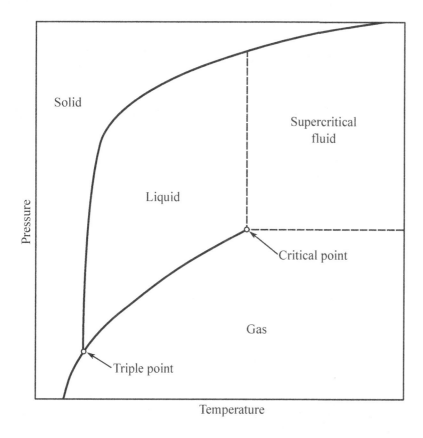

Figure 7.2   Phase diagram typical for CO_2 on the temperature-pressure plane.

The diagram is particularly useful for showing isobaric and isothermal processes. Such processes are represented with either vertical straight lines for isothermal processes or horizontal straight lines for isobaric processes. However, processes which involve phase change degenerate to a single point on the evaporation line. Since many processes in energy conversion systems are just of this character, the phase diagram is not particularly useful to represent them.

To resolve details of phase-change processes, two other types of diagrams are particularly useful. When isothermal processes prevail, T-s diagram is most proper (shown for water in Fig. 7.3), whereas when isenthalpic processes are involved, i-s diagram can be used.

The thermodynamic processes that are taking place in power plants, where water is used as a working fluid, can be represented on the i-s plane as shown in Fig. 7.4. Particularly important are three regions shown in the figure:

- liquid region,
- two-phase mixture region,
- vapor region.

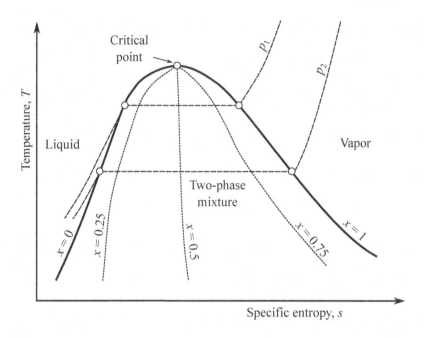

Figure 7.3 Plot of temperature T versus specific entropy s for water: x—mass fraction of the vapor phase.

An isenthalpic process in any of these regions is represented by a horizontal line ($i = const$), whereas an isentropic process by a vertical line ($s = const$). Any process in which losses are present and the entropy increases is called an irreversible process.

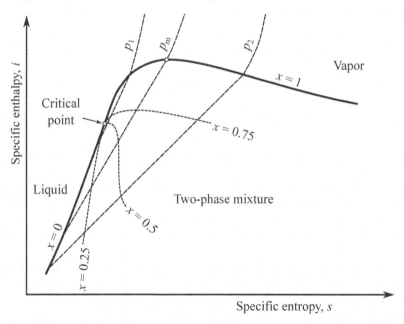

Figure 7.4 Plot of specific enthalpy i versus specific entropy s for water; x—mass fraction of the vapor phase.

The two-phase mixture region is located between curves connected to each other at the critical point and described by $x = 0$ and $x = 1$. Here x is the quality of the mixture determined as $x = (i - i_f)/i_{fg}$, where i_f is the specific enthalpy of the saturated liquid and i_{fg} is the latent heat. Two parameters which uniquely determine the state of two-phase mixture are the pressure and the quality. Once these parameters are known, all other quantities, such as mixture specific enthalpy, mixture density or mixture entropy, can be found.

The curve representing saturated vapor ($x = 1$) exhibits non-monotonic behavior. For pressures $p_2 < p_m$, the curve slope is negative (the specific enthalpy of the saturated vapor decreases with increasing specific entropy), whereas for pressures $p_1 > p_m$, the curve slope is positive. This means that any isenthalpic irreversible process for high pressure ($p > p_m$) dry saturated steam will lead to a wet steam, whereas for low pressure ($p < p_m$)—to superheated dry steam.

7.4   THERMODYNAMIC PROCESSES IN HEAT ENGINES

The classic thermodynamics allows to perform calculations only for equilibrium states or for processes which a realized by a consecutive series of equilibrium states. A process in which a system is infinitesimally close to thermodynamic equilibrium at every point in time is called a **quasi-equilibrium process**. All processes considered in this chapter are assumed to be quasi-equilibrium processes.

To explain the various thermodynamic processes with clear practical importance, we consider simple heat engines that employ a confined gas, known as a **working fluid**, to absorb or release heat in the process of doing work. In a **closed-cycle engine**, the working fluid is kept in the engine and is not replaced between the cycles. In an **open-cycle engine**, the working fluid is entirely or partly replaced during each cycle.

Any heat-engine cycle consists of a series of thermodynamic processes, during which heat is added or extracted, or work is performed. For example, we can consider a process during which the working fluid is expanding, and simultaneously heat is added to maintain a constant temperature. This process is known as isothermal expansion. In the following subsections we discuss the most common thermodynamic processes that take place in heat engines and thus are essential in energy transformation systems that use heat to generate electricity.

7.4.1   ISOTHERMAL PROCESS

An isothermal process is such thermodynamic process during which the temperature of a system is constant. During such process, the system can either expand or contract. Substituting $T = T_1 = T_2 = const$ to Clapeyron's equation, the following **Boyle-Mariotte's Law** is obtained:

$$\frac{p_1}{p_2} = \frac{V_2}{V_1}. \tag{7.35}$$

The process is shown in Fig. 7.5 using both the pressure-volume, and the temperature-entropy planes.

In the isothermal expansion of an ideal gas, the temperature T is constant and thus the thermal energy of the system does not change: $d\mathscr{U} = 0$. Thus from the first law of thermodynamics we have $đQ = đL$, and the ideal gas law yields $pV = const$. When an ideal gas expands from V_1 to V_2, the added heat is as follows:

$$\Delta Q = \Delta L = \int_{V_1}^{V_2} p\,dV = \mathscr{N} k_B T \int_{V_1}^{V_2} \frac{dV}{V} = \mathscr{N} k_B T \ln \frac{V_2}{V_1} = nRT \ln \frac{V_2}{V_1}, \tag{7.36}$$

and the associated entropy change is

$$\Delta S = \frac{\Delta Q}{T} = \mathscr{N} k_B \ln \frac{V_2}{V_1} = nR \ln \frac{V_2}{V_1}. \tag{7.37}$$

Here n is the number of moles, R is the universal gas constant, $\mathscr{N}$ is the number of molecules in the gas and k_B is the Boltzmann constant.

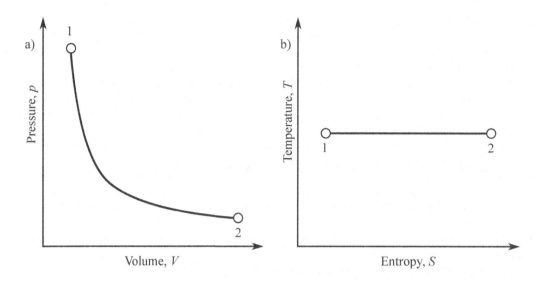

Figure 7.5 An isothermal process 1-2 in the (a) pressure-volume plane and (b) temperature-entropy plane.

7.4.2 ISOCHORIC PROCESS

An **isochoric process**, also referred to as **isometric process**, is such thermodynamic process during which the volume of a system is constant. Heating under constant volume is referred to as **isometric heating**. During this process, the working fluid is gradually heated from an initial temperature T_1 to a final temperature T_2. The initial and final values of pressure and temperature of a closed system that is undergoing isometric heating follow **Gay-Lussac's Law** as follows

$$\frac{p_1}{p_2} = \frac{T_1}{T_2}. \tag{7.38}$$

From the first law of thermodynamics we have

$$\mathrm{d}Q = \mathrm{d}\mathcal{U} + p\mathrm{d}V = \mathrm{d}\mathcal{U}. \tag{7.39}$$

Thus, integrating the above equation from state 1 to state 2 yields the following heat increase,

$$\Delta Q = \mathcal{U}_2 - \mathcal{U}_1 = C_V(T_2 - T_1), \tag{7.40}$$

and performed work

$$\Delta L = p(V_2 - V_1) = 0. \tag{7.41}$$

The corresponding entropy change is

$$\Delta S = \int_{T_1}^{T_2} \frac{C_V \mathrm{d}T}{T} = C_V \ln \frac{T_2}{T_1}. \tag{7.42}$$

The isochoric process is shown in Fig. 7.6 using both the pressure-volume, and the temperature-entropy planes.

Example 7.3:

A tank with volume $V = 5$ m^3 contains a monoatomic ideal gas at the following initial conditions: $p_1 = 0.13$ MPa, $T_1 = 288$ K. Due to added heat, the gas pressure increases to $p_2 = 0.18$ MPa. Calculate the amount of heat absorbed by the gas.

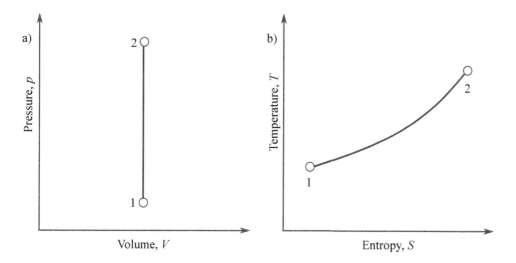

Figure 7.6 An isochoric process 1-2 in the (a) pressure-volume plane and (b) temperature-entropy plane.

Solution

From Clapeyron's equation, the number of moles of the gas in the tank is found as follows

$$n = \frac{p_1 V}{R T_1} = 271 \text{ mol.} \tag{7.43}$$

The final temperature after adding heat is

$$T_2 = T_1 \frac{p_2}{p_1} = 399 \text{ K} \tag{7.44}$$

For monoatomic ideal gas, the specific molar heat capacity at constant volume is found as

$$c_{MV} = \frac{R}{\kappa - 1} \simeq \frac{8.314 \text{ J/(mol} \cdot \text{K)}}{1.667 - 1} = 12.471 \text{ J/(mol} \cdot \text{K)} \tag{7.45}$$

For isochoric process, the amount of added heat is found as

$$\Delta Q = \mathscr{U}_2 - \mathscr{U}_1 = n c_{MV}(T_2 - T_1) = 271 \times 12.471(399 - 288) \simeq 375.14 \text{ kJ.} \tag{7.46}$$

Thus, the amount of heat added to the gas in tank is ~ 375 kJ.

7.4.3 ISOBARIC PROCESS

An isobaric process is such thermodynamic process during which the pressure of the system is constant. The Clapeyron's state equation for two thermodynamic states 1 and 2 of a certain system undergoing an isobaric expansion yields the following **Charles's Law**,

$$\frac{V_1}{V_2} = \frac{T_1}{T_2}. \tag{7.47}$$

Since during the expansion $p = p_1 = p_2 = const$, the work performed by the system is

$$\Delta L = \int_{V_1}^{V_2} p dV = p(V_2 - V_1) = nR(T_2 - T_1). \tag{7.48}$$

To maintain constant pressure in the system, heat ΔQ must be added during the expansion. From the first law of thermodynamics we get

$$\text{d}Q = \text{d}\mathscr{U} + \text{d}L = \text{d}\mathscr{U} + p\text{d}V = \text{d}I. \tag{7.49}$$

Integration from state point 1 to 2 yields the following amount of heat that must be transferred to the system

$$\Delta Q = I_2 - I_1 = C_p(T_2 - T_1), \tag{7.50}$$

and the corresponding entropy increase is found as

$$\Delta S = \int_{T_1}^{T_2} \frac{C_p \text{d}T}{T} = C_p \ln \frac{T_2}{T_1}. \tag{7.51}$$

Here C_p is the heat capacity at constant pressure of the working fluid.

The isobaric process is shown in Fig. 7.7 using both the pressure-volume, and the temperature-entropy planes.

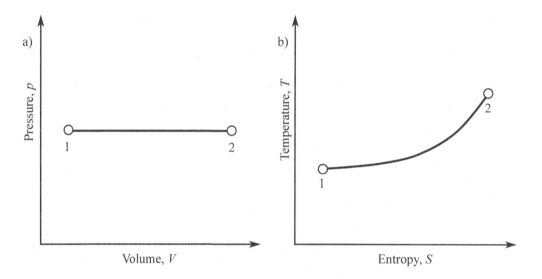

Figure 7.7   An isobaric process 1-2 in the (a) pressure-volume plane and (b) temperature-entropy plane.

Example 7.4:

A vertical cylinder, closed from the top with a piston moving without friction, contains nitrogen at initial parameters $V_1 = 0.05$ m³, $p_1 = 0.12$ MPa and $T_1 = 288$ K. Due to added heat, the volume of the gas increased to $V_2 = 0.07$ m³. Using the ideal gas law and isobaric process, calculate the amount of heat absorbed by the nitrogen.

Solution

From Charles's Law we get

$$T_2 = T_1 \frac{V_2}{V_1} = 403 \text{ K}. \tag{7.52}$$

From Clapeyron's equation, the number of moles of the gas in the tank is found as follows

$$n = \frac{p_1 V}{R T_1} \simeq 2.51 \text{ mol}. \tag{7.53}$$

For diatomic ideal gas, the specific molar heat capacity at constant pressure is found as

$$c_{Mp} = \frac{\kappa R}{\kappa - 1} \simeq 1.4 \frac{8.314 \text{ J}/(\text{mol} \cdot \text{K})}{1.4 - 1} = 29.1 \text{ J}/(\text{mol} \cdot \text{K}) \tag{7.54}$$

For isobaric process, the amount of added heat is found as

$$\Delta Q = I_2 - I_1 = nc_{Mp}(T_2 - T_1) = 2.51 \times 29.1(403 - 288) \simeq 8.4 \text{ kJ}. \tag{7.55}$$

Thus, the amount of heat added to the gas in cylinder is 8.4 kJ.

7.4.4 ADIABATIC PROCESS

An adiabatic process is such a thermodynamic process during which there is no heat exchange. Let us first consider reversible adiabatic expansion. Such process can be realized by slowly reducing the external pressure in a perfectly insulated system, so that $dQ = 0$. As the gas expands, its volume increases, while its pressure and temperature decrease. Since $đQ = 0$, $dS = đQ/T = 0$, and the entropy is constant. Thus, such process is also called *isentropic expansion*.

During the process the gas does work on the moving walls (piston) $L = \int p dV$. At the same time, the thermal energy decreases by the amount of work done, given in a differential form as $d\mathcal{U} = -p dV$, where p is not constant. Actually, the pressure changes according to the following differential expression: $C_V dT = d\mathcal{U} = -p dV$. From the ideal gas law we have $p dV + V dp = \mathcal{N} k_B dT$, so we get

$$(C_V + \mathcal{N} k_B)p dV + C_V V dp = 0. \tag{7.56}$$

Using the ideal gas equation $C_p = C_V + \mathcal{N} k_B$, we finally obtain

$$pV^\kappa = const. \tag{7.57}$$

Here $\kappa = C_p/C_V$ is the **adiabatic index**, also known as **isentropic index**. This parameter is quite important in the study of engines, and for an ideal gas it can be found from the following formula

$$\kappa = 1 + \frac{2}{n_f}, \tag{7.58}$$

where n_f is the number of degrees of freedom of gas molecules. Thus, for an ideal monoatomic gas (such as noble gases: helium, neon, argon, krypton, and xenon) $n_f = 3$ and $\kappa = 5/3 \simeq 1.667$, for an ideal diatomic gas (oxygen, nitrogen) $n_f = 5$ and $\kappa = 7/5 = 1.4$, and for an ideal gas with molecules containing three or more atoms $n_f = 6$ and $\kappa = 8/6 \simeq 1.333$.

The work done in adiabatic expansion is now found by integration of $p dV$ while keeping $pV^\kappa = const$,

$$L = \Delta E_I = \int_{V_1}^{V_2} p dV = \frac{1}{\kappa - 1}(P_1 V_1 - p_2 V_2) = \frac{1}{\kappa - 1} \mathcal{N} k_B (T_1 - T_2). \tag{7.59}$$

An adiabatic compression can be analyzed in a similar way.

Example 7.5:

A vertical cylinder with inner diameter $d = 60$ mm is closed from the top with a piston with mass $m_p = 0.5$ kg. The cylinder contains a combustion mixture of gas and air, which can be treated as an ideal gas. The NTP[1] standard density of the mixture is $\rho_m = 0.539$ kg/m^3 and its specific heat capacity is $c_p = 2386$ J/(kg·K). The initial parameters of the mixture are $V_1 = 0.002$ m^3 and

---

[1] Normal Temperature and Pressure.

$T_1 = 288$ K, and the ambient pressure is $p_a = 0.1$ MPa. The autoignition temperature of the mixture is $T_{ai} = 773$ K. Calculate a height above the piston H from which a hammer with mass $m_h = 5$ kg should be released to cause ignition of the mixture. Assume a free fall of the hammer and neglect the air resistance.

Solution

We find the specific gas constant of the mixture from the given density at NTP standard ($T_{NTP} = 293.15$ K, $p_{NTP} = 101325$ Pa) as

$$R_m = \frac{p_{NTP}}{\rho_m T_{NTP}} = \frac{101325}{0.539 \cdot 293.15} = 641.3 \text{ J}/(\text{kg} \cdot \text{K}). \tag{7.60}$$

The gas specific heat at constant volume is found as

$$c_V = c_p - R_m = 2386 - 641.3 = 1744.7 \text{ J}/(\text{kg} \cdot \text{K}).$$

The mass of gas mixture in the cylinder is found from Clapeyron's equation as

$$m_m = \frac{p_1 V_1}{R_m T_1} = \frac{101735 \cdot 0.002}{641.3 \cdot 288} = 0.0011 \text{kg},$$

where p_1 is found as

$$p_1 = p_a + \frac{m_p g}{\pi d^2/4} = 1 \times 10^5 + \frac{0.5 \cdot 9.81}{\pi (0.06)^2/4} = 101735 \text{ Pa}.$$

The required increase of the thermal energy of gas to attain the autoignition temperature can be found as

$$\Delta \mathscr{U} = \mathscr{U}_2 - \mathscr{U}_1 = m_m c_V (T_{ai} - T_1) = 0.0011 \cdot 1744.7 \times (773 - 288) = 930.8 \text{ J}.$$

Since the process of mixture compression is quite violent, we assume an adiabatic compression and the first law of thermodynamics can be written as

$$\Delta E_T = -L,$$

where ΔE_T is the total energy change and L is the work performed on the system. The total energy change of the system consists of the thermal energy increase $\Delta \mathscr{U}$ and the potential energy gained by the system from the piston: $E_{Pp} = m_p g H_p$, and the potential energy gained from the hammer: $E_{Ph} = m_h g (H_p + H)$, where H_p is the distance that the piston will move down in the cylinder from the initial position to the position of autoignition due to the hammer impact. Assuming an adiabatic and isentropic process, we find the gas mixture volume at the ignition time as

$$V_2 = V_1 \left(\frac{T_1}{T_2} \right)^{\frac{1}{\kappa-1}} = 0.002 \left(\frac{288}{773} \right)^{\frac{1}{1.367-1}} = 1.37 \times 10^{-4} \text{ m}^3,$$

where the specific heat ratio is found as $\kappa = c_p/c_V = 2386/1744.7 \simeq 1.367$. The piston travel distance is found now as

$$H_p = \frac{4(V_1 - V_2)}{\pi d^2} = \frac{4(0.002 - 1.37 \times 10^{-4})}{\pi (0.06)^2} = 0.659 \text{ m}.$$

From the first law of thermodynamics we have

$$\Delta E_T = \mathscr{U}_2 - (\mathscr{U}_1 + E_{Pp} + E_{Ph}) = -L = p_a (V_1 - V_2).$$

The height H can now be found as

$$H = \frac{\Delta E_I - H_p g (m_p + m_h) - p_a (V_1 - V_2)}{m_h g} \simeq 14.45 \text{ m}.$$

Thus, the height from which the hammer should be released to cause an autoignition of the gas mixture is equal to 14.45 m.

7.4.5 POLYTROPIC PROCESS

In many types of real thermodynamic processes very efficient heat exchangers are required to get as close as possible to an ideal isothermal process. However, due to economical and technical limitations, the ideal isothermal process is not possible and thus, the isothermal process model is not adequate to describe the system parameter change. A better model is a *polytropic process* in which $pV^\alpha = const$, where $1 < \alpha < \kappa$ is the **polytropic index**. Using Clapeyron's equation, the polytropic process can be expressed with three different pairs of independent parameters,

$$pV^\alpha = const, \ \ TV^{\alpha-1} = const, \ \ \frac{T}{p^{\frac{\alpha-1}{\alpha}}} = const. \tag{7.61}$$

These expressions lead to the following relationships between system parameters at two states 1 and 2,

$$\frac{p_1}{p_2} = \left(\frac{V_2}{V_1}\right)^\alpha, \ \ \frac{T_1}{T_2} = \left(\frac{V_2}{V_1}\right)^{\alpha-1}, \ \ \frac{T_1}{T_2} = \left(\frac{p_1}{p_2}\right)^{\frac{\alpha-1}{\alpha}}. \tag{7.62}$$

The absolute work performed by gas, undergoing a frictionless polytropic process in a closed system, can be found as

$$L_{12} = \int_1^2 pdV = \frac{p_1 V_1}{\alpha - 1}\left[1 - \left(\frac{p_2}{p_1}\right)^{\frac{\alpha-1}{\alpha}}\right] = \frac{mR}{\alpha - 1}(T_1 - T_2), \tag{7.63}$$

where m is the mass of the gas in the closed system.

Example 7.6:

An ideal gas is compressed in a polytropic process in a compressor transferring heat to the environment from pressure $p_1 = 0.1$ MPa and temperature $T_1 = 300$ K to $p_2 = 0.4$ MPa and $T_2 = 370$ K. Find the polytropic index of the process.

Solution

The pressure and temperature follow the following relationship,

$$T_1/T_2 = (p_1/p_2)^{(\alpha-1)/\alpha}$$

Thus, we find

$$\alpha = 1/[1 - \ln(T_1/T_2)/\ln(p_1/p_2)] = 1/[1 - \ln(300/370)/\ln(0.1/0.4)] \simeq 1.178.$$

The polytropic index of the compression process is $\alpha \simeq 1.178$.

7.5 THERMODYNAMIC CYCLES

Closed thermodynamic cycles play a very important role in energy conversion systems. The most important cycles and the involved thermodynamic processes are shown in Table 7.1.

7.5.1 CARNOT CYCLE

The Carnot cycle is a theoretical thermodynamic cycle that has the highest possible efficiency as determined by the second law of thermodynamics. As shown in Fig. 7.8, the cycle starts with working fluid at the lowest temperature and pressure, as described by point 1 in the figure. The cycle is realized in four processes as follows.

Table 7.1

Selected Thermodynamic Closed Cycles

Cycle	Compression	Heat Addition	Expansion	Heat Rejection
Carnot	adiabatic	isothermal	adiabatic	isothermal
Rankine	adiabatic	isobaric	adiabatic	isobaric
Brayton	adiabatic	isobaric	adiabatic	isobaric

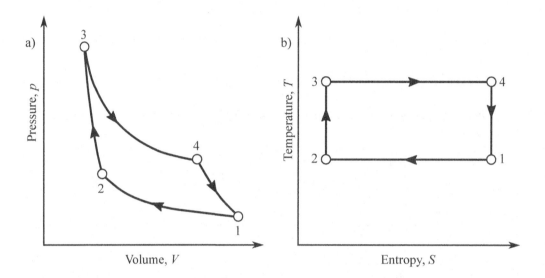

Figure 7.8 Carnot cycle shown in the (a) pressure-volume plane and (b) temperature-entropy plane. 1–2 heat rejection at isothermal compression, 2–3 isentropic compression, 3–4 heat addition at isothermal expansion, 4–1 isentropic expansion.

- Isothermal compression process from state 1 to state 2 at constant temperature $T_1 = T_2 = T_e$. During that process a work on the working fluid is done and heat is extracted from the system and dumped to the low-temperature reservoir. For this purpose the Carnot engine must be coupled to a heat reservoir at low temperature T_e.
- Adiabatic compression process from state 2 to state 3, doing more work on the working fluid with no heat transfer, rising pressure to p_3 and temperature to T_3.
- Isothermal expansion process from state 3 to state 4 at temperature $T_3 = T_4 = T_a$. During that process the system performs work and heat is added to the system from the high-temperature reservoir.
- Adiabatic expansion process from state 4 back to state 1, doing more work by the system with no heat transferred, and ending at low temperature $T_1 = T_e$.

Since the internal energy of the system at the beginning of the cycle is the same as at the end of the cycle, according to the first law of thermodynamics, the net added heat to the system, $Q_{34} - Q_{12}$, must be equal to the net work performed by the system L. The heat input is $Q_{34} = T_a(S_4 - S_3) = Q_a = T_a \Delta S$, where $\Delta S = S_4 - S_3$ is the entropy entering the system from the high-temperature reservoir. Since the same amount of entropy must be dumped to the low-temperature reservoir, the dumped heat in process 1–2 is $Q_{12} = T_e(S_1 - S_2) = Q_e = T_e \Delta S$. Thus the efficiency of the Carnot cycle is as follows,

$$\eta = \frac{L}{Q_a} = \frac{Q_a - Q_e}{Q_a} = \frac{T_a - T_e}{T_a}. \tag{7.64}$$

Despite its high efficiency, Carnot cycle is impractical, since relatively little work can be extracted for realistic volume and pressure ratios.

7.5.2 RANKINE CYCLE

The Rankine cycle is an idealized thermodynamic cycle of a heat engine that converts heat into mechanical work while adding and rejecting heat during phase change. During the cycle, cold working fluid is pressurized reversibly to a high pressure by a pump. Next, the working fluid is heated and evaporated in a boiler in an isobaric process. High-temperature and pressure vapor is used in a turbine to perform a mechanical work where the vapor is de-pressurized reversibly from the high pressure to a low pressure in a condenser. The low pressure vapor is condensed in the condenser by rejecting the latent heat. The condensate is next returning to the pump inlet closing in that way the cycle.

The Rankine cycle efficiency is lower than the Carnot efficiency since heat addition is partly taking place in a non-isothermal process. To improve the efficiency, it is desirable to bring the working fluid as close as possible to the saturation point while entering the boiler. This process is realized in a regenerative Rankine cycle that is commonly used in real power stations. To this end bleed steam is sent from a turbine to a regenerative heat exchanger (known as a feedwater heater), which has on the second side the working fluid returning to the boiler. Regenerative Rankine cycle shows improved efficiency, approaching that of Carnot cycle.

Figure 7.9 shows an example of a system in which Rankine cycle can be realized. The system consists of a boiler, in which heat is added to a working fluid (usually steam). The high pressure and temperature steam is next passed to a turbine, where a work is performed and the steam is expanding. Low pressure steam is condensed in condenser, and finally, the condensate is pumped back to the boiler.

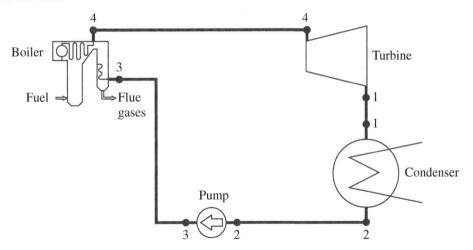

Figure 7.9 Schematic of a Rankine cycle system.

An ideal Rankine cycle can be represented in the T–S plane, as depicted in Fig. 7.10, where the labels of various state points correspond to the labels used in the schematic of Rankine-cycle system shown in Fig. 7.9.

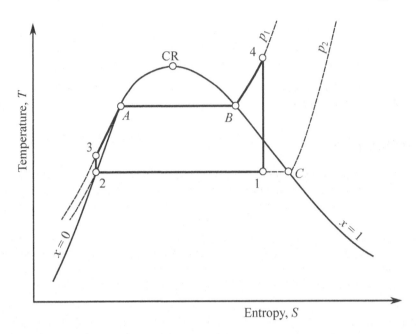

Figure 7.10 An ideal Rankine cycle: 1–2 isobaric heat rejection, 2–3 isentropic compression, 3–4 isobaric heat addition, 4–1 isentropic expansion, A—saturated liquid at pressure p_1, B—saturated vapor at pressure p_1, C—saturated vapor at pressure p_2, CR—critical point.

In real Rankine cycle systems the compression and expansion processes are irreversible. Thus, the compression 2–3 and expansion 4–1 processes are better represented by irreversible adiabatic processes, as shown in Fig. 7.11.

7.5.3 BRAYTON CYCLE

An ideal Brayton cycle, as shown in Fig. 7.12, comprises an isentropic process in which working fluid (usually gas) is compressed. The compressed fluid runs then through a heater (combustion chamber or a nuclear reactor) at constant pressure. In the next step the high pressure and temperature fluid is expanding in a turbine giving up the thermal energy in an isentropic process. After leaving the turbine, the fluid is giving up heat in a heat exchanger in an isobaric process and finally, as low-pressure low-temperature fluid, returns back to the inlet of the compressor.

The efficiency of the ideal Brayton cycle is,

$$\eta = \frac{T_a - T_e}{T_a} = 1 - \left(\frac{p_e}{p_a} \right)^{(\kappa - 1)/\kappa}, \tag{7.65}$$

where κ is the heat capacity ratio, p_a is the high pressure, at which heat is added, and p_e is the low pressure in the cycle, at which heat extracted.

Brayton cycle can also be realized in an open cycle, when air from the atmosphere is used as a working fluid. This type of a cycle is present in jet engines. A Brayton cycle with irreversible compression and expansion processes is shown in Fig. 7.13.

7.5.4 STIRLING CYCLE

Stirling engines follow a cycle similar to the Carnot cycle; however, the adiabatic expansion and compression processes are replaced with isochoric heating and cooling. The Stirling cycle contains the following steps, depicted in Fig. 7.14:

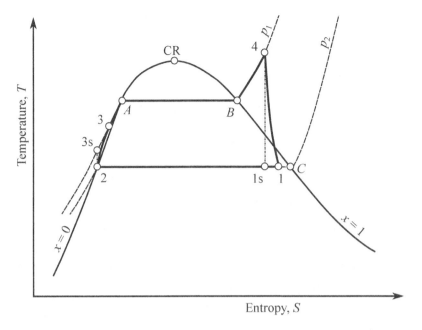

Figure 7.11   Rankine cycle with real compression and expansion processes: 1–2 isobaric heat rejection, 2–3 adiabatic compression, 3–4 isobaric heat addition, 4–1 adiabatic expansion, A—saturated liquid at pressure p_1, B—saturated vapor at pressure p_1, C—saturated vapor at pressure p_2, CR—critical point.

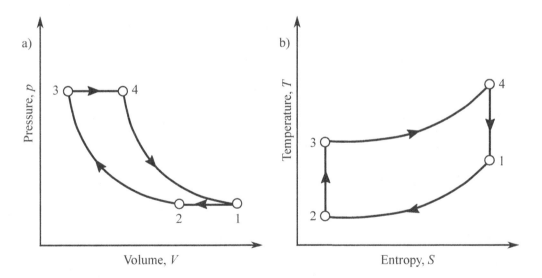

Figure 7.12   An ideal Brayton cycle in the (a) pressure-volume plane and (b) temperature-entropy plane. 1–2 isobaric heat rejection, 2–3 isentropic compression, 3–4 isobaric heat addition, 4–1 isentropic expansion.

- Isothermal compression at low temperature $T_1 = T_e$, heat $Q_{12} = Q_e$ extraction and work L_{12} input.
- Isochoric heating from $T_2 = T_e$ to $T_3 = T_a$, with heat input Q_{23} and no work done.
- Isothermal expansion at temperature $T_3 = T_4 = T_a$, with heat input Q_{34} and work output L_{34}.
- Isochoric cooling from $T_4 = T_a$ to $T_1 = T_e$ with heat output $Q_{41} = Q_e$ and no work done.

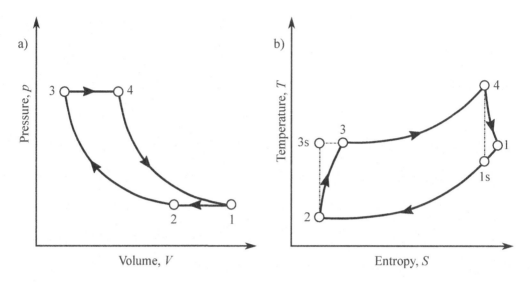

Figure 7.13 Brayton cycle with real compression and expansion processes presented in the (a) pressure-volume plane and (b) temperature-entropy plane. 1–2 isobaric heat rejection, 2–3 adiabatic compression, 3–4 isobaric heat addition, 4–1 adiabatic expansion.

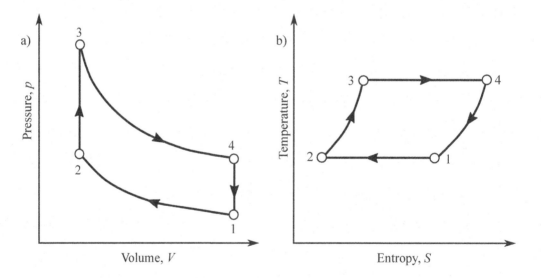

Figure 7.14 Stirling cycle: 1–2 isothermal compression and heat rejection to an external sink, 2–3 isochoric heat transfer from the regenerator to the gas, 3–4 isothermal expansion and heat addition from external source, 4–1 isochoric heat transfer from the gas to the regenerator.

The advantage of the Stirling engine is that it has the same theoretical efficiency as the Carnot engine, but it can provide more work once operating in the same range of pressures and volumes.

7.5.5 KALINA CYCLE

The Kalina cycle employs as a working fluid a solution of two liquids with different boiling points. Thanks to that the working fluid boils over a range of temperatures, rather than at a single temperature. As a result, more heat can be added to the working fluid from the high-temperature reservoir than in the case of a single fluid. This provides a higher cycle efficiency, comparable with a combined cycle. The temperature-entropy diagram of an ideal Kalina cycle is shown in Fig. 7.15.

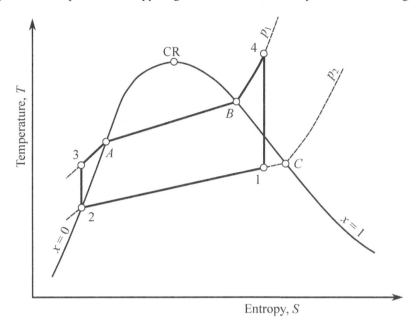

Figure 7.15 An ideal Kalina cycle: 1–2 isobaric heat rejection, 2–3 isentropic compression, 3–4 isobaric heat addition, 4–1 isentropic expansion, A—saturated liquid at pressure p_1, B—saturated vapor at pressure p_1, C—saturated vapor at pressure p_2, CR—critical point.

7.5.6 COMBINED CYCLE

A combined cycle usually consists of two heat engines, which use the same heat source. The engines operates in series, where the temperature of working fluid leaving the first engine is high enough for the second engine to extract additional work. As a result, the over-all efficiency of the combined cycle can be as high as 60%.

In stationary plants usually a gas turbine, which is operating in the Brayton cycle, is followed by a steam turbine, operating in the Rankine cycle. This arrangement is called Combined Cycle Gas Turbine (CCGT). Once operating in base load, the CCGT plant can achieve 62% efficiency. Figure 7.16 depicts CCGT in the temperature-entropy plane.

7.6 ENTROPY BALANCE

As an example of entropy balance, we consider a system shown in Fig. 7.17. The system consists of boiler, turbine, condenser and pump. We perform entropy balance for a control volume indicated with a dashed line. The control volume has four material ports, two outlets indicated with numbers 2 and 4, and two inlets, indicated with numbers 1 and 3. In addition, the control volume has a non-material port through which mechanical power N_e is extracted from the turbine.

Mass conservation applied to the control volume yields,

$$W_3 - W_4 + W_1 - W_2 = 0. \tag{7.66}$$

This equation is always satisfied, since during a cyclic steady-state operation of the system we have $W_1 = W_2 = W_3 = W_4$.

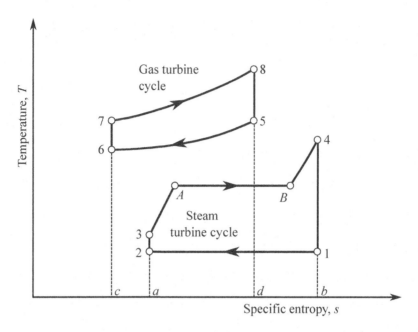

Figure 7.16 An ideal Combined Cycle Gas Turbine (CCGT). The area 5–6–c–d is equal to the area 3–A–B–4–b–a when all heat rejected in the gas turbine cycle is added to the steam turbine cycle.

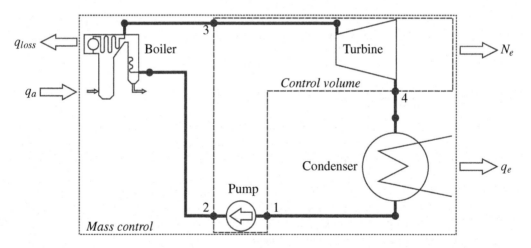

Figure 7.17 Entropy balance in a Rankine cycle system: q_a—added thermal power, q_e—extracted thermal power, N_e—extracted mechanical power, q_{loss}—thermal power losses.

Energy conservation equation applied to the system gives,

$$q_a + q_{loss} + q_e - N_e = 0, \tag{7.67}$$

where $q_a = W_3 i_3 - W_2 i_2$ is the added heat to the system, $q_e = W_1 i_1 - W_4 i_4$ is the extracted heat from the system, and $q_{loss} < 0$ represents thermal losses of the systems.

Entropy balance equation reads,

$$W_3 s_3 - W_4 s_4 + W_1 s_1 - W_2 s_2 + \left(\frac{dS}{dt}\right)_{loss} + \left(\frac{dS}{dt}\right)_{gen} = 0, \tag{7.68}$$

where $(dS/dt)_{loss}$ and $(dS/dt)_{gen}$ is external and internal entropy generation, respectively.

The entropy balance equation can be written as,

$$W_b(s_3 - s_2) + W_c(s_1 - s_4) + \left(\frac{dS}{dt}\right)_{loss} + \left(\frac{dS}{dt}\right)_{gen} = 0. \tag{7.69}$$

Here $W_b = W_3 = W_4$ is the mass flow rate through boiler, and $W_c = W_4 = W_1$ is the mass flow rate through condenser. At steady-state conditions we have in addition $W_b = W_c$.

From the energy balance for boiler we have

$$W_b(i_3 - i_2) = q_a \Rightarrow W_b = \frac{q_a}{i_3 - i_2}, \tag{7.70}$$

$$W_c(i_1 - i_4) = q_e \Rightarrow W_c = \frac{q_e}{i_1 - i_4}. \tag{7.71}$$

Substituting Eqs. (7.70) and (7.71) into Eq. (7.69) yields,

$$q_a \frac{s_3 - s_2}{i_3 - i_2} + q_e \frac{s_1 - s_4}{i_1 - i_4} + \left(\frac{dS}{dt}\right)_{loss} + \left(\frac{dS}{dt}\right)_{gen} = 0. \tag{7.72}$$

We introduce now the following average thermodynamic temperatures for heat addition in the boiler,

$$\overline{T}_a \equiv \left|\frac{i_3 - i_2}{s_3 - s_2}\right| \tag{7.73}$$

and for heat extraction in the condenser,

$$\overline{T}_e \equiv \left|\frac{i_4 - i_1}{s_4 - s_1}\right|. \tag{7.74}$$

Using Eqs. (7.73) and (7.74) in Eq. (7.72) yields,

$$\frac{q_a}{\overline{T}_a} + \frac{q_e}{\overline{T}_e} + \left(\frac{dS}{dt}\right)_{loss} + \left(\frac{dS}{dt}\right)_{gen} = 0. \tag{7.75}$$

Now we can eliminate q_e from the above entropy balance equation using the energy conservation (7.67) and we get,

$$\frac{q_a}{\overline{T}_a} - \frac{q_a}{\overline{T}_e} - \frac{q_{loss}}{\overline{T}_e} + \frac{N_e}{\overline{T}_e}\left(\frac{dS}{dt}\right)_{loss} + \left(\frac{dS}{dt}\right)_{gen} = 0, \tag{7.76}$$

from which we derive the following expression for the extracted power

$$N_e = q_a\left(1 - \frac{\overline{T}_e}{\overline{T}_a}\right) + q_{loss} - \overline{T}_e\left(\frac{dS}{dt}\right)_{loss} - \overline{T}_e\left(\frac{dS}{dt}\right)_{gen}. \tag{7.77}$$

We can now recognize that the first term on the right-hand-side of Eq. (7.77) represents the extracted power from an ideal Carnot cycle. Since all the remaining terms on the right-hand-side of Eq. (7.77) are negative (remember that $q_{loss} < 0$), the actual power extracted from the cycle is less than the ideal Carnot power. As we can see, the entropy analysis of a thermodynamic cycle provides important information on the efficiency losses and gives a theoretical indication of the origins of those losses. The maximum efficiency of a thermodynamic cycle is closely related to a maximum work theorem, which is discussed in the next section.

7.7 PRINCIPLE OF MAXIMUM WORK

For a given thermodynamic cycle, the *maximum work* can be determined from the first and second law of thermodynamics. A derivation of the *maximum work theorem* was first given by Gibbs, who considered a primary system, connected to reversible heat and reversible work sources, as shown in Fig. 7.18.

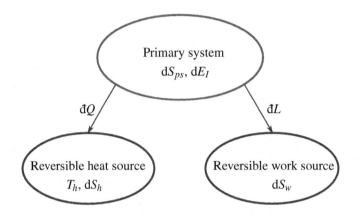

Figure 7.18 A primary system with connected reversible heat and work sources used for a derivation of the maximum work theorem.

The first law for the primary system can be written as follows,

$$dE_I = đQ - đL. \tag{7.78}$$

An internal energy change of the primary system is equal to a heat exchanged with the reversible heat source (RHS), minus a work exchanged with the reversible work source (RWS). We follow here the convention that heat added to a system and work extracted from it are positive. According to the second law of thermodynamics, the sum of entropy changes in the primary system and reversible sources is always non-negative, thus

$$dS_h + dS_w + dS_{ps} \geq 0. \tag{7.79}$$

Reversibility of the work source dictates that $dS_w = 0$, whereas reversibility of the heat source gives $-đQ = T_h dS_h$. The minus sign indicates that heat $đQ$ is extracted from the primary system and added to the reversible heat source, causing and increase of its entropy by dS_h. Thus, the first and second law of thermodynamics combined together yield,

$$-\frac{dE_I + đL}{T_h} + dS_{ps} \geq 0, \tag{7.80}$$

or

$$đL \leq T_h dS_{ps} - dE_I. \tag{7.81}$$

When the primary system is reversible, the equality sign in Eq. (7.81) will hold and the work term will correspond to the maximum work that the system can perform,

$$đL_{max} = T_h dS_{ps} - dE_I, \tag{7.82}$$

The above equation states that the maximum (positive) work that can be extracted from a system is equal to the internal energy change $(-dE_I)$ reduced by the term $T_h dS_{ps} < 0$. The maximum energy principle gives an explicit proof that it is not possible to use internal energy entirely to perform work.

Integrating Eq. (7.82) between two states (1) and (2), the total maximum work is obtained as

$$L_{max} = \int_1^2 dL_{max} = E_{I1} - E_{I2} + T_h(S_{ps2} - S_{ps1}). \tag{7.83}$$

It should be noted that this expression is not valid for cyclic processes, where states (1) and (2) are identical, which would give $L_{max} = 0$.

To derive an expression for the maximum work of a cyclic process, we need to consider a system as shown in Fig. 7.19.

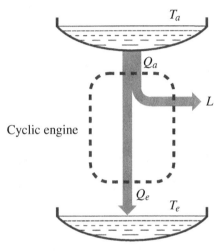

Figure 7.19 A cyclic engine with connected heat reservoirs used for a derivation of the maximum work for cyclic processes.

The system corresponds to an engine that operates between two heat reservoirs with temperatures T_a and T_e. During a cycle, heat Q_a is added to the system from reservoir with temperature T_a, and heat Q_e is extracted from the system and flows to reservoir with temperature T_e. During the same cycle, the engine delivers work L. Our goal is to determine the maximum work L_{max} that can be delivered.

Note that from the point of view of the cyclic engine, $Q_a > 0$ (system gains heat), $L > 0$ (engine performs work), and $Q_e < 0$ (heat is extracted from the system).

The first law of thermodynamics applied to the engine operating during one cycle is as follows,

$$\Delta E_I = Q_a + Q_e - L = 0, \tag{7.84}$$

thus,

$$L = Q_a + Q_e. \tag{7.85}$$

The reservoir with temperature T_a loses heat $Q_{Ra} = -Q_a < 0$ and its entropy changes as $\Delta S_a = Q_{Ra}/T_a = -Q_a/T_a < 0$. Similarly, entropy of the reservoir with temperature T_e changes as $\Delta S_e = Q_{Re}/T_e = -Q_e/T_e > 0$. During a cycle, the entropy change for the engine is $\Delta S_{engine} = 0$. The total entropy change is then

$$\Delta S_{total} = \Delta S_{engine} + \Delta S_a + \Delta S_e = \frac{Q_{Ra}}{T_a} + \frac{Q_{Re}}{T_e} = -\frac{Q_a}{T_a} - \frac{Q_e}{T_e} > 0. \tag{7.86}$$

Combining Eqs. (7.85) and (7.86) yields

$$-\frac{Q_a}{T_a} - \frac{L - Q_a}{T_e} > 0, \tag{7.87}$$

or

$$Q_a \left(1 - \frac{T_e}{T_a} \right) - L > 0. \tag{7.88}$$

Thus for a cyclic engine the following limitation for the work is valid

$$L < Q_a \left(1 - \frac{T_e}{T_a} \right) = Q_a \eta_C, \tag{7.89}$$

where

$$\eta_C \equiv \left(1 - \frac{T_e}{T_a} \right), \tag{7.90}$$

is the Carnot cycle efficiency.

7.8 EXERGY BALANCE

Once deriving the expressions for maximum work we noticed that only a fraction of the heat transferred to an engine is available (has potential) to do a useful work. Energy that is available to be used for work is called **exergy**, **available energy**, or **available work**. One of the first goals of thermodynamics was to determine exergy. Application of exergy analyses in chemical plants was partially responsible for their huge grow in 20th century.

The expression for the fundamental difference between energy and exergy is that the former is conserved, whereas the latter is not. According to the first law of thermodynamics, the energy input and energy output in a thermodynamic system at equilibrium will always balance. Exergy output will not balance the exergy input, since a fraction of exergy input will be destroyed according to the second law of thermodynamics.

An *energy efficiency* (or *first law efficiency*) determines the most efficient process, where as little as possible of the input energy is wasted. An *exergy efficiency* (or *second law efficiency*) determines the most efficient process, where as little as possible of the available work, from a given input of available work, is destroyed. These concepts are discussed further in §10.2.

In most general words, exergy, often denoted by B, is defined as the maximum amount of usable work that can be provided by a system or a material as it is brought into mechanical and thermodynamic equilibrium with the environment. There are various forms of exergy of a substance, as listed below:

- *Kinetic exergy*, B_K, which is calculated in terms of the absolute velocity of the substance, using the environment as the system of reference.
- *Potential exergy*, B_P, which is determined against this reference level in the environment, at which the substance is largely available.
- *Physical exergy*, B_{ph}, which is zero for a substance at the ambient temperature and pressure, and which is positive when the substance's temperature and pressure are different from the ambient reference values.
- *Chemical exergy*, B_{ch}, which is zero for a substance with the same chemical composition as commonly present in the environment. Otherwise, when the chemical composition is different, the substance has (usually) positive chemical exergy.
- *Nuclear exergy*, B_n, which is zero for a substance with Q-value zero when its nuclei undergo fission or fusion. Otherwise, when $Q > 0$, the nuclear exergy is positive.

Exergy can be interpreted as the amount of useful work that can be provided by a system as it is brought into thermodynamic equilibrium with its environment. A system is said to be in the **dead state** when it is in equilibrium with its environment, that is, it has the same temperature, pressure, composition, etc. No work can be extracted from a system that is in the dead state.

7.8.1 MECHANICAL AND ELECTRICAL EXERGY

Mechanical and electrical energy can be transformed into any other form with practically no losses. In particular, they can be transformed efficiently into useful work. The exergy of mechanical or electrical system is thus equal to its stored energy, using a proper frame of reference. A stream of substance passing through a stationary control boundary in that frame of reference contains two exergy components. The first one is the **kinetic exergy** $B_K = mv^2/2$, where v is the absolute velocity of the substance measured against the environment. The second energy component is the **potential exergy** $B_P = mgH$, where H is a level difference between the center of mass of the substance and a reference level, at which a large amount of the substance can be accumulated.

Similarly, the exergy of stored electromagnetic energy depends on the relevant value of the zero of electrostatic potential.

7.8.2 THERMAL EXERGY

Thermal exergy, B_{th} is defined as a sum of the physical exergy and the chemical exergy, thus,

$$B_{th} = B_{ph} + B_{ch}. \qquad (7.91)$$

The expression for the maximum work of a thermodynamic system was derived in §7.7 and is given by Eq. (7.83). We can use this equation to determine the amount of useful work that can be done by the system when operating between some initial state (1) to a state (2), which corresponds to the ambient pressure and temperature. We also include the term pV and thus replace the internal energy with enthalpy, since we are assuming an open system. The resulting maximum available work is equivalent to the thermal exergy of the system, and is given as follows,

$$\Delta B_{th} = I - I_0 - T_0(S - S_0). \qquad (7.92)$$

According to Eq. (7.92), the *specific thermal exergy* can be defined as,

$$b_{ph} = i_{ph} - T_0 s_{ph}, \qquad (7.93)$$

where i_{ph} and s_{ph} are physical specific enthalpy and entropy, respectively, calculated in reference to the state at ambient conditions, T_0, p_0. The physical exergy of an ideal gas is expressed with the following equation,

$$b_{ph} = c_p \left(T - T_0 - T_0 \ln \frac{T}{T_0} \right) + T_0 R \ln \frac{p}{p_0}. \qquad (7.94)$$

The first term of Eq. (7.94) takes into account the influence of temperature and the second, of pressure. The physical exergy of liquid and solid bodies can be expressed as

$$b_{ph} = \int_{T_0}^{T} \frac{T - T_0}{T} d i_{ph} + \overline{v}(p - p_0), \qquad (7.95)$$

where $i_{ph}(T)$ is the specific enthalpy at pressure p, and $\overline{v}$ is the mean specific volume at temperature T_0 for pressure range from p_0 to p.

The specific physical exergy of steam consists of two terms. The first term is equal to an isentropic enthalpy drop from the current state to ambient pressure and temperature. The second term results from an assumption that the remaining steam at ambient condition is first condensed and next the resulting liquid is compressed to the ambient pressure. Based on these assumptions, the specific physical exergy of steam at temperature T and pressure p is

$$b_{ph} = i(T,s) - i(T_0,s) - v_f(p_0 - p_{s0}), \tag{7.96}$$

where $s = s(T,p)$ is the steam specific entropy, v_f is the specific volume of saturated liquid at ambient pressure, and p_{s0} is the pressure of saturated steam at ambient temperature.

7.8.3 CHEMICAL EXERGY

Calculation of the specific chemical exergy of various substances can be complicated and requires application of specific procedures, depending of the type of a substance. If the substance contains the main components of air, the following equation can be applied,

$$b_{ch} = T_0 \sum_i g_i R_i \ln \frac{z_i}{z_{i0}}, \tag{7.97}$$

where z_i, z_{i0} is the molar fraction of i-th component of the substance and in the surrounding air, respectively; g_i is the mass fraction of i-th component in the substance and R_i is the specific gas constant of that component.

The specific chemical exergy of water vapor is equal to the exergy of liquid water at ambient pressure and temperature. It should be assumed that this water is decompressed to the saturation pressure p_{s0} and it evaporates completely (without exergy losses) to obtain the saturated vapor, which is next decompressed to the partial pressure of moisture at ambient pressure, p_{p0}. The resulting formula is as follows,

$$b_{ch} = v_f(p_0 - p_{s0}) + RT_0 \ln \frac{p_{s0}}{p_{p0}}. \tag{7.98}$$

Here $p_{p0} = \varphi_0 p_{s0}$, where φ_0 is the humidity of the atmospheric air representing a ratio of the partial pressure of moisture at ambient pressure to the pressure of saturated vapor at the same temperature.

7.8.4 TOTAL EXERGY OF SUBSTANCE

When a substance with mass m moves with a center-of-mass speed v at height H (both measured against a reference system), and has thermal exergy B_{th}, its total exergy is given as,

$$B = \frac{1}{2}mv^2 + mgH + B_{th} = m\left(\frac{1}{2}v^2 + gH + b_{th}\right), \tag{7.99}$$

where $b_{th} = B_{th}/m$ is the specific thermal exergy. In technical applications we usually consider a stream of substance flowing with a given mass flow rate $W = \dot{m}$. The corresponding rate of change of exergy is as follows,

$$\dot{B} = W\left(\frac{1}{2}v^2 + gH + b_{th}\right). \tag{7.100}$$

Both mechanical (potential and kinetic) and electrical energy can in principle be converted completely into useful work. Thus this type of exergy is equal to the stored energy. As can be expected, this is not the case for other types of exergy.

7.8.5 EXERGY OF HEAT RESERVOIRS

When energy is available only as heat transferred from a heat reservoir, only a fraction of the heat can be transformed into useful work. For example, a geothermal energy can be treated as "hot" heat reservoir at a certain temperature $T > T_0$, where T_0 is the ambient temperature. Assuming that temperature T doesn't change when extracting a certain amount of geothermal heat Q, the maximum amount of useful work that can be extracted is determined by the Carnot efficiency, η_C. Thus,

$$B_{gth} = Q\eta_C = Q\left(1 - \frac{T_0}{T}\right). \tag{7.101}$$

Sometime we have to consider the heat reservoir as a "cold" reservoir, as it is the case for the *ocean thermal energy conversion* (OTEC) systems, where cold water at depth plays the role of a cold reservoir. In this case, since $T < T_0$, and assuming that $T = const$ when adding amount of heat Q to the cold reservoir, we have,

$$B_{OTEC} = Q\left(\frac{T_0}{T} - 1\right), \tag{7.102}$$

where T is the cold water temperature at depth and T_0 is the ambient temperature.

7.8.6 EXERGY LOSSES

The exergy losses are present during various physical processes. One of the reasons for exergy losses can be friction and associated with it heat generation. If the friction heat dQ_f is absorbed by a body with temperature T, the exergy loss due to friction is determined as,

$$\Delta B = T_0 \sum \Delta S = T_0 \int_1^2 \frac{dQ_f}{T}. \tag{7.103}$$

Another process associated with exergy loss is heat transfer with a finite temperature drop. This is one of the most important irreversible processes in thermal engineering. The sum of entropy increases during irreversible heat flow per unit time, q, between two fluids at constant temperatures T_1 and T_2 is as follows,

$$\sum \Delta \dot{S} = \frac{q}{T_2} - \frac{q}{T_1} \tag{7.104}$$

The corresponding exergy loss per unit time is

$$\Delta \dot{B} = q\left(\frac{1}{T_2} - \frac{1}{T_1}\right)T_0 = q\frac{T_1 - T_2}{T_1 T_2}T_0 = \frac{1}{\mathscr{R}}\frac{(T_1 - T_2)^2}{T_1 T_2}T_0. \tag{7.105}$$

Here $\mathscr{R} \equiv (T_1 - T_2)/q$ is the heat transfer resistance and q is the heat flow per unit time.

Another common source of the exergy loss is the isenthalpic throttling. Even though the energy stream is conserved, such thermodynamic process is disadvantageous, since it is causing a significant loss of exergy, given as,

$$\Delta \dot{B} = WT_0(s_2 - s_1), \tag{7.106}$$

where W is a mass flow rate of the fluid, and s_1, s_2 is the specific entropy of the fluid before and after the throttling, respectively.

PROBLEMS

PROBLEM 7.1

A closed and thermally insulated tank contains gas that is mixed with a blender, which is powered from outside. Due to mixing during a certain period of time, the gas temperature increases from 293 to 303 K. The ratio of the internal energy increase to the gas temperature increase is constant and equal to $\Delta E_I/\Delta T = 1$ kJ/K. Find the mixing work added to the system.

PROBLEM 7.2

A closed but thermally uninsulated tank contains gas that is mixed with a blender, which is powered from outside. Due to mixing during a certain period of time, the gas temperature increases from 293 to 303 K. The ratio of the internal energy increase to the gas temperature increase is constant and equal to $\Delta E_I / \Delta T = 1$ kJ/K. Find the mixing work added to the system knowing that the heat transferred from the environment and added to the system during the same period of time was $Q = 2.1$ kJ.

PROBLEM 7.3

Mercury with mass 0.2 kg falls from 10 m height into a calorimeter containing 0.04 kg of water with temperature 293 K. Calculate the temperature of the mercury and water after long enough time to bring them to a thermodynamic equilibrium. Assume that the potential energy of mercury is entirely converted into internal energy of mercury and water. Temperature of mercury, when entering water, is 293 K. Specific heat at constant pressure for mercury and water is 0.14 kJ/(kg·K) and 4.186 kJ/(kg·K), respectively.

PROBLEM 7.4

Calculate the volume of 1 kmol of ideal gas at standard temperature and pressure according to the IUPAC (international Union of Pure and Applied Chemistry): $p = 10^5$ Pa, $T = 273.15$ K.

PROBLEM 7.5

The volume ratio of air to methane (CH_4) in a mixture is 1 to 0.31. Calculate the volume fractions of gases in the mixture, the mixture efective molecular mass and the mixture gas constant.

PROBLEM 7.6

A closed tank with volume $V = 100$ m^3 contains $n = 20$ kmol of methane (CH_4). Calculate the mass of the gas and its volume at international standard metric condition for natural gas: $p = 101325$ Pa, $T = 288.15$ K.

PROBLEM 7.7

Nitrogen gas (N_2) flows through a pipe with inner diameter $d = 0.6$ m. The pipe cross-section average gas velocity is $U = 35$ m/s. Assuming that the gas density is equal to $\rho = 2.3$ kg/m^3, calculate: (a) gas mass flow rate W, kg/s; (b) gas molar flow rate W_M, kmol/h; (c) gas normal volumetric flow rate Q_n, sm^3/min. Use international standard metric condition for natural gas: $p = 101325$ Pa, $T = 288.15$ K.

PROBLEM 7.8

A stream of ideal gas with $\kappa = 1.33$ and initial parameters $p_1 = 1.5$ MPa and $T_1 = 305$ K is heated at constant pressure in a heater and next it expands in a turbine in an reversible adiabatic process to pressure $p_2 = 0.12$ MPa. The turbine has power 12 kW and the heat supplied to the heater is 16.5 kW. Calculate the mass flow rate of the gas through the system.

PROBLEM 7.9

A perfectly insulated tank with volume $V = 18$ m^3 contains saturated steam at pressure $p_1 = 2.4$ MPa and quality $x_1 = 0.96$. Water at temperature 300 K is injected to the tank and, as a result, the pressure in the tank decreases to $p_2 = 0.75$ MPa. Find the amount of water injected to the tank.

PROBLEM 7.10

Using exergy balance, find the minimum work which is needed to produce 1 kmol of industrial oxygen in which the molar fraction of oxygen is 0.94 and the remaining gas is nitrogen at ambient pressure and temperature, $p_a = 101325$ Pa and $T_a = 293$ K.

8 Fluid Flow in Energy Systems

Wind and flowing water carry kinetic energy that can be captured by turbines and subsequently can be converted into electricity in generators. Hot water in geothermal resources can be used for heating, whereas dry-steam power plants use steam directly from a geothermal reservoir to turn turbines. Water at high pressure is used in nuclear reactors to remove fission energy from a core and transfer the heat to high-pressure steam in steam generators. All these applications are only few examples showing the important role of fluid flows in energy systems.

A fluid, unlike a solid, can not support a shear stress in mechanical equilibrium, since all fluid molecules are free to move. Thus the local properties ascribed to fluids include not only such parameters as density, temperature, and pressure, but also a velocity vector, which is actually an average velocity of the molecules in the fluid at the location. This particular feature requires that full local characterization of fluid dynamics includes principles of conservation of mass, momentum, and energy.

The dynamics of fluids can be very complicated, since the conservation equations turn out to be nonlinear and their solution for practical purposes can be obtained only numerically. In fact, the existence and uniqueness of the solution of equation that governs fluid flow, so called Navier-Stokes equation, has not been proven yet[1]. Fortunately, practically useful results can be obtained when simplifying assumptions to the conservation laws are adopted. In many cases it suffices to study ideal fluid at steady-state, using Bernoulli's equation.

In this chapter we explain how these simplified approaches can be derived from more rigorous formulations. To this end we first introduce mass, energy and momentum conservation in various types of systems. The fundamental physical laws that we invoke are Newton's second law of dynamics, the principle of conservation of mass, and the first law of thermodynamics. We employ the integral and differential approaches and derive the governing equations, such as the Navier-Stokes, continuity, and energy equations. Simplifications and solutions of these equations for particular applications are shown.

8.1 GENERALIZED CONSERVATION LAW

Conservation laws of mass, momentum and energy constitute the basis for fluid mechanics and heat transfer. Usually the laws are derived and analyzed separately, since they concern physical quantities that have quite distinct properties. However, the laws have many common aspects, that can be treated in a more general way.

Let Ψ be an arbitrary extensive property of a system with mass m and volume V for which,

$$m = \iiint_V \rho \, dV, \tag{8.1}$$

where ρ is a local mass density in the volume V. The conservation law for Ψ contained in any open system (that is a system which is exchanging mass with the surroundings) can be formulated as follows,

$$\frac{d\Psi}{dt} = \Gamma, \tag{8.2}$$

---

[1]This proof is one of the seven Millennium Prize Problems announced by the Clay Mathematics Institute in May 2000.

DOI: 10.1201/9781003036982-8

where Γ represents all sources and sinks per unit time of the property Ψ in the system under consideration. Since Ψ is an extensive property, it can be represented in an integral form as follows,

$$\Psi = \iiint_{V(t)} \rho \psi dV. \tag{8.3}$$

The integration is over the volume $V(t)$ to indicate that in general the volume can be time-dependent.

Taking $\psi = 1$ in Eq. (8.3) leads to $\Psi = m$, which is the mass contained in volume V. In a similar manner, taking $\psi = \mathbf{v}$ gives the momentum, and $\psi = e_I + \frac{1}{2}v^2$—the total energy of the system. Here $\mathbf{v}$ is the local velocity vector, $\frac{1}{2}v^2 \equiv \frac{1}{2}(\mathbf{v} \cdot \mathbf{v})$ is the specific kinetic energy, and e_I is the specific internal energy of the fluid. Thus by a proper choice of ψ we obtain various physical quantities describing the system.

To close the conservation equation given by Eq. (8.2), the source term on the right-hand-side has to be determined. In general, the source term Γ can be decomposed into two distinct parts,

$$\Gamma = \Gamma_v + \Gamma_s, \tag{8.4}$$

where Γ_v results from volumetric sources, distributed in volume $V(t)$, and Γ_s results from sources distributed over surface $S(t)$, which is surrounding the volume $V(t)$. Thus,

$$\Gamma_v = \iiint_{V(t)} \rho \phi dV, \tag{8.5}$$

where ϕ is a source or sink of Ψ per unit mass and unit time. Not all conserved physical quantities have non-zero source term. For example, in the framework of classical mechanics the mass cannot be created nor destroyed, which dictates $\phi = 0$. However, there are volumetric sources of both the momentum and the total energy. Any bulk force per unit mass $\mathbf{b}$, such as the gravity acceleration $\mathbf{g}$, will lead to a creation of the momentum. In a similar manner, any source of heat per unit volume q''', or a friction work done by the bulk force $\mathbf{b} \cdot \mathbf{v}$ contribute to the total energy change in the system.

The surface source Γ_s, in turn, can result from convection and diffusion of the property Ψ through surface S. Thus,

$$\Gamma_s = \Gamma_{sc} + \Gamma_{sd}, \tag{8.6}$$

where the convective part is given as,

$$\Gamma_{sc} = -\iint_{S(t)} \rho \psi \mathbf{v}_r \cdot \mathbf{n} dS, \tag{8.7}$$

and the diffusive part is given as,

$$\Gamma_{sd} = -\iint_{S(t)} \mathbf{n} \cdot \mathbf{J} dS. \tag{8.8}$$

Here $\mathbf{n}$ is an outward-pointing unit vector normal to surface S, $\mathbf{v}_r = \mathbf{v} - \mathbf{v}_s$ is the relative fluid velocity at the surface, $\mathbf{v}$ is the fluid velocity, $\mathbf{v}_s$ is the surface velocity, and $\mathbf{J}$ is the diffusive flux of Ψ per unit area and unit time. The minus signs in Eqs. (8.7) and (8.8) are used since the positive surface sources are assumed when the property Ψ is convected into volume V, that is in the opposite direction than the vector $\mathbf{n}$ is pointing to.

The diffusive flux of mass is equal to zero $\mathbf{J} = \mathbf{0}$, since mass is not transported through a diffusion process. However, both the momentum and the energy can be transported through diffusion. The momentum is transported by the total stress in flowing fluids, which for the incompressible media is as follows,

$$\mathbf{T} = -p\mathbf{I} + \tau. \tag{8.9}$$

Here $-p\mathbf{I}$ represents the isotropic pressure stress and τ is the anisotropic viscous stress. When operating on a surface, the total stress tensor produces traction $\mathbf{n} \cdot \mathbf{T}$. Thus, for the momentum equation we have $\mathbf{J} = -\mathbf{T}$.

Table 8.1

Definitions of Terms in the Generalized Conservation Law

Conserved Quantity	ψ	$\mathbf{J}$	ϕ
Mass	1	0	0
Linear momentum	$\mathbf{v}$	$-\mathbf{T}$	$\mathbf{b}$
Angular momentum	$\mathbf{r} \times \mathbf{v}$	$-(\mathbf{r} \times \mathbf{T})^T$	$\mathbf{r} \times \mathbf{b}$
Energy	$e_I + \frac{1}{2}v^2$	$\mathbf{q}'' - \mathbf{T} \cdot \mathbf{v}$	$\mathbf{b} \cdot \mathbf{v} + \frac{1}{\rho}q'''$
Entropy	s	$\frac{1}{T}\mathbf{q}''$	$\frac{1}{\rho}\Delta_s$

In a similar manner we obtain the terms ψ, $\mathbf{J}$, and ϕ for the remaining conservation equations, such as the angular momentum conservation and entropy balance, as summarized in Table 8.1.

The vector $\mathbf{r}$ in Table 8.1 describes position of a point in the system with respect to a selected origin of the coordinate system. The form of diffusion vector $\mathbf{J}$ for the angular momentum equation contains the assumption that the shear stress tensor τ is symmetric, as for Newtonian fluids. A symbol $()^T$ denotes a transpose of a tensor. In the energy balance, e_I is the internal specific energy, $\mathbf{q}''$ is the heat flux vector and q''' is the volumetric heat source. In the entropy balance, s is the specific entropy, T is the absolute temperature and Δ_s is the local entropy source per unit volume and time.

8.1.1 GENERAL INTEGRAL CONSERVATION EQUATION

Using terms given by Eqs. (8.3) through (8.8), the conservation equation for the property Ψ in an open system with volume $V(t)$ and surface $S(t)$ can be written as follows,

$$\frac{d}{dt}\left(\iiint_{V(t)} \rho\psi dV\right) = -\iint_{S(t)} [\rho\psi \mathbf{v}_r \cdot \mathbf{n} + \mathbf{n} \cdot \mathbf{J}] dS + \iiint_{V(t)} \rho\phi dV. \qquad (8.10)$$

This equation is valid for any additive (extensive) property Ψ, which in general can be either a scalar (for mass and energy conservation) or a vector (for linear and angular momentum conservation). Since the surface flux term $\mathbf{J}$ is dot-multiplied with the unit vector $\mathbf{n}$, its tensor rank must be higher by 1 compared to the tensor rank of Ψ.

The left-hand-side term represents time change of property Ψ due to its changes inside the volume $V(t)$ and caused by changes of the volume. Since volume $V(t)$ is a function of time, the differentiation operator d/dt cannot be moved under the integration operator.

The right-hand-side of Eq. (8.10) describes sources and sinks of property Ψ resulting from the convection and diffusion through the surface $S(t)$, and from sources which are present in the volume $V(t)$.

8.1.2 STATIONARY CONTROL VOLUME

For a stationary control volume, when both the volume V and the surface S are time-independent ($\mathbf{v}_s = \mathbf{0}$), the integral conservation equation given by Eq. (8.10) is as follows,

$$\frac{d}{dt}\left(\iiint_{V} \rho\psi dV\right) = -\iint_{S} [\rho\psi \mathbf{v} \cdot \mathbf{n} + \mathbf{n} \cdot \mathbf{J}] dS + \iiint_{V} \rho\phi dV. \qquad (8.11)$$

Since now $V = const$, the differentiation on the left-hand-side can be moved under the integration operator to obtain,

$$\iiint_{V} \frac{\partial(\rho\psi)}{\partial t} dV = -\iint_{S} [\rho\psi \mathbf{v} \cdot \mathbf{n} + \mathbf{n} \cdot \mathbf{J}] dS + \iiint_{V} \rho\phi dV. \qquad (8.12)$$

The partial differentiation is used in Eq. (8.12) since the product $\rho \psi$ is in general a function of the time and space coordinates.

8.1.3 MOVING CONTROL VOLUME

When the control volume is moving in reference to a stationary coordinate system, it is necessary to describe the nature of the movement, that is, to specify the velocity (for inertial systems) and the acceleration (for non-inertial systems) of the control volume in some reference inertial coordinate system.

For a control volume $V(t)$ with a surface $S(t)$ moving with velocity $\mathbf{v}_s$ Eq. (8.10) is valid, thus,

$$\frac{\mathrm{d}}{\mathrm{d}t} \left(\iiint_{V(t)} \rho \psi \mathrm{d}V \right) = - \iint_{S(t)} [\rho \psi \mathbf{v}_r \cdot \mathbf{n} + \mathbf{n} \cdot \mathbf{J}] \, \mathrm{d}S + \iiint_{V(t)} \rho \phi \mathrm{d}V. \tag{8.13}$$

where $\mathbf{v}_r = \mathbf{v} - \mathbf{v}_s$ is the relative velocity of fluid at the surface $S(t)$ and $\mathbf{v}$ is the fluid absolute velocity.

8.1.4 MATERIAL VOLUME

A material volume is characterized by the fact that every fluid particle remains in the volume. The boundary of the material volume can alter its shape but the fluid particles that are on the boundary remain on it. In that way the boundary always consists of the same fluid particles. Such boundary is called the material surface. Assuming that the volume $V(t)$ and the surface $S(t)$ in Eq. (8.10) are the material volume and the material surface, respectively, the general conservation equation is as follows,

$$\frac{D}{Dt} \left(\iiint_{V_m(t)} \rho \psi \mathrm{d}V \right) = - \iint_{S_m(t)} \mathbf{n} \cdot \mathbf{J} \mathrm{d}S + \iiint_{V_m(t)} \rho \phi \mathrm{d}V, \tag{8.14}$$

where notation $V_m(t)$ and $S_m(t)$ is used to indicate that integration is over time-dependent material volume and material surface. Since the time derivative on the left-hand-side concerns the material volume, the Stokes derivative, called also the **substantial** or **material derivative**, D/Dt, is used. It should be also noted that for a substantial surface there is no convective component in the surface source term on the right-hand-side of Eq. (8.14), since the relative velocity between the fluid and the surface is zero.

Applying Reynolds' transport theorem, Eq. (8.14) can be written as,

$$\iiint_{V_m(t)} \frac{\partial (\rho \psi)}{\partial t} \mathrm{d}V + \iint_{S_m(t)} \rho \psi \mathbf{v} \cdot \mathbf{n} \mathrm{d}S$$
$$= - \iint_{S_m(t)} \mathbf{n} \cdot \mathbf{J} \mathrm{d}S + \iiint_{V_m(t)} \rho \phi \mathrm{d}V \tag{8.15}$$

or, after rearrangements,

$$\iiint_{V_m(t)} \frac{\partial (\rho \psi)}{\partial t} \mathrm{d}V =$$
$$- \iint_{S_m(t)} [\rho \psi \mathbf{v} \cdot \mathbf{n} + \mathbf{n} \cdot \mathbf{J}] \mathrm{d}S + \iiint_{V_m(t)} \rho \phi \mathrm{d}V \tag{8.16}$$

8.1.5 LOCAL DIFFERENTIAL FORMULATION

The surface integrals in Eqs. (8.15) and (8.16) can be replaced with the volume integrals using the Gauss theorem and the following is obtained,

$$\iiint_{V_m(t)} \frac{\partial (\rho \psi)}{\partial t} \mathrm{d}V + \iiint_{V_m(t)} \nabla \cdot (\rho \psi \mathbf{v} + \mathbf{J}) \mathrm{d}V - \iiint_{V_m(t)} \rho \phi \mathrm{d}V = 0. \tag{8.17}$$

Since all integrals are over the same, arbitrary volume, they can be combined to a single integral, which can be zero only when the expression under the integration operator is zero. Thus, the following instantaneous, local differential balance equation is obtained,

$$\frac{\partial (\rho \psi)}{\partial t} + \nabla \cdot (\rho \psi \mathbf{v} + \mathbf{J}) - \rho \phi = 0. \tag{8.18}$$

We can now substitute terms ψ, $\mathbf{J}$, and ϕ from Table 8.1 to obtain the following local differential balance equations:

Mass

$$\frac{\partial \rho}{\partial t} + \nabla \cdot (\rho \mathbf{v}) = 0. \tag{8.19}$$

Linear momentum

$$\frac{\partial (\rho \mathbf{v})}{\partial t} + \nabla \cdot (\rho \mathbf{v} \mathbf{v} - \mathbf{T}) - \rho \mathbf{b} = 0. \tag{8.20}$$

Angular momentum

$$\frac{\partial (\rho \mathbf{r} \times \mathbf{v})}{\partial t} + \nabla \cdot \left[\mathbf{r} \times \rho \mathbf{v} \mathbf{v} - (\mathbf{r} \times \mathbf{T})^T \right] - \mathbf{r} \times \rho \mathbf{b} = 0. \tag{8.21}$$

Energy

$$\frac{\partial \left[\rho \left(e_I + \frac{1}{2} v^2 \right) \right]}{\partial t} + \nabla \cdot \left[\rho \left(e_I + \frac{1}{2} v^2 \right) \mathbf{v} + \mathbf{q}'' - \mathbf{T} \cdot \mathbf{v} \right] - \rho \mathbf{b} \cdot \mathbf{v} = 0. \tag{8.22}$$

Entropy inequality

$$\frac{\partial (\rho s)}{\partial t} + \nabla \cdot \left(\rho s \mathbf{v} + \frac{1}{T} \mathbf{q}'' \right) - \Delta_s \geq 0. \tag{8.23}$$

Equations (8.19) through (8.23) constitute a general system of equations that describe local behavior of fluid. The system needs additional closure relationships for $\mathbf{T}, \mathbf{q}''$ and Δ_s to be solved. Not all equations need to be considered for particular cases, however. For example, most isothermal cases are uniquely described by the mass and momentum equations only.

8.2 CLOSURE RELATIONSHIPS

A general problem of fluid flow with heat transfer includes five equations describing the conservation of mass, linear momentum (which has three components) and energy. These equations have seventeen unknowns: density ρ, velocity vector $\mathbf{v}$ with three unknown components, total stress tensor $\mathbf{T}$ with nine unknown components, internal energy e_I, and heat flux vector $\mathbf{q}''$ with three unknown components. The typical closure relationships for the set of conservation equations are described below.

8.2.1 TOTAL STRESS TENSOR

The fluid stress tensor for incompressible fluids is given by Eq. (8.9). It can be decomposed as follows,

$$\mathbf{T} = -p_h \mathbf{I} - p_d \mathbf{I} + \tau_n + \tau_s, \tag{8.24}$$

where the first term on the right-hand side represents the hydrostatic pressure stress. This is the only pressure stress present in stationary fluids. We can find traction due to the hydrostatic pressure on a surface with outer normal vector $\mathbf{n}$ as,

$$\mathbf{f_h} = \mathbf{n} \cdot (-p_h \mathbf{I}) = -p_h \mathbf{n}. \tag{8.25}$$

Thus the absolute value of the traction is equal to the hydrostatic pressure, and it is directed normal to the surface.

In moving fluid all four terms on the right-hand side of Eq. (8.24) become non-zero, but the two last ones are present only when the moving fluid is viscous. The viscous normal stress, τ_n, is due to accelerating and decelerating perpendicular motion toward the surface, and is proportional to the viscosity of the fluid and to the velocity gradient. The viscous shear stress, τ_s, is caused by shearing motion of fluid tangential to the surface.

To close the system of conservation equations, each stress component is expressed in terms of the velocity field and fluid properties.

It has been found empirically that for all gases and all liquids with molecular weight of less than about 5000 the resistance to flow is proportional to velocity gradient,

$$\tau \sim \mu \frac{du}{dy}, \tag{8.26}$$

where μ is the dynamic viscosity of the fluid. Newtonian fluids have no "memory" and the stress can be in general given as,

$$\tau = \phi\left(\mathbf{v}, \nabla \mathbf{v}, \mathbf{D}\right). \tag{8.27}$$

Here $\mathbf{D}$ is the **deformation tensor** (symmetric rate-of-strain tensor),

$$\mathbf{D} = \frac{1}{2} \begin{bmatrix} 2\dfrac{\partial u}{\partial x} & \left(\dfrac{\partial u}{\partial y} + \dfrac{\partial v}{\partial x}\right) & \left(\dfrac{\partial u}{\partial z} + \dfrac{\partial w}{\partial x}\right) \\[2mm] \left(\dfrac{\partial v}{\partial x} + \dfrac{\partial u}{\partial y}\right) & 2\dfrac{\partial v}{\partial y} & \left(\dfrac{\partial v}{\partial z} + \dfrac{\partial w}{\partial y}\right) \\[2mm] \left(\dfrac{\partial w}{\partial x} + \dfrac{\partial u}{\partial z}\right) & \left(\dfrac{\partial w}{\partial y} + \dfrac{\partial v}{\partial z}\right) & 2\dfrac{\partial w}{\partial z} \end{bmatrix}, \tag{8.28}$$

and the velocity gradient tensor is as follows,

$$\nabla \mathbf{v} = \begin{bmatrix} \dfrac{\partial u}{\partial x} & \dfrac{\partial v}{\partial x} & \dfrac{\partial w}{\partial x} \\[2mm] \dfrac{\partial u}{\partial y} & \dfrac{\partial v}{\partial y} & \dfrac{\partial w}{\partial y} \\[2mm] \dfrac{\partial u}{\partial z} & \dfrac{\partial v}{\partial z} & \dfrac{\partial w}{\partial z} \end{bmatrix}. \tag{8.29}$$

It can be shown that

$$\nabla \mathbf{v} = \mathbf{D} + \mathbf{S}, \tag{8.30}$$

where $\mathbf{S}$ is the antisymmetric **vorticity tensor**,

$$\mathbf{S} = \frac{1}{2} \begin{bmatrix} 0 & -\left(\dfrac{\partial v}{\partial x} - \dfrac{\partial u}{\partial y}\right) & \left(\dfrac{\partial u}{\partial z} - \dfrac{\partial w}{\partial x}\right) \\[2mm] \left(\dfrac{\partial v}{\partial x} - \dfrac{\partial u}{\partial y}\right) & 0 & -\left(\dfrac{\partial w}{\partial y} - \dfrac{\partial v}{\partial z}\right) \\[2mm] -\left(\dfrac{\partial u}{\partial z} - \dfrac{\partial w}{\partial x}\right) & \left(\dfrac{\partial w}{\partial y} - \dfrac{\partial v}{\partial z}\right) & 0 \end{bmatrix}. \tag{8.31}$$

For Newtonian fluids, the local stress is entirely due to the local deformation tensor and is well-approximated with the following equation

$$\tau = 2\mu \mathbf{D} = \mu \left[\nabla \mathbf{v} + (\nabla \mathbf{v})^T\right]. \tag{8.32}$$

This expression is sometime generalized as proposed by Stokes,

$$\tau = \mu \left[\nabla \mathbf{v} + (\nabla \mathbf{v})^T \right] + \left(\kappa - \frac{2}{3}\mu \right) (\nabla \cdot \mathbf{v}) \mathbf{I} \tag{8.33}$$

where κ is called the **bulk coefficient of viscosity**. For incompressible flows $\nabla \cdot \mathbf{v} = 0$ and the second term on the right-hand side of Eq. (8.33) is zero.

8.2.2 HEAT FLUX

The heat flux vector can be expressed according to **Fourier's law** as

$$\mathbf{q}'' = -\lambda \nabla T, \tag{8.34}$$

where λ is the thermal conductivity and ∇T is the temperature gradient.

8.2.3 ENTROPY GENERATION

There are various sources of entropy in irreversible processes caused by fluid viscosity and heat conduction. For a Newtonian fluid, the entropy source due to both these effects is given as,

$$\Delta_s = \mathbf{q}'' \cdot \nabla \left(\frac{1}{T} \right) + \frac{1}{T} \tau : \mathbf{D} \geq 0, \tag{8.35}$$

where T is the temperature, $\mathbf{q}''$ is the heat flux vector, τ is the shear stress tensor and $\mathbf{D}$ is the deformation tensor.

8.3 SPACE-AVERAGED FLOW IN A TUBE

We will consider a fixed tube aligned with Oz axis, with constant cross-section area A. A fixed volume V is cut by two planes perpendicular to Oz, located at a distance Δz from each other. Thus the volume of the pipe segment that is limited by the planes and the tube wall is $V = A\Delta z$. The first plane, representing the inflow to the control volume, is located at coordinate z and the second, at which the fluid leaves the control volume, at $z + \Delta z$. The unit normal vector at the inlet to the control volume is thus $\mathbf{n}_i = -\mathbf{e}_z$ and at the outlet $\mathbf{n}_o = \mathbf{e}_z$. Here $\mathbf{e}_z$ is the base unit vector along z-axis. Further we assume that the flow is rectilinear in the direction of z-axis, so that the velocity vector is $\mathbf{v} = w(x,y,z,t)\mathbf{e}_z$ everywhere in the tube. For such defined stationary control volume, the generalized integral conservation equation (8.11) can be written as,

$$\frac{\mathrm{d}}{\mathrm{d}t} \left(\iiint_V \rho \psi \mathrm{d}V \right) = - \iint_S \rho \psi \mathbf{v} \cdot \mathbf{n} \mathrm{d}S - \iint_S \mathbf{n} \cdot \mathbf{J} \mathrm{d}S + \iiint_V \rho \phi \mathrm{d}V. \tag{8.36}$$

We make an assumption that the fluid density is constant, thus

$$\frac{\mathrm{d}}{\mathrm{d}t} \left(\iiint_V \psi \mathrm{d}V \right) = - \iint_S \psi \mathbf{v} \cdot \mathbf{n} \mathrm{d}S - \frac{1}{\rho} \iint_S \mathbf{n} \cdot \mathbf{J} \mathrm{d}S + \iiint_V \phi \mathrm{d}V. \tag{8.37}$$

The volume integral on the left-hand side can be written as,

$$\iiint_V \psi \mathrm{d}V = \int_z^{z+\Delta z} \iint_A \psi \mathrm{d}A \mathrm{d}z. \tag{8.38}$$

Introducing the following definition of the area-averaged ψ

$$\overline{\psi}^A \equiv \frac{1}{A} \iint_A \psi \mathrm{d}A, \tag{8.39}$$

the volume integral in Eq. (8.38) can be written as

$$\iiint_V \psi dV = A \int_z^{z+\Delta z} \overline{\psi}^A dz. \tag{8.40}$$

In a similar manner, the third integral on the right-hand side of Eq. (8.37) can be written as,

$$\iiint_V \phi dV = A \int_z^{z+\Delta z} \overline{\phi}^A dz, \tag{8.41}$$

where now $\overline{\phi}^A$ is the area-averaged source of quantity ψ in the tube cross-section. It should be mentioned that both averaged parameters are now functions of z and t only.

The surface integrals in Eq. (8.37) consists of three parts: the integration over the inlet surface at z, with area A_i and surface normal unit vector $\mathbf{n}_i = -\mathbf{e}_z$, the integration over the outlet surface at $z + \Delta z$, with area A_o and surface normal unit vector $\mathbf{n}_o = \mathbf{e}_z$, and the integration over the tube wall surface with area $\Delta A_w = P_w \Delta z$ and with surface normal unit vector $\mathbf{n}_w$. Here P_w is the perimeter of the tube. Thus, we can write

$$\iint_S \psi \mathbf{v} \cdot \mathbf{n} dS = \iint_{A_i} \psi w \mathbf{e}_z \cdot (-\mathbf{e}_z) dA_i + \iint_{A_o} \psi w \mathbf{e}_z \cdot \mathbf{e}_z dA_o$$
$$+ \iint_{A_w} \psi w \mathbf{e}_z \cdot \mathbf{n}_w dA_w \tag{8.42}$$

The third term on the right-hand side is zero, since vectors $\mathbf{e}_z$ and $\mathbf{n}_w$ are perpendicular to each other, thus $\mathbf{e}_z \cdot \mathbf{n}_w = 0$. Since $\mathbf{e}_z \cdot \mathbf{e}_z = 1$, we get

$$\iint_S \psi \mathbf{v} \cdot \mathbf{n} dS = -\iint_{A_i} \psi w dA_i + \iint_{A_o} \psi w dA_o. \tag{8.43}$$

We can introduce here the area-averaged and velocity weighted quantity ψ as follows

$$\overline{\psi}^{Aw} \equiv \frac{\iint_A \psi w dA}{\iint_A w dA} = \frac{\iint_A \psi w dA}{A \overline{w}^A}, \tag{8.44}$$

to get

$$\iint_S \psi \mathbf{v} \cdot \mathbf{n} dS = A_o \overline{\psi}^{Aw}(z+\Delta z, t) \overline{w}^A(z+\Delta z, t) - A_i \overline{\psi}^{Aw}(z, t) \overline{w}^A(z, t). \tag{8.45}$$

Here we introduced the area-averaged flow velocity as follows,

$$\overline{w}^A \equiv U \equiv \frac{1}{A} \iint_A w dA. \tag{8.46}$$

Thus, Eq. (8.45) becomes,

$$\iint_S \psi \mathbf{v} \cdot \mathbf{n} dS = A \left[\overline{\psi}^{Aw}(z+\Delta z, t) U(z+\Delta z, t) - \overline{\psi}^{Aw}(z, t) U(z, t) \right]. \tag{8.47}$$

Here the assumption on the constancy of the cross-section area was used, thus $A_o = A_i = A$.

The second term on the right-hand side of Eq. (8.37) can be written as,

$$\iint_S \mathbf{n} \cdot \mathbf{J} dS = -\iint_{A_i} \mathbf{e}_z \cdot \mathbf{J} dA_i + \iint_{A_o} \mathbf{e}_z \cdot \mathbf{J} dA_o + \iint_{A_w} \mathbf{n}_w \cdot \mathbf{J} dA_w. \tag{8.48}$$

Further operations depend on the tensor rank of $\mathbf{J}$. In case it is a vector (a tensor rank 1), which is the case for the energy and entropy balance equations, the dot product $\mathbf{n} \cdot \mathbf{J}$ is a scalar, so all terms in Eq. (8.48) will be scalars. For the linear and angular momentum $\mathbf{J}$ is a tensor of rank 2, so the dot product $\mathbf{n} \cdot \mathbf{J}$ will result in a vector. That is, Eq. (8.48) is a vector equation with three components. However, since the flow is rectilinear and we are interested in flow evolution only in the direction of Oz axis, all terms of the equation can be dot multiplied with base unit vector $\mathbf{e}_z$. Thus, in this case the final equation will be a scalar equation as well.

To further facilitate the analysis, let us assume a circular tube and that the flow is described in cylindrical polar coordinates (r, θ, z). Then we have $\mathbf{e}_z = [0\ 0\ 1]$ and $\mathbf{n}_w = [1\ 0\ 0]$. For $\mathbf{J} = [J_r\ J_\theta\ J_z]^T$ we have,

$$\mathbf{e}_z \cdot \mathbf{J} = J_z, \qquad \mathbf{n}_w \cdot \mathbf{J} = J_r. \tag{8.49}$$

Equation (8.48) gives now,

$$\iint_S \mathbf{n} \cdot \mathbf{J} dS = -\iint_{A_i} J_z dA_i + \iint_{A_o} J_z dA_o + \iint_{A_w} J_r dA_w. \tag{8.50}$$

For the first two integrals on the right-hand side of Eq. (8.50) we use the area-averaged quantities as defined by Eq. (8.39). For the third integral we have,

$$\iint_{A_w} J_r dA_w = \int_z^{z+\Delta z} \int_{P_w} J_r dP_w dz = P_w \int_z^{z+\Delta z} \overline{J_r}^P dz, \tag{8.51}$$

where we introduced the perimeter-averaged flux term,

$$\overline{J_r}^P \equiv \frac{1}{P_w} \int_{P_w} J_r dP_w. \tag{8.52}$$

Using the definitions of averaged-flux terms, Eq. (8.50) becomes,

$$\iint_S \mathbf{n} \cdot \mathbf{J} dS = A \left[\overline{J_z}^A (z + \Delta z, t) - \overline{J_z}^A (z, t) \right] + P_w \int_z^{z+\Delta z} \overline{J_r}^P dz. \tag{8.53}$$

Substituting Eqs. (8.40), (8.41), (8.47) and (8.53) into (8.37) and taking $\Delta z \to 0$ we obtain the following one-dimensional general conservation equation

$$\frac{\partial \overline{\psi}^A}{\partial t} = -\frac{\partial \left(\overline{\psi}^{Aw} U \right)}{\partial z} - \frac{1}{\rho} \frac{\partial \overline{J_z}^A}{\partial z} - \frac{1}{\rho} \frac{P_w}{A} \overline{J_r}^P + \overline{\phi}^A. \tag{8.54}$$

The obtained equation is a generic conservation equation for flow of constant-density fluid in a circular tube with a constant cross section area. The equation is valid when the flux term $\mathbf{J}$ is a vector. A similar derivation can be performed for $\mathbf{J}$ as a tensor, and the following equation is obtained,

$$\frac{\partial \overline{\psi}^A}{\partial t} = -\frac{\partial \left(\overline{\psi}^{Aw} U \right)}{\partial z} - \frac{1}{\rho} \frac{\partial \overline{J_{zz}}^A}{\partial z} - \frac{1}{\rho} \frac{P_w}{A} \overline{J_{rz}}^P + \overline{\phi}^A. \tag{8.55}$$

Here we assume that $\mathbf{J}$ has the following form,

$$\mathbf{J} = \begin{bmatrix} J_{rr} & J_{r\theta} & J_{rz} \\ J_{\theta r} & J_{\theta\theta} & J_{\theta z} \\ J_{zr} & J_{z\theta} & J_{zz} \end{bmatrix}. \tag{8.56}$$

In the following, we will use the generic equations to formulate the mass and momentum conservation equation for flows in channels.

8.3.1 AVERAGED MASS CONSERVATION EQUATION

For mass conservation, we take $\psi = 1$, $\mathbf{J} = \mathbf{0}$, $\phi = 0$ and substitute to Eq. (8.54). As a result, we have,

$$\frac{\partial U}{\partial z} = 0. \tag{8.57}$$

This equation indicates that the mean flow velocity in the tube is independent of the location. Integration along the tube length gives $U = f(t)$. In particular, if the velocity at the inlet to the tube is a known function of time $U_{in}(t)$, then at any location in the tube the velocity is also known and equal to $U(z,t) = U_{in}(t)$. It should be remembered that this result is valid for constant-density fluids flowing in channels with constant cross section area.

8.3.2 AVERAGED MOMENTUM CONSERVATION EQUATION

For the momentum equation we have $\psi = w\mathbf{e}_z$, $\mathbf{J} = -\mathbf{T} = p\mathbf{I} - \tau$, $\phi = \mathbf{g}$ and we get from Eq. (8.55) the following

$$\frac{\partial U}{\partial t} = -\frac{\partial \left(\overline{w}^{Aw}U\right)}{\partial z} - \frac{1}{\rho}\frac{\partial p}{\partial z} + \frac{1}{\rho}\frac{\partial \overline{\tau}_{zz}^{A}}{\partial z} - \frac{1}{\rho}\frac{P_w}{A}\overline{\tau}_{rz}^{P} + \mathbf{g}\cdot\mathbf{e}_z. \tag{8.58}$$

To solve the equation, the closure relationships for $\overline{w}^{Aw}$, $\overline{\tau}_{zz}^{A}$ and $\overline{\tau}_{rz}^{P}$ are needed. In general, for non-uniform velocity distribution in the tube cross section we have $\overline{w}^{Aw} \neq \overline{w}^{A} = U$. However, it is a common practice, as a first approximation, to take $\overline{w}^{Aw} = U$. For turbulent flows, when an empirical $\frac{1}{7}$ power law velocity profile prevails, the assumption introduces an error of about 2%.

The normal averaged viscous shear stress $\overline{\tau}_{zz}^{A}$ is very small in most cases of interest for energy systems and can be neglected.

The most significant closure issue is to express the perimeter-averaged wall shear stress $\tau_w \equiv \overline{\tau}_{rz}^{P}$ in terms of the flow velocity, fluid properties and geometry characteristics.

After the above-mentioned assumptions, the momentum equation is as follows,

$$\frac{\partial U}{\partial t} = -\frac{\partial \left(U^2\right)}{\partial z} - \frac{1}{\rho}\frac{\partial p}{\partial z} - \frac{1}{\rho}\frac{P_w\tau_w}{A} + \mathbf{g}\cdot\mathbf{e}_z. \tag{8.59}$$

Since in view of Eq. (8.57)

$$\frac{\partial \left(U^2\right)}{\partial z} = 2U\frac{\partial U}{\partial z} = 0,$$

the momentum equation becomes,

$$\frac{\partial U}{\partial t} = -\frac{1}{\rho}\frac{\partial p}{\partial z} - \frac{1}{\rho}\frac{P_w\tau_w}{A} + \mathbf{g}\cdot\mathbf{e}_z. \tag{8.60}$$

When the inlet velocity is constant, we have

$$-\frac{\mathrm{d}p}{\mathrm{d}z} = \frac{P_w\tau_w}{A} - \rho\mathbf{g}\cdot\mathbf{e}_z. \tag{8.61}$$

For a tube with length L, the total pressure drop is as follows

$$\Delta p = p_o - p_i = -\frac{LP_w\tau_w}{A} + L\rho\mathbf{g}\cdot\mathbf{e}_z. \tag{8.62}$$

8.4 INTERNAL FLOWS

Flows bounded by solid surfaces are called internal or duct flows. The internal flow of liquids, where there is a free surface, is termed open-channel flow. Common examples of internal flows include flow through fluid machines and pipes, whereas open-channel flow exists in rivers, aqueducts, and irrigation ditches. The basic parameters, which are describing a duct, or a channel, are:

- flow cross-section area, A,
- wetted perimeter, P_w,
- wall roughness, k,
- hydraulic diameter, D_h,
- length, L.

The hydraulic diameter is defined as,

$$D_h \equiv \frac{4 \times \text{cross-section area}}{\text{wetted perimeter}} = \frac{4A}{P_w}. \tag{8.63}$$

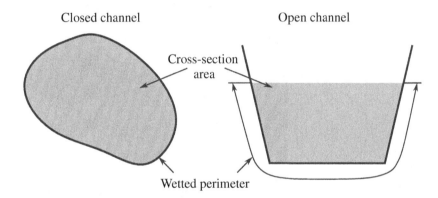

Figure 8.1 The channel cross-section area and the wetted perimeter used in a definition of the hydraulic diameter.

For a circular channel with a diameter D we have,

$$D_h \equiv \frac{4A}{P_w} = \frac{4 \cdot \pi D^2 / 4}{\pi D} = D. \tag{8.64}$$

A hydraulic diameter can be defined for a channel with a perimeter containing a rigid stationary wall (such as for the closed channel shown in Fig. 8.1), and also for more complex channels, in which a solid wall constitutes only a fraction of the perimeter. In such cases care must be exercised when defining the wetted perimeter. As an example, let us consider such more complex channels.

Figure 8.2 shows a cross-section of a typical fuel-rod array for a light water reactor. The array is characterized with two parameters: a *fuel rod diameter*, D, and a fuel rod centerline-to-centerline spacing, commonly called the *fuel rod pitch*, P. The total flow area can be divided into parts called **subchannels**. As shown in the figure, different definitions of subchannels are possible.

Subchannel S_1 has the following hydraulic diameter,

$$D_{h,S_1} = \frac{4A}{P_w} = \frac{4(P^2 - \pi D^2 / 4)}{\pi D} = D \left[\frac{4}{\pi} \left(\frac{P}{D} \right)^2 - 1 \right]. \tag{8.65}$$

In a similar manner we can obtain a hydraulic diameter for subchannel S_2. Since both subchannels have identical cross-section areas A and wetted perimeters P_w, we conclude immediately that hydraulic diameters of S_1 and S_2 are equal. For subchannel S_3 we get (noting that here the cross-section area and the wetted perimeter are both equal to 1/4-th of values for subchannel S_1),

$$D_{h,S_3} = \frac{4A}{P_w} = \frac{4(P^2 - \pi D^2/4)/4}{\pi D/4} = D\left[\frac{4}{\pi}\left(\frac{P}{D}\right)^2 - 1\right]. \tag{8.66}$$

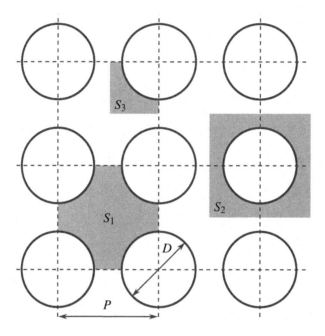

Figure 8.2 Fuel rod array with various subchannel definitions.

Thus, the conclusion is that for all three subchannel definitions shown in Fig. 8.2, the hydraulic diameters remain the same.

For a channel with a hydraulic diameter D_h and a mean in cross-section flow velocity U, a characteristic non-dimensional number called **Reynolds number** is defined as,

$$\mathrm{Re} = \frac{U D_h}{\nu} = \frac{\rho U D_h}{\mu}, \tag{8.67}$$

where ν is a kinematic viscosity of fluid and $\mu = \rho\nu$ is the dynamic viscosity of the fluid flowing in the channel.

As first established by Osborne Reynolds, for a specific channel and fluid flowing in it, there exists a certain value of the cross-section mean velocity that the flow transits from a well-ordered, **laminar flow**, to a chaotic and disordered **turbulent flow**. The transition condition is determined by the Reynolds number and for a pipe it is given as,

$$\mathrm{Re} \approx 2300. \tag{8.68}$$

Most flows of industrial interest have much higher Reynolds numbers and thus are turbulent.

For laminar flows in long circular pipes, when flow is fully developed and the pipe entrance effects are not present, the local velocity distribution is axisymmetric and has a parabolic shape as a function of the pipe radius. When flow is turbulent, the shape of velocity distribution changes significantly and becomes almost flat in the channel cross-section area. The flow is then turbulent everywhere in the channel except for a very thin region close to the wall, known as a **viscous sublayer**. The characteristic features of this layer are that the shear stress within the layer is virtually constant and the turbulent effects are essentially absent. The thickness δ of the viscous sublayer in a pipe with a diameter D can be approximately determined from the following equation,

$$\frac{\delta}{D} = 62\mathrm{Re}^{-7/8}. \tag{8.69}$$

Since this viscous sublayer is very thin, it is commonly accepted to approximate the velocity distribution for turbulent flow in the entire cross-section area of a circular pipe according to the so-called $\frac{1}{n}$ formula as follows,

$$\frac{w(r)}{w_c} = \left(\frac{R-r}{R}\right)^{1/n}, \tag{8.70}$$

where w_c is the flow velocity at the channel centerline and n is an exponent that depends on the Reynolds number. For $\mathrm{Re} \sim 10^5$, $n = 7$.

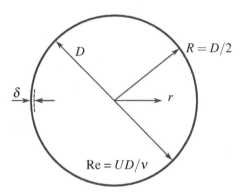

Figure 8.3   A viscous sublayer for turbulent flow in a pipe cross section.

8.4.1   AVERAGE FLOW PARAMETERS

Pressure losses in channels can be expressed in terms of cross-section averaged flow parameters, such as an averaged flow velocity, volumetric flow rate, mass flux or mass flow rate.

The area-averaged flow velocity can be in general calculated as,

$$U \equiv \frac{1}{A} \iint_A w \, dA, \tag{8.71}$$

where w is the local flow velocity in the cross section area. Taking a circular pipe with a radius R shown in Fig. 8.3 as an example, and employing the polar cylindrical coordinates system, the average flow velocity can be found as,

$$U = \frac{1}{\pi R^2} \int_0^R w(r) 2\pi r \, dr \tag{8.72}$$

Taking a velocity profile given by Eq. (8.70) we get

$$U = \frac{2n^2}{(1+n)(1+2n)} w_c, \tag{8.73}$$

where w_c is the local velocity at the center of the pipe. Substituting $n = 7$, we have $U \simeq 0.817 w_c$.

In many applications the parameter of interest to describe a fluid flow in a channel is the rate of flow of the fluid volume. This parameter is called the **volumetric flow rate** and is defined as,

$$Q = \iint_A w \, dA. \tag{8.74}$$

Combining Eqs. (8.71) and (8.74) we get

$$Q = UA. \tag{8.75}$$

Thus the volumetric flow rate can be found from the average flow velocity by multiplying it with the channel cross-section area. Similarly we define the **mass flux** G in a channel as,

$$G \equiv \frac{1}{A} \iint_A \rho w \, dA, \tag{8.76}$$

and a **mass flow rate** W as

$$W \equiv \iint_A \rho w \, dA = GA, \tag{8.77}$$

where ρ is the fluid density.

8.4.2 WALL SHEAR STRESS AND FRICTION PRESSURE LOSS

We will consider a stationary momentum balance for a differential pipe volume as shown in Fig. 8.4. The pipe is assumed to have a constant cross-section area $A = \pi R^2$, with walls parallel everywhere to the pipe axis Oz. That is, we have everywhere $\mathbf{n}_w \cdot \mathbf{e}_z = 0$, where $\mathbf{n}_w$ is a vector normal to the wall and $\mathbf{e}_z$ is the unit base vector in the pipe axis direction. The momentum balance for the differential volume is as follows:

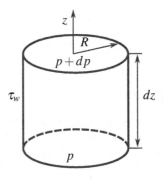

Figure 8.4 Differential pipe element and forces acting on the fluid in the axial direction.

$$p \pi R^2 - (p + dp) \pi R^2 - \tau_w \, dz \, 2\pi R + \pi R^2 \, dz \, \rho g_z = 0, \tag{8.78}$$

where R is the pipe radius, τ_w is the wall shear stress, p is pressure, ρ is the fluid density and g_z is a gravity acceleration vector projected on axis of the pipe. Equation (8.78) leads to the following differential equation,

$$-\frac{dp}{dz} = \frac{2\tau_w}{R} - \rho g_z. \tag{8.79}$$

As can be seen, we obtained an equation equivalent to that derived in §8.3, where space averaging of flow in a tube was used. Assuming a constant shear stress along the pipe and vertical upward flow, we can integrate the differential equation between cross sections 1 and 2 to obtain,

$$p_1 - p_2 = \frac{2\tau_w}{R}(z_2 - z_1) + \rho g(z_2 - z_1). \tag{8.80}$$

The above equation shows that the pressure drop along the channel consists of two parts: the friction loss part and the gravity head part. To find the friction pressure loss it is necessary to know the value of the wall shear stress.

Assuming stationary, fully developed, axisymmetric laminar flow of a Newtonian fluid in a circular tube, in which pressure is a function of the axial distance only, the momentum conservation equation can be expressed in the polar cylindrical coordinate system as follows,

$$0 = -\frac{dp}{dz} - \rho g + \frac{\mu}{r}\frac{d}{dr}\left(r\frac{dw}{dr}\right), \tag{8.81}$$

where $w(r)$ is a radial distribution of the flow velocity in the axial direction. Since $dp/dz = const$, we find the following solution of the above differential equation,

$$w(r) = 2U\left[1 - \left(\frac{r}{R}\right)^2\right], \tag{8.82}$$

where U is the cross-section averaged flow velocity:

$$U \equiv \frac{1}{\pi R^2}\int_0^R w(r)2\pi r\,dr = -\frac{dp/dz + \rho g}{8\mu}R^2. \tag{8.83}$$

Since the velocity distribution in the pipe is known, the wall shear stress can be calculated as

$$\tau_w = -\tau(r)|_{r=R} = -\mu\frac{dw}{dr}|_{r=R} = \frac{4\mu U}{R}. \tag{8.84}$$

The equation shows that for laminar flows the wall shear stress is proportional to the area-averaged flow velocity and inversely proportional to the tube radius.

For turbulent flows no analytical expression for the wall shear stress exists. Experimental data suggest, however, that the wall shear stress for turbulent flow is proportional to the square of U, and the following general relationship is valid,

$$\tau_w = C_f\frac{1}{2}\rho U^2, \tag{8.85}$$

where C_f is the **Fanning friction factor**. Combining Eqs. (8.84) and (8.85) we get

$$C_f = \frac{4\mu U/R}{\rho U^2/2} = \frac{16}{Re}, \tag{8.86}$$

where $Re = \rho UD/\mu$ is the Reynolds number. The above equation gives the Fanning friction factor for laminar flows in circular tubes.

A general expression for the frictional pressure gradient obtained in Eq. (8.79) can be written in terms of the Fanning friction factor as,

$$-\left(\frac{dp}{dz}\right)_f = \frac{2\tau_w}{R} = \frac{4C_f}{D}\frac{\rho U^2}{2} = \frac{\lambda}{D}\frac{\rho U^2}{2}. \tag{8.87}$$

Here the **Darcy-Weisbach friction factor** (also called the **Moody friction factor**) is introduced:

$$\lambda = 4C_f. \tag{8.88}$$

Table 8.2
Roughness k for Pipes and Channels

Type of Wall Material	Roughness k (mm)
Glass	Smooth, < 0.0003
Drawn tubing	0.0015
Wrought iron	0.046
Asphalted cast iron	0.122
Galvanized iron	0.152
Cast iron	0.259
Asphalted steel	0.030
Commercial steel	0.046
Welded steel	0.091
Riveted steel	0.762
Rusted steel	1.524
Vibrated concrete	0.061
Smooth concrete	0.183
Cement plaster	0.457
Unfinished concrete	0.305
Old concrete	15.24
Planned wood	0.030
Rough wood	0.610
Old wood	1.524
Brick	0.610
Stones	122.0

Friction factors for turbulent flows in tubes have been obtained experimentally. When a tube has smooth walls, the **Blasius correlation** is applicable,

$$C_f = \frac{0.0791}{\mathrm{Re}^{0.25}}. \tag{8.89}$$

The correlation is valid for $4000 < \mathrm{Re} < 10^5$. For commercial tubes with non-zero wall roughness, the following formula was proposed by Coolebrook [18],

$$\lambda^{-0.5} = -2.0 \log\left(\frac{k/D}{3.7} + \frac{2.51}{\mathrm{Re}} \lambda^{-0.5}\right), \tag{8.90}$$

where k is the wall roughness height, given in Table 8.2 for various types of wall materials. The above equation is transcendental and a few iterations are required to find λ. An explicit formula, agreeing within 1% with the Coolbrook formula, was proposed by Haaland [38],

$$\lambda^{-0.5} = -1.8 \log\left[\left(\frac{k/D}{3.7}\right)^{1.11} + \frac{6.9}{\mathrm{Re}}\right]. \tag{8.91}$$

A friction pressure loss in a channel with length L and hydraulic diameter D_h can be found from integration of the pressure gradient given by Eq. (8.87). Assuming constant C_f, D_h and U along the pipe, the following expression is obtained,

$$-\Delta p_f = -\int_0^L \left(\frac{dp}{dz}\right)_f dz = \frac{4 C_f L}{D_h} \frac{\rho U^2}{2}. \tag{8.92}$$

8.4.3 MACROSCOPIC ENERGY BALANCE FOR ADIABATIC CHANNEL

For steady-state flow of incompressible fluid in a channel, the mass conservation requires that $W_1 = W_2$, thus, the mass flow rate along the channel is constant. This condition is satisfied irrespective of any energy losses in the channel.

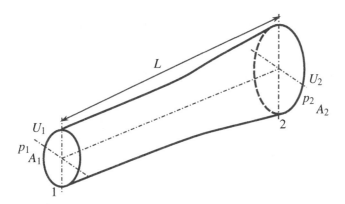

Figure 8.5 Flow in a channel with a variable cross-section area.

For a channel segment, as shown in Fig. 8.5, the steady-state energy conservation equation, given in a general form by Eq. (7.12), can be written as,

$$\frac{dE_T}{dt} = 0 = q - N + (i + e_P + e_K)_1 W_1 - (i + e_P + e_K)_2 W_2, \tag{8.93}$$

where E_T is the total energy, q is the thermal power, N is the mechanical power, i is the specific enthalpy of the fluid, e_K is the specific kinetic energy of the fluid and e_P is the specific potential energy of the fluid.

Assuming a steady and adiabatic flow, with no mechanical work performed, the energy conservation equation becomes,

$$\left(i_1 + \frac{U_1^2}{2} + gH_1\right) - \left(i_2 + \frac{U_2^2}{2} + gH_2\right) = 0. \tag{8.94}$$

Since $i = e_I + p/\rho$, the above equation can be written as,

$$\left(\frac{p_1}{\rho} + \frac{U_1^2}{2} + gH_1\right) - \left(\frac{p_2}{\rho} + \frac{U_2^2}{2} + gH_2\right) - (e_{I2} - e_{I1}) = 0. \tag{8.95}$$

Both above equations are equivalent to each other. However, the latter explicitly shows a possible difference between fluid internal energies at cross sections 1 and 2. If this difference is greater than zero ($e_{I2} > e_{I1}$), the equation requires that $p_1 - p_2 = \rho(e_{I2} - e_{I1}) > 0$, assuming further that channel cross section area doesn't change ($U_1 = U_2$) and the elevations at the inlet and outlet are the same ($H_1 = H_2$). Equation (8.95) shows that any increase of the internal energy in the considered system must be accompanied with a corresponding pressure loss. Such **energy degradation** (that is transformation of the mechanical energy into the internal energy) takes place in all systems where a viscous dissipation exists.

Assuming an ideal system with no viscous dissipation, we obtain the following equality:

$$\left(\frac{p_1}{\rho} + \frac{U_1^2}{2} + gH_1\right) = \left(\frac{p_2}{\rho} + \frac{U_2^2}{2} + gH_2\right) = const. \tag{8.96}$$

The obtained equation represents a principle of conservation of mechanical energy in an ideal inviscid flow. The equation is sometime called the **engineering Bernoulli equation** and is applied to pressure drop calculations in channels.

When Eq. (8.96) is applied to flows in channels with non-uniform distribution of flow velocity in the cross section, the following form of equation should be used

$$\left(\frac{p_1}{\rho} + \frac{K_1^{(3)}U_1^2}{2} + gH_1 \right) = \left(\frac{p_2}{\rho} + \frac{K_2^{(3)}U_2^2}{2} + gH_2 \right) = const, \tag{8.97}$$

where $K^{(r)}$ is the **average velocity coefficient** of r-th order, defined as

$$K^{(r)} \equiv \frac{\overline{w^r}^A}{(\overline{w}^A)^r} = \frac{\overline{w^r}^A}{U^r}. \tag{8.98}$$

Here, $U \equiv \overline{w}^A$ is the averaged flow velocity in the cross-section area and w is the local flow velocity in the cross-section.

Assuming $\frac{1}{n}$-th turbulent velocity profile given by Eq. (8.70), the average velocity coefficient of 3-rd order is as follows,

$$K^{(3)} = \left(\frac{w_c}{U} \right)^3 \frac{2n^2}{(3+n)(3+2n)} = \frac{(1+n)^3(1+2n)^3}{4n^4(3+n)(3+2n)}. \tag{8.99}$$

Using $n = 7$, we get $K^{(3)} \simeq 1.0584$. Since $K^{(3)}$ for turbulent flows is very close to 1, it is common practice to omit it in Eq. (8.97). However, for laminar flows in circular pipes we have $K^{(3)} = 2$ and the coefficient shouldn't be omitted.

8.4.4   LOCAL PRESSURE LOSSES

Additional pressure losses for flows in channels can occur due to local obstacles, such as bends, local cross-section area contractions, local cross-section area expansions, as well as inflows or outflows from a channel to or from a vessel.

Let us consider flow through a sudden channel area expansion, as shown in Fig. 8.6.

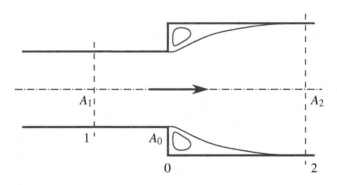

Figure 8.6   Flow through a sudden expansion of a channel.

The mass conservation equation between cross sections 1 and 2 yields

$$\rho A_1 U_1 - \rho A_2 U_2 = 0, \quad \text{thus,} \quad U_2 = U_1 \frac{A_1}{A_2}. \tag{8.100}$$

The momentum conservation equation gives

$$p_1 A_1 + W_1 U_1 + p_0 (A_2 - A_1) - p_2 A_2 - W_2 U_2 = 0. \tag{8.101}$$

Here term $p_0(A_2 - A_1)$ gives a force resulting from the pressure exerted by the expansion (washer-shaped) area $A_2 - A_1$ on the fluid. Since there are no significant pressure losses between 1 and 0 (we assume that cross section 1 is located close to the sudden expansion cross section 0), it can be assumed that $p_0 = p_1$, which yields the following pressure drop through the expansion region,

$$\Delta p = p_2 - p_1 = \rho U_2 (U_1 - U_2). \tag{8.102}$$

Since from mass conservation $U_1 > U_2$, the pressure increases through the sudden expansion, as $\Delta p > 0$. The found pressure increase Δp combines two effects: a pressure increase due to flow deceleration from velocity U_1 to $U_2 < U_1$; and pressure irreversible loss due to energy dissipation at the cross-section area expansion. The first term can be obtained from the Bernoulli equation as follows,

$$\Delta p_R = (p_2 - p_1)_R = \rho \left(\frac{U_1^2}{2} - \frac{U_2^2}{2} \right). \tag{8.103}$$

Anticipating that $\Delta p = \Delta p_R + \Delta p_I$, where Δp_I is the irreversible pressure loss, we obtain,

$$-\Delta p_I = \rho \frac{U_1^2}{2} \left[1 + \left(\frac{U_2}{U_1} \right)^2 - 2 \left(\frac{U_2}{U_1} \right) \right] = \rho \frac{U_1^2}{2} \left[1 - \left(\frac{A_1}{A_2} \right) \right]^2. \tag{8.104}$$

The above irreversible pressure loss can be written in general form as,

$$-\Delta p_I = \zeta_{loc} \frac{\rho U_{A=min}^2}{2} = \zeta_{loc} \frac{G_{A=min}^2}{2\rho}, \tag{8.105}$$

where

$$\zeta_{loc} = \zeta_{exp} = \left[1 - \left(\frac{A_1}{A_2} \right) \right]^2 \tag{8.106}$$

is the local loss coefficient, $U_{A=min} = U_1$ is the characteristic loss velocity taken at the minimum cross-section area at the obstacle location, and $G_{A=min} = G_1 = \rho U_1$ is the corresponding mass flux. We have introduced here a convention that for local pressure loss calculations, the velocity and the mass flux used in calculations should be defined at the minimum flow cross section at the obstacle location.

Using this convention, the irreversible pressure loss at sudden contraction, shown in Fig. 8.7, is found as,

$$-\Delta p_I = \zeta_{con} \frac{\rho U_2^2}{2}, \tag{8.107}$$

where

$$\zeta_{con} = \left(\frac{A_2}{A_c} - 1 \right)^2, \tag{8.108}$$

and

$$\frac{A_c}{A_2} = 0.62 + 0.38 \left(\frac{A_2}{A_1} \right)^3. \tag{8.109}$$

Here A_c is the flow cross-section in *vena contracta* created just downstream of the sudden contraction of the pipe cross-section area.

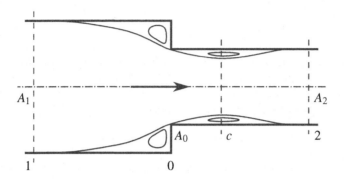

Figure 8.7 Flow through a sudden contraction of channel.

8.5 EXTERNAL FLOWS

External flows are flows over bodies immersed in an unbounded fluid. This concept includes, e.g., flow over a semi-infinite flat plate and a flow around an airfoil. The most important phenomena that occur in external flows include creation of forces acting on immersed bodies. By virtue of these forces, fluid can perform work on the bodies. These mechanisms are behind harvesting of kinetic energy contained in moving air masses (winds) and ocean currents.

The total force experienced by an object immersed in a fluid is decomposed into two components called the *drag force* and the *lift force*.

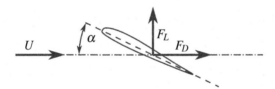

Figure 8.8 Drag and lift forces acting on an airfoil oriented at an angle of attack α with respect to unperturbed air flow velocity U.

The magnitude of the forces experienced by the airfoil shown in Fig. 8.8 depends on the relative velocity U, the airfoil shape and the *angle of attack*. For a given airfoil and velocity U, the lift force increases with increasing angle of attack until a so-called *critical angle of attack* is reached. After that point a stall occurs and the lift force decreases rapidly. The physical processes that lead airfoils to stall are quite complicated and for practical applications, such as the wind turbine dynamics, it is sufficient to parametrize lift and drag forces in terms of empirically measured lift and drag coefficients.

The *drag coefficient*, C_D, is defined as,

$$C_D = \frac{F_D}{\frac{1}{2}\rho U^2 A}, \tag{8.110}$$

where F_D is the drag force, A is the reference area of the immersed body, usually taken as the cross-section area of the body in a plane perpendicular to the velocity vector $U\mathbf{e}_x$, U is the velocity, and ρ is the fluid density. Similarly, the *lift coefficient*, C_L, is defined as,

$$C_L = \frac{F_L}{\frac{1}{2}\rho U^2 A}. \tag{8.111}$$

For certain immersed body shapes (such as, e.g., airfoils) area A depends on the angle of attack. Consequently, the maximum projected area of the body is used to define lift and drag coefficients.

Drag and lift coefficients for airfoils are provided in a graphical form as functions of the angle of attack. Drag coefficient data for selected objects are shown in Fig. 8.9.

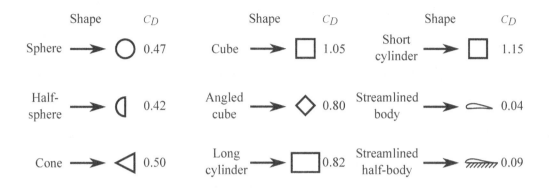

Figure 8.9 Drag coefficient data for selected objects (Re $> 10^3$).

8.6 MULTIPHASE FLOWS

Multiphase flows are flows in which multiple phases, for example, liquid, gas, and/or solid, flow simultaneously in the same common space. Such flows are more complex than single phase flows, since the phases interact with each other, and these interactions are in general very complex and flow dependent. In this section we will consider some common approaches to predict the main characteristics of such flows. In particular, we will focus on prediction of pressure drops and phase content during multiphase flows in channels.

8.6.1 NOTATION AND NOMENCLATURE

The main parameters describing the mass, momentum, and energy balances for multiphase flows are represented with the same notation as employed for single phase flows. However, to distinguish various components or phases of the flow, proper indices are used. The standard approach is to use index k to refer to k-phase. For example $\mathbf{v}_k$ represents velocity vector field for phase k. In particular, $k = l$ indicates the liquid phase and $k = v$ refers to the vapor (gas) phase. When a two-phase mixture of saturated liquid and vapor phases of the same species is considered, $k = f$ is used to indicate the saturated liquid phase and $k = g$ indicates the saturated vapor phase.

The two-phase mixture composition can be expressed in terms of volumetric flow rates of each of the components or phases. For liquid-vapor flow as shown in Fig. 8.10, the volumetric flow rates are defined as,

$$Q_l = \iint_{A_l} w_l dA_l, \quad Q_v = \iint_{A_v} w_v dA_v, \tag{8.112}$$

where w_l, w_v—local velocity of liquid and gas, respectively, in the channel cross-section area, and A_l, A_v—instantaneous cross-section area of the channel occupied by the liquid and gas phase, respectively. In a similar way, area averaged velocities for each phase are defined as,

$$U_l = \frac{1}{A_l} \iint_{A_l} w_l dA_l, \quad U_v = \frac{1}{A_v} \iint_{A_v} w_v dA_v. \tag{8.113}$$

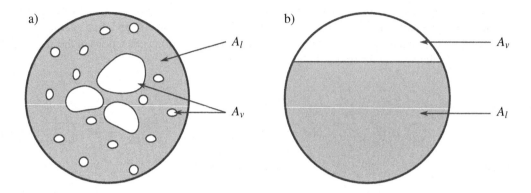

Figure 8.10 An instantaneous distribution of liquid and vapor phases in channel cross section during a) dispersed two-phase flow and b) stratified two-phase flow.

We call these velocities as "phase-area average" velocities, or simply **phase average velocities**, to distinguish them from the channel-area averaged velocities, commonly referred to as **superficial velocities**, defined as,

$$J_l = \frac{1}{A} \iint_{A_l} w_l \mathrm{d}A_l, \quad J_v = \frac{1}{A} \iint_{A_v} w_v \mathrm{d}A_v. \tag{8.114}$$

For any component k, the relation between the superficial velocity and the phase average velocity is as follows,

$$J_k = \frac{1}{A} \iint_{A_k} w_k \mathrm{d}A_k = \frac{A_k}{A} \frac{1}{A_k} \iint_{A_k} w_k \mathrm{d}A_k = \alpha_k U_k, \tag{8.115}$$

where $\alpha_k = A_k/A$ is the **volume fraction** of phase k. It should be noted that, since $\sum_{k=1}^{K} A_k = A$, then $\sum_{k=1}^{K} \alpha_k = 1$. Here K is the total number of phases or components in the flow. In particular, for liquid-vapor two-phase flow we have

$$\alpha_l + \alpha_v = 1. \tag{8.116}$$

Here it is a common practice to drop the index for the vapor volume fraction and use α for the vapor phase and $1 - \alpha$ for the volume fraction of the liquid phase. Further, in nuclear applications, the volume fraction of vapor is called the **void fraction**.

Mass flux of phase k is defined as,

$$G_k = \rho_k J_k, \tag{8.117}$$

where ρ_k is the mass density of phase k. For liquid-vapor two-phase flow we have,

$$G_l = \rho_l J_l = (1 - \alpha)\rho_l U_l, \quad G_v = \rho_v J_v = \alpha \rho_v U_v. \tag{8.118}$$

Mass flow rate of phase k is found as,

$$W_k = G_k A = \rho_k J_k A. \tag{8.119}$$

Flow quality or **actual quality** of liquid-vapor two-phase flow is traditionally defined as the vapor mass fraction in the flow,

$$x_a = \frac{W_v}{W_v + W_l} = \frac{W_v}{W} = \frac{G_v}{G_v + G_l} = \frac{G_v}{G}, \tag{8.120}$$

where G is the total mass flux and W is the total mass flow rate of the two-phase mixture.

Using the expressions for the phase mass fluxes expressed in terms of α, we obtain the following quality-volume fraction relationship,

$$x_a = \frac{G_v}{G} = \frac{\rho_v \alpha U_v}{\rho_v \alpha U_v + \rho_l (1 - \alpha) U_l} = \frac{1}{1 + \frac{(1-\alpha)}{\alpha} \frac{\rho_l}{\rho_v} \frac{U_l}{U_v}}. \tag{8.121}$$

Resolving the above equation for α in terms of x_a yields,

$$\alpha = \frac{1}{1 + \frac{(1-x_a)}{x_a} \frac{\rho_v}{\rho_l} \frac{U_v}{U_l}}. \tag{8.122}$$

8.6.2 FLOW PATTERNS

One of the major difficulties from the modeling point of view is that a multiphase flow behavior is very sensitive to the spatial distribution or topology of the components of the flow. Through observations of multiphase flows, it has been established that several typical component distributions, commonly called **flow patterns**, can be discerned. The taxonomy of flow patterns is based on visual observations and strict physically-based expressions to distinguish between various flow patterns, even though very desirable, are still not well-developed. The main reason for this is that flow patterns tend to change, depending on the channel geometry (for example, pipe diameter), channel orientation (vertical, horizontal, inclined), flow conditions (flow rates of each of the components), and component properties.

Figure 8.11 illustrates typical flow patterns for co-current two-phase flow in vertical, middle-size (with a diameter in a range from 5 to 50 mm) channels. **Bubbly flow** consists of a *continuous*

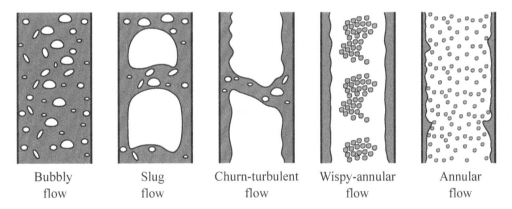

| Bubbly | Slug | Churn-turbulent | Wispy-annular | Annular |
| flow | flow | flow | flow | flow |

Figure 8.11 Two-phase flow patterns for upward cocurrent flow in a vertical pipe.

liquid phase and a *dispersed gas phase* flowing as bubbles. The bubbles usually have a certain size distribution, and thus, they are not monodisperse. However, their diameters are always significantly smaller than the channel diameter.

With increasing fraction of the gas phase, the bubbles coalesce and form large bubbles that concentrate at the center of the channel. These large bubbles coalesce with each-other until they form channel-wide bubbles called **Taylor bubbles**. The Taylor bubbles are separated from channel walls with a liquid film. A flow pattern consisting of Taylor bubbles, separated with liquid slugs, is called **slug flow**.

With increasing momentum of the gas phase in the channel, the slug flow becomes unstable and transits into **churn flow**. This flow pattern is an intermediate flow regime that exists between the slug flow and the annular flow, and in small-diameter channels it may not develop at all.

When liquid bridges in churn flow break up into droplet clouds, a **wispy-annular flow** is created. Since the interfacial shear dominates over gravity, the liquid phase is expelled from the channel center and forms a thin liquid film flowing along channel walls. The gas phase mixed with disperse droplets fills the central regions of the channel.

At high gas flow rates, droplets are entirely dispersed in the gas core and the **annular flow** pattern is formed. The name of the flow pattern stems from fact that the gas core is surrounded by an annular ring of a thin liquid film flowing along channel walls. The characteristic feature of annular flow is existence of the liquid phase both as the disperse phase (droplets) and as the continuous phase (liquid film).

8.6.3 HOMOGENEOUS EQUILIBRIUM MODEL

One of the simplest models to describe two-phase flows is the **homogeneous equilibrium model**. This model employs the following assumptions:

- the phases are in the thermodynamic equilibrium,
- the phases have the same velocity everywhere,
- the phases behave as a single fluid with effective mixture properties.

With these assumptions, single-phase balance equations can be used; however, definitions of effective properties, such as mixture density and mixture viscosity, are required.

One immediate consequence of the assumptions is that $U_v \equiv U_g = U_l \equiv U_f = U_m$, where U_m is the mixture center-of-mass velocity. The relationship between the volume fraction α and quality x_a becomes,

$$\alpha = \frac{1}{1 + \frac{(1-x_a)}{x_a} \frac{\rho_g}{\rho_f}}. \tag{8.123}$$

From the mass conservation equation for the mixture we have

$$G = G_f + G_g = (1-\alpha)\rho_f U_m + \alpha \rho_g U_m = \rho_m U_m, \tag{8.124}$$

where

$$\rho_m = (1-\alpha)\rho_f + \alpha \rho_g \tag{8.125}$$

is the effective mixture density. Similarly we can derive an effective mixture density that satisfies the momentum conservation by finding the total momentum flow through the channel cross-section area,

$$\iint_A (\rho w) w \, dA = \iint_{A_f} \rho_f w_f^2 \, dA_f + \iint_{A_g} \rho_g w_g^2 \, dA_g = U_m^2 A \left(\rho_f \frac{A_f}{A} + \rho_g \frac{A_g}{A} \right). \tag{8.126}$$

Here we use the assumption that $w_f = w_g = U_m = const$. Since $A_f/A = 1 - \alpha$ and $A_g/A = \alpha$, we get

$$\iint_A (\rho w) w \, dA = U_m^2 A \left[(1-\alpha)\rho_f + \alpha \rho_g \right] = \rho_m U_m^2 A = \frac{G^2}{\rho_m} A. \tag{8.127}$$

Hence the mixture momentum flow rate in the cross-section is correctly represented with the mixture density ρ_m and the mixture velocity U_m.

Enthalpy balance in a cross-section provides the following result,

$$\iint_A (\rho i) w \, dA = \iint_{A_f} \rho_f i_f w_f \, dA_f + \iint_{A_g} \rho_g i_g w_g \, dA_g = \rho_m i_m U_m A. \tag{8.128}$$

where the effective mixture enthalpy is defined as,

$$i_m = \frac{(1-\alpha)\rho_f i_f + \alpha\rho_g i_g}{\rho_m}. \tag{8.129}$$

Similar expressions can be obtained for other mixture quantities, such as entropy and internal energy.

Friction Pressure Loss

It is postulated that the friction pressure gradient can be found from the same relationship as used for single phase flow, which, according to Eq. (8.87), gives

$$-\left(\frac{dp}{dz}\right)_{f,tp} = \frac{4C_{f,tp}}{D_h}\frac{\rho_m U_m^2}{2} = \frac{4C_{f,tp}}{D_h}\frac{G^2}{2\rho_m}, \tag{8.130}$$

where index tp indicates a value for two-phase flow. As can be seen we need to devise a proper relationship for the Fanning friction factor, $C_{f,tp}$, valid for two-phase flows. Clearly, this factor should be obtained from experimental data, as was the case for single-phase flows. However, such correlation would be inevitably flow-pattern dependent. Since the homogeneous equilibrium model is flow-pattern independent, we rather deduce $C_{f,tp}$ from the analogy to single-phase flows. Based on the Blasius correlation, and assuming that it should be valid for two-phase flows as well, we get,

$$C_{f,tp} = A\mathrm{Re}_{tp}^{-a}, \tag{8.131}$$

where A and a are constants and $\mathrm{Re}_{tp} = GD_h/\mu_{tp}$ is an "effective" Reynolds number for two-phase flow with total mass flux G, in a channel with hydraulic diameter D_h. Assume now that we have liquid-only (that is, a saturated water) flow in the same channel, with the same mass flux as for the two-phase flow. The friction gradient becomes,

$$-\left(\frac{dp}{dz}\right)_{f,lo} = \frac{4C_{f,lo}}{D_h}\frac{G^2}{2\rho_f}, \tag{8.132}$$

where $C_{f,lo}$ is now the Fanning friction factor as defined and valid for single phase flow, given as,

$$C_{f,lo} = A\mathrm{Re}_{lo}^{-a}, \tag{8.133}$$

where $\mathrm{Re}_{lo} = GD_h/\mu_f$ is the Reynolds number for liquid flow with mass flux G, in a channel with hydraulic diameter D_h, and having dynamic viscosity μ_f.

Combining Eqs. (8.130) and (8.132) yields,

$$-\left(\frac{dp}{dz}\right)_{f,tp} = -\left(\frac{dp}{dz}\right)_{f,lo}\phi_{lo}^2, \tag{8.134}$$

where a two-phase **friction multiplier** is introduced as follows,

$$\phi_{lo}^2 \equiv \frac{(dp/dz)_{f,tp}}{(dp/dz)_{f,lo}} = \frac{C_{f,tp}}{C_{f,lo}}\frac{\rho_f}{\rho_m} = \left(\frac{\mu_{tp}}{\mu_f}\right)^a\frac{\rho_f}{\rho_m}. \tag{8.135}$$

Defining the effective two-phase mixture dynamic viscosity as

$$\frac{1}{\mu_{tp}} = \frac{x}{\mu_g} + \frac{1-x}{\mu_f}, \tag{8.136}$$

and using the derived expressions for mixture density and volume fraction, the two-phase flow multiplier is obtained as,

$$\phi_{lo}^2 = \left[1 + \left(\frac{\mu_f}{\mu_g}\right)x\right]^{-0.25}\left[1 + \left(\frac{\rho_f}{\rho_g} - 1\right)x\right].$$ (8.137)

where x is the mixture quality and it is assumed that $a = 0.25$.

The two-phase friction pressure loss in a channel with length L can be now found as,

$$-\Delta p_{f,tp} = -\int_0^L \left(\frac{dp}{dz}\right)_{f,tp} dz = -\Delta p_{lo}\frac{1}{L}\int_0^L \phi_{lo}^2 dz = -r_3\Delta p_{lo}.$$ (8.138)

Here the **integral friction multiplier** is introduced as,

$$r_3 \equiv \frac{1}{L}\int_0^L \phi_{lo}^2 dz.$$ (8.139)

Assuming that the quality of mixture at the pipe inlet is $x_{in} = 0$, the integral friction multiplier can be represented as a function of the exit quality x_{ex} and pressure p, as shown in Fig. 8.12.

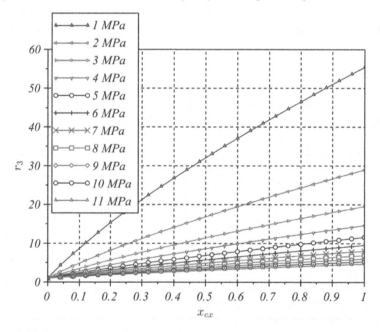

Figure 8.12 Integral friction multiplier r_3 for water-steam flow with linear variation of quality from $x_{in} = 0$ to x_{ex}.

Gravity Pressure Drop

The pressure gradient due to gravity was derived in §8.4.2 and for the homogeneous two-phase flow it can be written as,

$$\left(\frac{dp}{dz}\right)_{gr} = \rho_m g_z,$$ (8.140)

where g_z is a projection of the gravity acceleration vector on the channel axial direction. The total pressure drop due to gravity for two-phase flow in a pipe with length L, inlet quality $x_{in} = 0$ and exit quality x_{ex} can be obtained as,

$$\Delta p_{gr,tp} = \int_0^L \left(\frac{dp}{dz} \right)_{gr} dz = \left(\frac{1}{L} \int_0^L \frac{\rho_m}{\rho_f} dz \right) \rho_f g_z L = r_4 \Delta p_{lo}, \qquad (8.141)$$

where $\Delta p_{lo} = \rho_f g_z L$ is the gravity pressure drop in the same channel for liquid-only flow, and the **integral gravity multiplier** r_4 has been defined as,

$$r_4 \equiv \frac{1}{L} \int_0^L \frac{\rho_m}{\rho_f} dz. \qquad (8.142)$$

Figure 8.13 shows r_4-multiplier calculated for water-steam flows at various pressures in channels with a linear variation of quality from $x_{in} = 0$ to x_{ex}.

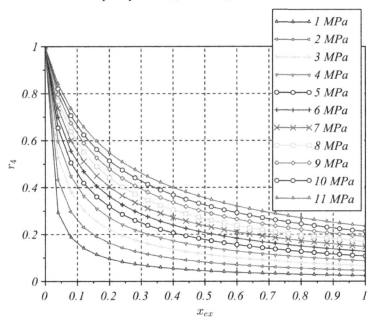

Figure 8.13 Integral gravity multiplier r_4 for water-steam flow with linear variation of quality from $x_{in} = 0$ to x_{ex}.

Acceleration Pressure Drop

The multiphase channel flow can accelerate or decelerate for the following three reasons: (i) due to a variable channel cross-section area, (ii) due to changes of mixture density caused by changes in flow quality, (iii) due to changes of mixture density caused by pressure variations. In the continuation we will assume that the channel has a constant cross-section area and that the pressure variations are not significant. Under such conditions, the only remaining effect causing flow acceleration is due to changes of the flow quality, which typically result from boiling or condensation.

The acceleration pressure gradient can be obtained from the momentum balance in a channel with a differential length dz and constant cross-section area A, for which we have,

$$pA + \rho_m U_m U_m A - (p+dp)A - [\rho_m U_m U_m A + d(\rho_m U_m U_m A)]A = 0. \qquad (8.143)$$

Thus, reducing terms and dividing the equation by dz, we obtain,

$$-\left(\frac{dp}{dz} \right)_{ac} = \frac{d\left(\rho_m U_m^2 \right)}{dz} = \frac{d}{dz} \left(\frac{G^2}{\rho_m} \right) = G^2 \frac{d\upsilon_m}{dz}, \qquad (8.144)$$

where $v_m = 1/\rho_m$ is the mixture specific volume.

The acceleration pressure drop is found as,

$$-\Delta p_{ac} = -\int_0^L \left(\frac{dp}{dz}\right)_{ac} = r_2 \frac{G^2}{\rho_f}, \tag{8.145}$$

where we introduce here the **integral acceleration multiplier** r_2 defined as,

$$r_2 \equiv \rho_f \left(v_{m,ex} - v_{m,in}\right). \tag{8.146}$$

Here $v_{m,in}$ and $v_{m,ex}$ are the mixture specific volume at the channel inlet and outlet, respectively. Assuming two-phase flow in a channel with inlet quality $x_{in} = 0$ and exit quality x_{ex}, the integral acceleration multiplier becomes,

$$r_2 = \rho_f \left(v_{m,ex} - v_f\right) = \left(\frac{\rho_f}{\rho_g} - 1\right) x_{ex}. \tag{8.147}$$

Figure 8.14 shows r_2-multiplier calculated for water-steam flows at various pressures in channels with a variation of quality from $x_{in} = 0$ to x_{ex}.

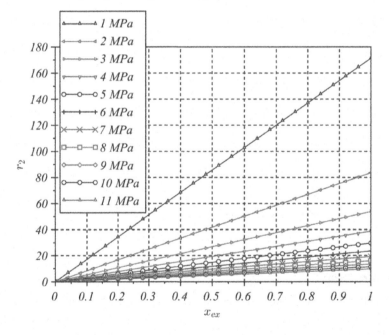

Figure 8.14 Integral gravity multiplier r_2 for water-steam flow with variation of quality from $x_{in} = 0$ to x_{ex}.

Local Pressure Loss

In homogeneous equilibrium model, the local pressure loss due to flow obstacle for two-phase flow is expressed in a similar way as for a single-phase flow, given by Eq. (8.105). Replacing the liquid density with the mixture density in the equation, we have,

$$-\Delta p_{l,tp} = \zeta \frac{G^2}{2\rho_m} = \zeta \frac{\rho_f}{\rho_m} \frac{G^2}{2\rho_f} = \phi_{lo,d}^2 \zeta \frac{G^2}{2\rho_f} = -\phi_{lo,d}^2 \Delta p_{l,lo}. \tag{8.148}$$

Here we introduce the **two phase local loss multiplier** $\phi_{lo,d}^2$. In homogeneous two-phase model, it can be calculated as,

$$\phi_{lo,d}^2 = \frac{\rho_f}{\rho_m} = 1 + x\left(\frac{\rho_f}{\rho_g} - 1\right), \tag{8.149}$$

where x is the thermodynamic equilibrium quality at the obstacle location.

Total Pressure Drop

The total pressure drop in a channel with a constant cross-section area, a constant mass flow rate of saturated liquid at the inlet, and any number of local flow obstacles can be found from the following equation,

$$-\Delta p = \left[r_3 \frac{4C_{f,lo}L}{D_h} + 2r_2 + \left(\sum_{i=1}^{N} \phi_{lo,di}^2 \zeta_i\right)\right]\frac{G^2}{2\rho_f} + r_4 L \rho_f g \sin\varphi, \tag{8.150}$$

where $\phi_{lo,di}^2 = \left[1 + x_i\left(\rho_f/\rho_g - 1\right)\right]$ is the two-phase local loss multiplier, x_i is the quality at location i, ζ_i is the loss coefficient at the same location, and φ is the channel inclination angle to the horizontal direction.

Differential Balance Equations

Performing a similar derivations as for the single-phase flow (see §8.3), the conservation equations of mass, momentum and energy for a homogeneous equilibrium transient flow in a channel are obtained as follows:

mass conservation

$$\frac{\partial \rho_m}{\partial t} + \frac{\partial G}{\partial z} = 0, \tag{8.151}$$

momentum conservation

$$\frac{\partial G}{\partial t} + \frac{\partial}{\partial z}\left(\frac{G^2}{\rho_m}\right) + \frac{\partial p}{\partial z} + \left[\frac{4C_f}{D_h} + \sum_{j=1}^{N} \zeta_j \delta(z - z_j)\right]\frac{|G|G}{2\rho_m} + \rho_m g_z = 0, \tag{8.152}$$

energy conservation

$$\frac{\partial(\rho i_m)}{\partial t} + \frac{\partial(G i_m)}{\partial z} = \frac{q'}{A} + \frac{\partial p}{\partial t}. \tag{8.153}$$

Here all variables describe the two-phase mixture and in particular: ρ_m is the mixture density, G is the mass flux, i_m is the specific enthalpy and p is the pressure. In addition t is the time, z is the distance along the channel, A is the channel cross-section area and q' is the linear power. The friction and the local pressure loss, as well as gravity pressure drop terms contain C_f—Fanning friction factor, ζ_j—local pressure loss coefficient at obstacle j, z_j—location of obstacle j, D_h—hydraulic diameter, g_z—gravity vector projected on the channel axis.

8.6.4   HOMOGENEOUS RELAXATION MODEL

Homogeneous relaxation model is derived with the same assumptions as the homogeneous equilibrium model. However, to make the model more realistic, a relaxation time is introduced in the modeling of the mixture quality. Thus, an additional equation is introduced as follows,

$$\frac{dx}{dz} = \frac{x - x_e}{\theta U_m}, \tag{8.154}$$

where x_e is the mixture quality calculated based on the thermodynamic equilibrium assumptions, x is the modeled mixture quality and θ is a proper relaxation time. Since a thermodynamic nonequilibrium is assumed, the equation of state of the mixture becomes,

$$i = i(T_v, T_l, p, x) = x i_v(p, T_v) + (1 - x) i_l(p, T_l). \tag{8.155}$$

Here T_v and T_l are temperatures of the vapor and the liquid phase, respectively. To close the model it is necessary to determine the phase temperatures T_v and T_l, and the relaxation time θ. The latter is rather difficult, since there is a lack of a good model for this parameter that would be valid and applicable in a wide range of operating conditions. However, in spite of these difficulties, the homogeneous relaxation model represents some success in modeling two-phase critical flows.

8.6.5 SEPARATED FLOW MODEL

Disperse multiphase flows are relatively well-described by the homogeneous equilibrium model. However, when phases or components are separated and do not mix with each other, the homogeneous model might fail. For example, for horizontal two-phase flows with moderate mass flow rates of components or phases, the gaseous phase flows above the liquid phase, and the phases are well-separated from each other. Similarly for vertical annular flows the liquid film phase is separated from the gas core. For such flows, a more adequate approach is to treat the phases separately in the corresponding portion of the channel cross-section area occupied by them.

The main differences between the **separate flow model**, in comparison to the homogeneous equilibrium model, are as follows,

- the velocities of phases are different,
- each phase is flowing separately and can be treated as a single phase flow with own set of properties, and can separately interact with channel walls or other phases or components flowing in the channel.

With these assumptions, relationships for pressure drops and losses, as well as for the volume fraction of the gas phase, can be derived. A detailed theory of separated two-phase flow can be found in [87].

8.6.6 DRIFT FLUX MODEL

The **drift flux model** represents a significant improvement over the homogeneous equilibrium model, since it allows for the relative motion of the phases. The model was first developed, among others, by Zuber and Findlay [91] and Wallis [87], to better describe experimental data.

The **drift flux** is a physical quantity representing the volumetric flux of a component relative to a surface moving with the average mixture velocity. To focus our attention, let us assume a rectilinear two-phase flow in z-direction. The local liquid and vapor phase velocities are w_l and w_v, respectively, and the local volume fraction of vapor is α. The following local drift velocities are defined:

drift velocity of vapor phase, also called the **void drift velocity**

$$w_{vj} \equiv w_v - j, \tag{8.156}$$

drift velocity of liquid phase

$$w_{lj} \equiv w_l - j, \tag{8.157}$$

where $j = j_v + j_l = \alpha w_v + (1 - \alpha) w_l$ is the average mixture velocity. In a similar manner, drift fluxes are defined as follows:

drift flux of vapor phase, which in the continuation will be called just drift flux,

$$j_{vl} \equiv \alpha(w_v - j) = \alpha w_{vj} = j_v - \alpha j, \tag{8.158}$$

drift flux of liquid phase,

$$j_{lv} \equiv (1 - \alpha)(w_l - j). \tag{8.159}$$

The drift flux can be transformed as follows,

$$j_{vl} = \alpha w_v - \alpha \left[\alpha w_v + (1 - \alpha)w_l \right] = \alpha(1 - \alpha)(w_v - w_l) = \alpha(1 - \alpha)w_{vl}, \tag{8.160}$$

where $w_v - w_l \equiv w_{vl}$ is the relative velocity. Thus, we can notice that the drift flux is proportional to the relative velocity and, in particular, it is equal to zero when the relative velocity is zero.

Drift Flux Void Correlation

Consider two-phase flow in a channel where all local parameters are certain functions of radial position in the channel cross-sectional area. Integration of the drift flux over channel area A gives,

$$\iint_A \alpha w_{vj} dA = \iint_A j_v dA - \iint_A \alpha j dA. \tag{8.161}$$

The area-averaged quantities are introduced as follows,

$$J_v \equiv \frac{1}{A} \iint_A j_v dA, \quad J \equiv \frac{1}{A} \iint_A j dA, \quad \overline{\alpha}^A \equiv \frac{1}{A} \iint_A \alpha dA. \tag{8.162}$$

In addition, the **average drift velocity** is defined as,

$$U_{vj} \equiv \frac{\iint_A \alpha w_{vj} dA}{\iint_A \alpha dA} = \frac{\iint_A \alpha w_{vj} dA}{\overline{\alpha}^A A} \tag{8.163}$$

The second term on the right-hand side of Eq. (8.161) contains an integral of the product of two functions, which can be represented as a product of two integrals as follows,

$$\frac{1}{A} \iint_A \alpha j dA \equiv \overline{\alpha j}^A = C_0 \overline{\alpha}^A \overline{j}^A = C_0 \overline{\alpha}^A J. \tag{8.164}$$

Here C_0 is the **distribution parameter**, sometime also called **concentration parameter**. This parameter quantifies the effect of the void and volumetric flux distribution in cross-sectional area A and is equal to the ratio of the average of the product of flux and void to the product of averages,

$$C_0 \equiv \frac{\overline{\alpha j}^A}{\overline{\alpha}^A \overline{j}^A}. \tag{8.165}$$

Substituting Eqs. (8.163) and (8.164) into (8.161) and re-arranging, yields,

$$\overline{\alpha}^A = \frac{J_v}{C_0 J + U_{vj}}. \tag{8.166}$$

The above equation is the **drift-flux void correlation** and represents a relationship to determine the cross-section averaged void fraction using the drift-flux model. It can be seen that for $C_0 = 1$ and $U_{vj} = 0$ the homogeneous equilibrium model is obtained.

Numerous correlations are provided for C_0 and U_{vj} in the literature and an interested reader should consult the book by Ishii and Hibiki [52] for a thorough coverage of this topic. Zuber and Findlay [91] provided a single set of parameters, which are valid for high-pressure steam water flows and for all flow patterns,

$$C_0 = 1.13, \quad U_{vj} = 1.41 \left[\frac{\sigma g (\rho_l - \rho_v)}{\rho_l^2} \right]^{0.25}, \tag{8.167}$$

where σ is the surface tension, ρ_l is liquid density, ρ_v is vapor density and g is gravity acceleration.

Balance Equations

The conservation equations in the drift flux model can be formulated with various assumptions. In the most general case, five conservation equations are obtained: two mass and energy conservation equations for each of the phases and one common momentum conservation equation for the mixture. The equations are as follows:

mass conservation equation for the liquid phase

$$\frac{\partial}{\partial t}\left[\rho_l\left(1-\alpha\right)A\right]+\frac{\partial}{\partial z}\left(\rho_l J_l A\right)=-\Gamma A, \tag{8.168}$$

mass conservation equation for the vapor phase

$$\frac{\partial}{\partial t}\left(\rho_v \alpha A\right)+\frac{\partial}{\partial z}\left(\rho_v J_v A\right)=\Gamma A, \tag{8.169}$$

energy conservation equation for the liquid phase

$$\frac{\partial}{\partial t}\left[\left(\rho_l i_l-p\right)\left(1-\alpha\right)A\right]+\frac{\partial}{\partial z}\left(\rho_l i_l J_l A\right)=q_l^{''}P_H, \tag{8.170}$$

energy conservation equation for the vapor phase

$$\frac{\partial}{\partial t}\left[\left(\rho_v i_v-p\right)\alpha A\right]+\frac{\partial}{\partial z}\left(\rho_v i_v J_v A\right)=q_v^{''}P_H, \tag{8.171}$$

momentum conservation equation for the mixture

$$\frac{\partial G}{\partial t}+\frac{1}{A}\frac{\partial}{\partial z}\left(\frac{G^2 A}{\rho_M}\right)+\frac{\partial p}{\partial z}+$$
$$\left[\frac{4C_f}{D_h}\phi_{lo}^2+\sum_{j=1}^{N}\zeta_j\phi_{lo,dj}^2\delta(z-z_j)\right]\frac{|G|G}{2\rho_l}+\rho_m g_z=0. \tag{8.172}$$

Here all variables describe the two-phase mixture and in particular: ρ_m is the "static" mixture density given by Eq. (8.173), ρ_M is the "dynamic" mixture density given by Eq. (8.174), G is the mixture mass flux, i_l is the specific enthalpy of the liquid phase, i_v is the specific enthalpy of the vapor phase, and p is the pressure of the mixture. In additions t is the time, z is the distance along the channel, A is the channel cross-sectional area and $q_l^{''}, q_v^{''}$ are the heat fluxes to liquid and vapor, respectively. The friction and the local pressure loss, as well as gravity pressure drop terms contain C_f—Fanning friction factor, ζ_j—local pressure loss coefficient at obstacle j, z_j—location of obstacle j, D_h—hydraulic diameter, g_z—gravity vector projected on the channel axis, ϕ_{lo}^2—two-phase pressure loss multiplier and $\phi_{lo,dj}^2$—two-phase local pressure loss multiplier at location z_j. To close the system of equations, the mixture densities are expressed as,

$$\rho_m = \rho_l\left(1-\alpha\right)+\rho_v \alpha, \tag{8.173}$$

$$\rho_M = \left[\frac{1-x^2}{\rho_l\left(1-\alpha\right)}+\frac{x^2}{\rho_v \alpha}\right]^{-1}. \tag{8.174}$$

8.6.7 TWO-FLUID MODEL

In two-fluid formulation of two-phase flow, each of the phases is described with a separate set of conservation equations for mass, energy and momentum. This formulation allows for a proper treatment of the thermodynamic non-equilibrium between phases, since the temperature for each phase is found from own energy balance equation.

The conservation equations for one-dimensional, two-fluid model are as follows:

mass conservation equation for phase k,

$$\frac{\partial}{\partial t}(\alpha_k \rho_k) + \frac{\partial}{\partial z}(\alpha_k \rho_k U_k) = \Gamma_k, \tag{8.175}$$

energy conservation equation for phase k,

$$\frac{\partial}{\partial t}(\alpha_k \rho_k i_k) + \frac{\partial}{\partial z}(\alpha_k \rho_k i_k U_k) = -\frac{\partial}{\partial z}\left[\alpha_k \left(q_k + q_k^t\right)_z\right] + \alpha_k \frac{Dp_k}{Dt}$$
$$+ \frac{P_H}{A} \alpha_{kw} q_{kw}'' + \Gamma_k i_{ki} + a_i q_{ki}'' + \Phi_k, \tag{8.176}$$

momentum conservation equation for phase k,

$$\frac{\partial}{\partial t}(\alpha_k \rho_k U_k) + \frac{\partial}{\partial z}(\alpha_k \rho_k U_k^2) = -\alpha_k \frac{\partial p_k}{\partial z} + \frac{\partial}{\partial z}\left[\alpha_k \left(\tau_{zz,k} + \tau_{zz,k}^t\right)\right]$$
$$- \frac{4\alpha_{kw} \tau_{kw}}{D_h} + \Gamma_k U_{ki} + (p_{ki} - p_k)\frac{\partial \alpha_k}{\partial z} - \alpha_k \rho_k g_z + M_k^d, \tag{8.177}$$

where: ρ_k—phase-k density, α_k—phase-k volume fraction, U_k—phase-k velocity, i_k—phase-k specific enthalpy, p_k—phase-k pressure. In additions t is the time, z is the distance along the channel, D_h—hydraulic diameter, Γ_k—phase-change rate for phase-k, q_{ki}''—heat flux from phase-k to the interface, g_z—gravity acceleration projected on the channel axis, Φ_k—volumetric heat sources in phase k, U_{ki}—phase-k velocity at the interface, M_k^d—interfacial momentum transfer term.

The set of $3 \times k$ equations given above requires closure relationships for interfacial mass, energy and momentum transfer. In general these closure relationships are flow-regime dependent. Further details on the two-fluid model can be found in [52].

PROBLEMS

PROBLEM 8.1

Water at temperature $T = 350$ K and pressure $p = 0.25$ MPa flows through a horizontal pipe with inner diameter $D = 150$ mm and length $L = 15$ m. Calculate the total pressure change $\Delta p = p_2 - p_1$ over the pipe length when the mass flow rate of water is $W = 10$ kg/s. Assume the same water properties everywhere. Neglect the local inlet and outlet losses and assume smooth pipe walls.

PROBLEM 8.2

Water at temperature $T = 403$ K and pressure $p = 1.0$ MPa flows upward through a vertical pipe with total length 7 m and inner diameter 25 mm that suddenly expands to 50 mm at 3.5 m distance from the inlet. Calculate the total pressure change $\Delta p = p_2 - p_1$ over the pipe length when the mass flow rate of water is $W = 4$ kg/s. Assume the same water properties everywhere. Neglect the local inlet and outlet losses and assume smooth pipe walls.

PROBLEM 8.3

Water at temperature $T = 438$ K and pressure $p = 2.0$ MPa flows downward through a vertical pipe with total length 5 m and inner diameter 50 mm that suddenly contracts to 38 mm at 2 m distance from the inlet. Calculate the total pressure change $\Delta p = p_2 - p_1$ over the pipe length when the mass flow rate of water is $W = 5$ kg/s. Assume the same water properties everywhere. Neglect the local inlet and outlet losses and assume smooth pipe walls.

PROBLEM 8.4

Water steam at temperature $T = 608$ K and pressure $p = 7.0$ MPa flows through a horizontal pipe with inner diameter $D = 38$ mm and length $L = 22$ m. Calculate the total pressure change $\Delta p = p_2 - p_1$ over the pipe length when the mass flow rate of steam is $W = 0.2$ kg/s. Assume the same steam properties everywhere. Neglect the local inlet and outlet losses and assume that pipe walls have absolute roughness $k = 0.091$ mm.

PROBLEM 8.5

A hydraulic turbine is supplied with water at temperature $T = 288$ K and atmospheric pressure from a mountain lake located $\Delta H = 250$ m above the turbine. The water flows through a supply pipe made of unfinished concrete, with diameter $D = 3$ m and length $L = 150$ m. Water leaves the pipe at atmospheric pressure. Calculate the discharge velocity. Neglect all local pressure losses.

PROBLEM 8.6

Saturated water and vapor mixture at temperature $T = 453$ K flows through a horizontal pipe with diameter $D = 100$ mm and length $L = 12.5$ m. Using the homogeneous equilibrium model, calculate the void fraction and the total pressure change $\Delta p = p_2 - p_1$ over the pipe length when the mass flow rate of water is $W_f = 8.7$ kg/s and the mass flow rate of vapor is $W_g = 0.049$ kg/s. Assume the same properties of water and vapor in the whole pipe. Assume rough walls with roughness $k = 0.05$ mm.

PROBLEM 8.7

Saturated water and vapor mixture at temperature $T = 638.9$ K flows upward through a vertical pipe with diameter $D = 100$ mm and length $L = 7$ m. Using the homogeneous equilibrium model, calculate the void fraction and the total pressure change $\Delta p = p_2 - p_1$ over the pipe length when the mass flow rate of water is $W_f = 2.89$ kg/s and the mass flow rate of vapor is $W_g = 3.0$ kg/s. Assume the same properties of water and vapor in the whole pipe. Assume rough walls with roughness $k = 0.05$ mm and do not include inlet and exit losses.

PROBLEM 8.8

Saturated water and vapor mixture at temperature $T = 485.53$ K flows downward through a vertical uniformly heated pipe with diameter $D = 100$ mm and length $L = 7$ m. Using the homogeneous equilibrium model, calculate the exit void fraction, the exit mixture speed, and the total pressure change $\Delta p = p_2 - p_1$ over the pipe length when the mass flow rate of saturated water at the pipe inlet is $W_f = 2$ kg/s and the exit thermodynamic equilibrium quality is $x_{ex} = 0.25$. Assume the same properties of water and vapor in the whole pipe. Assume rough walls with roughness $k = 0.05$ mm and do not include inlet and exit losses.

PROBLEM 8.9

Show that

$$\alpha = \frac{j_v}{j}\left(1 - \frac{j_{vl}}{j_v}\right)$$

where j_{vl} is a drift flux given by Eq. (8.160).

9 Heat Transfer in Energy Systems

Heat transfer plays a major role in energy transformation systems, in which the internal energy transformation is involved. A conversion of the internal energy into another form of energy requires some kind of heat transfer from one system to another. In this chapter we discuss heat transfer modes that frequently take place in energy systems, such as heat conduction, convection and radiation.

9.1 GOVERNING EQUATIONS

Heat conduction is governed by the energy conservation equation. The general differential energy conservation equation derived in §8.1.5 is as follows

$$\frac{\partial \left[\rho \left(e_I + \frac{1}{2}v^2\right)\right]}{\partial t} + \nabla \cdot \left[\rho \left(e_I + \frac{1}{2}v^2\right)\mathbf{v} + \mathbf{q}'' - \mathbf{T} \cdot \mathbf{v}\right] - \rho \mathbf{b} \cdot \mathbf{v} = 0. \tag{9.1}$$

Here e_I is the specific internal energy, $\mathbf{v}$ is the flow velocity vector, $v = |\mathbf{v}|$ is the velocity vector magnitude, $\mathbf{q}''$ is the heat flux vector, $\mathbf{T}$ is the total stress tensor, and $\mathbf{b}$ is the external bulk force.

The most practical form of the energy conservation equation depends on the unknown variable for which the equation should be solved. In most cases it is the temperature, sometime it is the specific enthalpy. Equation (9.1) needs to be further transformed to be useful. The equation in the current form contains, from left to right, the following five terms:

1. rate of change of kinetic and internal energy per unit volume,
2. rate of internal and kinetic energy change per unit volume by convective transport,
3. rate of heat transfer per unit volume by conduction,
4. rate of work done per unit volume by stresses,
5. rate of work done per unit volume by external bulk forces, e.g., by gravity.

Equation (9.1) is expressed in terms of the internal and kinetic energy $(e_I + v^2/2)$ rather than the total energy $(e_T = e_I + e_K + e_P)$ since the potential energy e_P is replaced with the work done by external bulk forces, e.g., $\rho \mathbf{b} \cdot \mathbf{v}$. However, since

$$\rho \mathbf{b} \cdot \mathbf{v} = -\nabla \cdot (\rho \mathbf{v} e_P) - \frac{\partial (\rho e_P)}{\partial t}, \tag{9.2}$$

the energy conservation equation can be written as,

$$\frac{\partial (\rho e_T)}{\partial t} + \nabla \cdot \left(\rho e_T \mathbf{v} + \mathbf{q}'' - \mathbf{T} \cdot \mathbf{v}\right) = 0, \tag{9.3}$$

where $e_T = e_I + v^2/2 + e_P$ is the total specific energy. When the total specific energy of fluid is of interest, this form of the energy conservation equation should be used. It is worth noting here that the chemical, nuclear and radiative energy forms can be included as an additional source term in Eq. (9.3), and the internal energy e_I consists of the thermal energy part only. This is because these energy forms are governed by separate equations (describing the nuclear or chemical reactions) and their net effect is to contribute to the thermal energy change only.

Equations (9.1) and (9.3) still include different effects, and sometimes it is desirable to separate these effects from each other. In particular, the mechanical energy (containing the kinetic and

DOI: 10.1201/9781003036982-9

potential energy) can be subtracted from the total energy conservation equation and a pure internal energy conservation equation can be obtained. The mechanical energy balance can be obtained from the linear momentum balance equation, Eq. (8.20), multiplied by velocity vector **v**. As a result, we get,

$$\frac{\partial}{\partial t}\left(\frac{1}{2}\rho v^2\right) + \nabla \cdot \left(\frac{1}{2}\rho v^2 \mathbf{v}\right) - \rho \mathbf{b} \cdot \mathbf{v} - \nabla \cdot (\mathbf{T} \cdot \mathbf{v}) + \mathbf{T} : \nabla \mathbf{v} = \mathbf{0}. \tag{9.4}$$

Subtracting the mechanical energy equation, Eq. (9.4), from the total energy equation, Eq. (9.1), yields the following internal energy balance equation,

$$\frac{\partial}{\partial t}(\rho e_I) + \nabla \cdot (\rho e_I \mathbf{v}) + \nabla \cdot \mathbf{q}'' - \mathbf{T} : \nabla \mathbf{v} = 0. \tag{9.5}$$

Two most frequently used energy equation forms can be derived from the above equation. First we express the specific internal energy (which is essentially the thermal energy) in terms of the specific enthalpy. Since $e_I = i - p/\rho$ and $\mathbf{T} = -p\mathbf{I} + \tau$, Eq. (9.5) becomes,

$$\frac{\partial}{\partial t}(\rho i - p) + \nabla \cdot (\rho i \mathbf{v}) - \mathbf{v} \cdot \nabla p + \nabla \cdot \mathbf{q}'' - \tau : \nabla \mathbf{v} = 0. \tag{9.6}$$

This equation can be written in a more compact way by using the substantial derivative and by combining it with the continuity equation, yielding the following **equation of change for specific enthalpy**,

$$\rho \frac{Di}{Dt} = -\nabla \cdot \mathbf{q}'' + \tau : \nabla \mathbf{v} + \frac{Dp}{Dt}. \tag{9.7}$$

The second useful form of the energy balance is expressed in terms of temperature. Since the material derivative of specific enthalpy can be expressed in terms of temperature and pressure as,

$$\rho \frac{Di}{Dt} = \rho c_p \frac{DT}{Dt} + \left[1 + \left(\frac{\partial \ln \rho}{\partial \ln T}\right)_p\right] \frac{Dp}{Dt}, \tag{9.8}$$

we obtain the following **equation of change for temperature**,

$$\rho c_p \frac{DT}{Dt} = -\nabla \cdot \mathbf{q}'' + \tau : \nabla \mathbf{v} - \left(\frac{\partial \ln \rho}{\partial \ln T}\right)_p \frac{Dp}{Dt}. \tag{9.9}$$

It should be noted that Eqs. (9.1) through (9.9) are exact, since no simplifying assumptions were made once deriving them. However, specific simplifying assumptions are frequently adopted, since not all terms in the equations are equally important in various applications.

9.2 CONDUCTION

Heat conduction is present in both solids and fluids, but only in solids it exists in pure form. The conduction in fluids is in most cases accompanied with fluid motion. Such form of heat transfer is called heat convection and will be discussed in §9.3. In this section, the pure form of heat conduction as present in solids, is considered.

Combining Eq. (9.9) with Fourier's law of heat conduction given by Eq. (8.34) yields,

$$\rho c_p \frac{\partial T}{\partial t} = \nabla \cdot \lambda \nabla T + q''', \tag{9.10}$$

where q''' is added to represent the heat source rate per unit volume. If the thermal conductivity can be assumed independent of spacial location and temperature, the equation becomes,

$$\frac{\partial T}{\partial t} = a\nabla^2 T + \frac{q'''}{\rho c_p}, \tag{9.11}$$

in which $a = \lambda/\rho c_p$ is the *thermal diffusivity* of the solid material. Equation (9.11) has been solved analytically for many practical applications. The book of Carslaw and Jaeger [15] contains many solutions with wide spectrum of geometries and boundary conditions. For more complex geometries and boundary conditions the equation needs to be solved numerically.

Solution of Eqs. (9.10) and (9.11) in an arbitrary volume V, with boundary S, requires specification of the initial and boundary conditions. The initial condition specifies the distribution of temperature in the whole volume at time $t = 0$. The boundary conditions can have different forms depending on the physical phenomena taking place at the boundary. The three types of boundaries that are particularly common are as follows: (i) the boundary S_1 on which the temperature is known, that is the boundary condition of the first kind can be specified; (ii) the boundary S_2 where heat flux is known and the boundary condition of the second kind (or the Neumann boundary condition) can be specified; (iii) the boundary S_3 where the convective heat transfer takes place with known heat transfer coefficient, h, and known fluid bulk temperature, T_b, and where the boundary condition of the third kind (or the Robin boundary condition) can be specified. The applicable initial and boundary conditions can be formulated as,

1. initial temperature distribution $T(x,y,z,0) = T_0(x,y,z)$ at any point $P(x,y,z) \in V$,
2. boundary temperature distribution $T(x,y,z,t)\,|_{S_1} = T_1(x,y,z,t)$ at boundary S_1,
3. boundary heat flux distribution $-\lambda \nabla T\,|_{S_2} \cdot \mathbf{n} = q''(x,y,z)\,|_{S_2}$ at boundary S_2,
4. Robin boundary condition $-\lambda \nabla T\,|_{S_3} \cdot \mathbf{n} = h[T(x,y,z) - T_b]\,|_{S_3}$ at boundary S_3.

9.2.1 STEADY-STATE HEAT CONDUCTION

For steady state conditions, the equation of change for temperature reduces to the so called *Poisson equation*,

$$\nabla^2 T + \frac{q'''}{\lambda} = 0. \tag{9.12}$$

We will consider solutions of this equation for several simple cases of practical importance.

Plane Wall

For steady-state one-dimensional heat conduction through a wall without internal heat sources, the temperature equation is as follows,

$$\nabla^2 T = \frac{d^2 T}{dx^2} = 0. \tag{9.13}$$

A general solution of the equation is then,

$$T(x) = Ax + B, \tag{9.14}$$

where A and B are constants that need to be determined from boundary conditions. Let us first consider a case as shown in Fig. 9.1, when wall surface temperatures are known: $T\,|_{x=0} = T_1$ and $T\,|_{x=L} = T_2$. Substituting the boundary conditions in Eq. (9.14) gives the following temperature distribution in the wall,

$$T(x) = T_1 + \frac{x}{L}(T_2 - T_1). \tag{9.15}$$

Heat flux through the wall is obtained as,

$$q'' = -\lambda \frac{dT(x)}{dx} = -\lambda \frac{T_2 - T_1}{L} = \frac{\lambda}{L}(T_1 - T_2). \tag{9.16}$$

As can be seen, $q'' > 0$ when $T_1 > T_2$, and q'' is decreasing with increasing wall thickness L. The total rate of heat transfer through a wall with area A is thus,

$$q = q''A = \frac{\lambda A}{L}(T_1 - T_2). \tag{9.17}$$

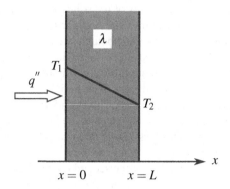

Figure 9.1 Heat conduction through a wall with given surface temperature.

We can find the *thermal resistance* of the wall as follows,

$$R_{th} \equiv \frac{\Delta T}{q} = \frac{T_1 - T_2}{q} = \frac{L}{\lambda A}. \tag{9.18}$$

Sometimes we know the wall thermal resistance and the temperature difference on both wall surfaces. The corresponding heat transfer rate can be found as,

$$q = \frac{\Delta T}{R_{th}} = \frac{T_1 - T_2}{R_{th}}. \tag{9.19}$$

The above equation can be thought as an analogy of Ohm's law, according to which the current flow (or q here) is equal to the ratio of the voltage difference (ΔT) to the electric resistance (R_{th}).

Let us now assume that the wall is dividing fluids at known far-field temperatures, $T_{1\infty}$, $T_{2\infty}$, and the heat is convected with known heat transfer coefficients on both sides, h_1, h_2, as shown in Fig. 9.2. The boundary conditions are as follows,

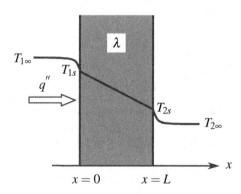

Figure 9.2 Heat conduction through a wall with convective heat transfer on both sides.

$$-\lambda \frac{dT}{dx}\Big|_{x=0} = h_1(T_{1\infty} - T_{1s}) = q'', \tag{9.20}$$

$$\lambda \frac{dT}{dx}\Big|_{x=L} = h_2(T_{2\infty} - T_{2s}) = q'', \tag{9.21}$$

and the solution in solid wall yields,

$$q'' = \lambda \frac{T_{1s} - T_{2s}}{L}. \tag{9.22}$$

Combining the above three equations gives the following expression for the heat flux flowing through the wall,

$$q'' = \frac{T_{1\infty} - T_{2\infty}}{\frac{1}{h_1} + \frac{L}{\lambda} + \frac{1}{h_2}}. \tag{9.23}$$

The corresponding thermal resistance for the wall is as follows,

$$R_{th} = \frac{T_{1\infty} - T_{2\infty}}{q} = \frac{1}{A} \left(\frac{1}{h_1} + \frac{L}{\lambda} + \frac{1}{h_2} \right). \tag{9.24}$$

Hollow Cylinder

Let us consider steady-state conduction in an infinite hollow cylinder, as shown in Fig. 9.3. As a first task, we will find the temperature distribution in the cylindrical wall, assuming that the inner and outer surface temperatures, T_i and T_o, are known. The equation of change for the temperature

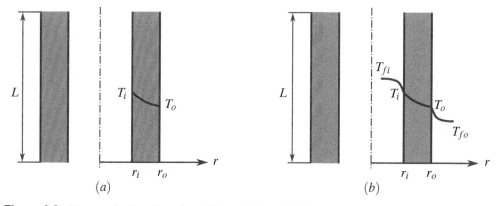

Figure 9.3 Heat conduction through a hollow cylinder with (a) given inner and outer surface temperature and (b) convective heat transfer on both sides.

can be written in the cylindrical polar coordinates as follows,

$$\frac{1}{r} \frac{\partial}{\partial r} \left(r \frac{\partial T}{\partial r} \right) + \frac{1}{r^2} \frac{\partial^2 T}{\partial \theta^2} + \frac{\partial^2 T}{\partial z^2} = 0. \tag{9.25}$$

Assuming further that the temperature has axisymmetric distribution (it does not depend on θ-coordinate) and it does not depend on z-coordinate, the following second order ordinary differential equation is obtained,

$$\frac{1}{r} \frac{d}{dr} \left(r \frac{dT}{dr} \right) = 0, \tag{9.26}$$

with boundary conditions,

$$T(r_i) = T_i, \quad T(r_o) = T_o. \tag{9.27}$$

A single integration of Eq. (9.26) yields,

$$r \frac{dT}{dr} = C, \tag{9.28}$$

where C is a constant. Integrating once more time, we obtain the following general solution,

$$T(r) = C \int \frac{dr}{r} = C \ln r + D, \tag{9.29}$$

where D is another constant. Employing the boundary conditions, the two constants C and D can be determined and the solution can be written as,

$$T(r) = T_o + (T_i - T_o) \frac{\ln \frac{r}{r_o}}{\ln \frac{r_i}{r_o}}. \tag{9.30}$$

We often are interested to know what is the heat flow rate through a cylindrical wall. To calculate this quantity, we can consider any control surface within the wall. Taking a cylindrical surface with a radius r, where $r_i < r < r_o$, and length L, and calculating the local heat flux from Fourier's law $q'' = -\lambda \frac{dT}{dr}$, the heat flow rate becomes,

$$q = -2\pi r L \lambda \frac{dT}{dr} = \frac{2\pi L \lambda}{\ln \frac{r_o}{r_i}} (T_i - T_o) \neq f(r). \tag{9.31}$$

As can be seen, the heat flow rate does not depend on the radial position and is a function of the cylinder dimensions (L, r_i, r_o), temperatures (T_o, T_i), and material property (λ). We can also find the *linear heat rate q'* as follows,

$$q' \equiv \frac{q}{L} = \frac{2\pi \lambda}{\ln \frac{r_o}{r_i}} (T_i - T_o). \tag{9.32}$$

Similarly as for the plane wall, the thermal resistance can be defined for the heat conduction in a cylindrical wall as follows,

$$q' = \frac{2\pi \lambda}{\ln \frac{r_o}{r_i}} (T_i - T_o) = \frac{T_i - T_o}{R_{th} L}, \tag{9.33}$$

where the thermal resistance for a cylindrical wall is defined as,

$$R_{th} \equiv \frac{\ln \frac{r_o}{r_i}}{2\pi \lambda L}. \tag{9.34}$$

Once the heat flow rate q is known, we can obtain heat fluxes on both the inner and outer surfaces as follows,

$$q''_o = \frac{q}{2\pi r_o L} = \frac{\lambda}{r_o \ln \frac{r_o}{r_i}} (T_i - T_o), \quad q''_i = \frac{q}{2\pi r_i L} = \frac{\lambda}{r_i \ln \frac{r_o}{r_i}} (T_i - T_o). \tag{9.35}$$

Thus, as it should be expected, $q''_o \neq q''_i$.

We can now extend our analysis to include convective heat transfer on both sides of the cylindrical wall. the expected temperature distribution for such case is shown in Fig. 9.3 (b). It is convenient to write expressions for the linear heat rate for convective heat transfer on both sides of the cylindrical wall. On the inner surface we have $q' = h_i 2\pi r_i (T_{fi} - T_i)$, thus,

$$T_{fi} = T_i + \frac{q'}{h_i 2\pi r_i}, \tag{9.36}$$

where T_{fi} is the bulk temperature of the inner fluid and h_i is the convective heat transfer coefficient between the inner fluid and the wall inner surface. In a similar way, using obvious notation, the bulk temperature of the outer fluid is found as,

$$T_{fo} = T_o - \frac{q'}{h_o 2\pi r_o}. \tag{9.37}$$

Combining the above two equations with Eq. (9.32) we have,

$$T_{fi} - T_{fo} = q' \left(\frac{1}{h_i 2\pi r_i} + \frac{\ln \frac{r_o}{r_i}}{2\pi\lambda} + \frac{1}{h_o 2\pi r_o} \right). \tag{9.38}$$

Thus for a hollow conducting cylinder with convective heat transfer on the inner and outer surfaces, the over-all thermal resistance is as follows,

$$R_{th} = \frac{1}{h_i 2\pi r_i L} + \frac{\ln \frac{r_o}{r_i}}{2\pi\lambda L} + \frac{1}{h_o 2\pi r_o L}. \tag{9.39}$$

Using the above expression for the thermal resistance, the total heat transferred through a hollow cylinder with length L can be found as,

$$q = \frac{T_{fi} - T_{fo}}{R_{th}}. \tag{9.40}$$

Composite Wall

Composite walls are made up of layers of various materials, which have different thermal properties. Such walls are frequently used in industry and building construction, where they have to simultaneously serve various purposes, such as to minimize the thermal losses and at the same time to satisfy the structure integrity requirements.

Our previous results obtained for single-property walls can be easily extended to composite walls. Figure 9.4 illustrates two composite walls: (a)—a planar wall, and (b)—a hollow cylinder wall. For

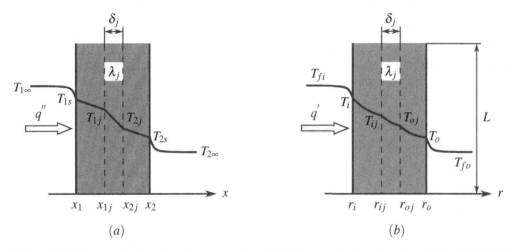

Figure 9.4 Heat conduction through a composite wall with convective heat transfer on both sides (a) planar wall (b) cylindrical wall.

a composite planar wall consisting of M-layers of different materials, placed between two fluid streams at temperatures $T_{1\infty}$ and $T_{2\infty}$, Eq. (9.23) can be extended as follows,

$$q'' = \frac{T_{1\infty} - T_{2\infty}}{\frac{1}{h_1} + \sum_{j=1}^{M} \frac{\delta_j}{\lambda_j} + \frac{1}{h_2}}. \tag{9.41}$$

Here δ_j and λ_j are j-th layer thickness and thermal conductivity, respectively. Introducing the thermal resistance R_{th}, we have,

$$q'' = \frac{T_{1\infty} - T_{2\infty}}{AR_{th}}, \quad R_{th} = \frac{1}{A}\left(\frac{1}{h_1} + \sum_{j=1}^{M}\frac{\delta_j}{\lambda_j} + \frac{1}{h_2}\right). \tag{9.42}$$

Here A is the composite wall surface area for which the thermal resistance is defined. Sometimes Eq. (9.41) is written in a form similar to Newton's equation of cooling

$$q'' = h_O(T_{1\infty} - T_{2\infty}), \quad h_O = \left(\frac{1}{h_1} + \sum_{j=1}^{M}\frac{\delta_j}{\lambda_j} + \frac{1}{h_2}\right)^{-1}, \tag{9.43}$$

where h_O is the so-called *over-all heat transfer coefficient*.

For a cylindrical composite wall shown in Fig. 9.4 (b), the linear heat rate is as follows,

$$q' = \frac{T_{fi} - T_{fo}}{\frac{1}{h_i 2\pi r_i} + \sum_{j=1}^{M}\frac{\ln r_{oj}/r_{ij}}{2\pi\lambda_j} + \frac{1}{h_o 2\pi r_o}} = \frac{T_{fi} - T_{fo}}{R_{th}L}, \tag{9.44}$$

where the thermal resistance of a composite cylindrical wall with length L is defined as,

$$R_{th} = \frac{1}{h_i 2\pi r_i L} + \frac{1}{2\pi L}\sum_{j=1}^{M}\frac{\ln\frac{r_{oj}}{r_{ij}}}{\lambda_j} + \frac{1}{h_o 2\pi r_o L}. \tag{9.45}$$

Critical Insulation Thickness

A composite wall containing an insulation material is often designed to minimize heat losses. It is thus important to know how the wall should be designed to satisfy this criterion. When a plane surface is insulated, the rate of heat transfer q always decreases with increasing thickness of the insulation. Thus the optimum insulation thickness can be determined purely on the cost calculation basis. For cylindrical or spherical surfaces, there is always a certain thickness of the insulation, for which the total thermal resistance gets its minimum. We call this thickness the *critical thickness of insulation*. This behavior can be explained by two contradicting effects, when increasing the insulation thickness,

- increasing conductance resistance in the insulation layer,
- decreasing convection resistance due to increasing convection surface area.

Let us consider a hollow cylinder with outer layer of insulation. We can neglect the resistance of the well-conducting solid wall and write the thermal resistance of the composite cylindrical wall as follows,

$$R_{th} = \frac{1}{h_i 2\pi r_i L} + \frac{1}{2\pi L\lambda}\ln\frac{r_o}{r_i} + \frac{1}{h_o 2\pi r_o L}. \tag{9.46}$$

Here r_o represents the outer radius of the insulation. Assuming that all other parameters in the above equation are constant, the minimum of the thermal resistance in function of r_o can be found as,

$$\frac{dR_{th}}{dr_o} = \frac{1}{2\pi L\lambda}\frac{1}{r_o} - \frac{1}{h_o 2\pi r_o^2 L} = 0, \quad \text{thus} \quad r_{o,cr} = \frac{\lambda}{h_o}. \tag{9.47}$$

The thermal resistance of an insulated wall will decrease when the outer radius of the insulation is $r_o < r_{cr}$. Usually under normally expected conditions the critical thickness of insulation is quite small and adding additional insulation will increase the thermal resistance. However, in case of electrical cables, when their efficient cooling is an objective, the critical thickness of insulation can be chosen to increase the amount of heat released to the surroundings.

Infinite Cylinder with Uniform Heat Sources

For an infinite cylinder with radius r_o, a known surface temperature $T(r_o) = T_o$, and with uniform heat sources q''', the temperature change equation is as follows,

$$\frac{1}{r}\frac{d}{dr}\left(r\frac{dT}{dr}\right) + \frac{q'''}{\lambda} = 0, \qquad (9.48)$$

and the boundary conditions are written as,

$$\left.\frac{dT}{dr}\right|_{r=0} = 0, \quad T(r)\,|_{r=r_o} = T_o. \qquad (9.49)$$

An integration of the temperature change equation gives,

$$r\frac{dT}{dr} = -\frac{q'''r^2}{2\lambda} + C, \quad \text{or} \quad \frac{dT}{dr} = -\frac{q'''r}{2\lambda} + \frac{C}{r}. \qquad (9.50)$$

To satisfy the boundary condition at $r = 0$, we need $C = 0$. The second integration yields,

$$T = -\frac{q'''r^2}{4\lambda} + D. \qquad (9.51)$$

Employing the boundary condition at $r = r_o$ we have,

$$D = T_o + \frac{q'''r_o^2}{4\lambda}, \quad \text{thus} \quad T = \frac{q'''r_o^2}{4\lambda}\left(1 - \frac{r^2}{r_o^2}\right) + T_o. \qquad (9.52)$$

Infinite Cylinder with Heat Sources from Nuclear Fission

Nuclear fuel rod can be considered as an infinite cylinder with internal heat sources, described by the modified Bessel function of the first kind and zero order, $q''' = AI_0(kr)$, where A and k are constants. Thus, the temperature change equation can be written as,

$$\frac{1}{r}\frac{d}{dr}\left(r\frac{dT}{dr}\right) + \frac{A}{\lambda}I_0(kr) = 0. \qquad (9.53)$$

We take the following boundary conditions,

$$\left.\frac{dT}{dr}\right|_{r=0} = 0, \quad T(r)\,|_{r=r_o} = T_o. \qquad (9.54)$$

The differential equation can be written as,

$$d\left(r\frac{dT}{dr}\right) = -\frac{A}{\lambda k^2}(kr)I_0(kr)d(kr). \qquad (9.55)$$

An integration of the equation yields,

$$r\frac{dT}{dr} = -\frac{A}{\lambda k^2}krI_1(kr), \quad \text{thus} \quad \frac{dT}{dr} = -\frac{A}{\lambda k}I_1(kr), \qquad (9.56)$$

where $I_1(kr)$ is the modified Bessel function of the first kind and first order. An integration of the obtained equation gives,

$$T = -\frac{A}{\lambda k^2}I_0(kr) + C. \qquad (9.57)$$

However, since $T(r)\,|_{r=r_o} = T_o$, we have,

$$T - T_o = \frac{A}{\lambda k^2}\left[I_0(kr_o) - I_0(kr)\right]. \tag{9.58}$$

The constant A can be found as,

$$q' = \int_0^{r_o} q''' 2\pi r dr = 2\pi A \int_0^{r_o} I_0(kr)dr, \tag{9.59}$$

thus,

$$A = \frac{q'k}{2\pi r_o I_1(kr_o)}, \tag{9.60}$$

and the temperature distribution can be written as,

$$T - T_o = \frac{q'}{2\pi\lambda kr_o}\frac{I_0(kr_o) - I_0(kr)}{I_1(kr_o)}. \tag{9.61}$$

9.2.2 TRANSIENT HEAT CONDUCTION

During transient heat transfer, the temperature is changing with time. We discuss here two popular approaches to solve transient heat transfer problems:

- a lumped thermal capacity model,
- a semi-infinite region model, which is applied to solve a generic one-dimensional transient problem.

Lumped Thermal Capacity Model

In the **lumped thermal capacity approach**, spatial temperature variations within the investigated body are neglected. For a body of an arbitrary shape with volume V and surface S, containing material with mass density ρ, specific heat at constant pressure c_p, submerged in a fluid with far-field temperature T_f, the energy conservation equation can be written as,

$$\rho V c_p \frac{dT}{dt} = -hS(T - T_f), \tag{9.62}$$

where h is the convective heat transfer coefficient on the surface of the body. Assuming that the initial temperature of the body is T_0, the solution of the differential energy equation gives,

$$\frac{T - T_f}{T_0 - T_f} = e^{-\frac{hS}{\rho V c_p}t}. \tag{9.63}$$

Let us introduce the following parameters,

$$L = \frac{V}{S}, \quad \text{equivalent length of the body}$$

$$\text{Bi} = \frac{hl}{\lambda}, \quad \text{Biot number, } l - \text{characteristic length}$$

$$\tau = \frac{\lambda t}{\rho c_p l^2} = \frac{at}{l^2}, \quad \text{non-dimensional time scale (Fourier number)}$$

Using the newly-defined parameters, the solution can be written as,

$$\Theta \equiv \frac{T - T_f}{T_0 - T_f} = e^{-\frac{l}{L}\text{Bi}\tau}. \tag{9.64}$$

The above equation can be used to approximately describe the temperature change for bodies with various shapes by a proper choice of the characteristic length. For example, for an infinite plate, $l = L = W/2$, where W is the plate width, for an infinite cylinder $l = R$, $L = R/2$, where R is the cylinder radius, and for a sphere $l = R$, $L = R/3$, where R is the sphere radius.

Generic One-Dimensional Transient Model

A space and time resolved one dimensional transient equation of change for temperature can be written as,

$$\frac{1}{\xi^n}\frac{\partial}{\partial\xi}\left(\xi^n\frac{\partial\Theta}{\partial\xi}\right) = \frac{\partial\Theta}{\partial\tau},$$ (9.65)

with the boundary condition

$$\frac{\partial\Theta}{\partial\xi} = 0 \quad \text{for} \quad \xi = 0, \qquad \frac{\partial\Theta}{\partial\xi} + \text{Bi}\Theta = 0 \quad \text{for} \quad \xi = 1,$$ (9.66)

and the initial condition

$$\Theta = 1 \quad \text{for} \quad \tau = 0.$$ (9.67)

The above generic non-dimensional system of equations is valid for an infinite symmetric plate ($n = 0$), an infinite axi-symmetric cylinder ($n = 1$) or a sphere ($n = 2$), where the following non-dimensional variables and parameters are introduced,

$$\Theta \equiv \frac{T - T_f}{T_0 - T_f}, \quad \tau \equiv \frac{at}{r_0^2}, \quad \xi \equiv \frac{r}{r_0}, \quad \text{Bi} \equiv \frac{hr_0}{\lambda}.$$ (9.68)

Here T is the temperature in the body, T_f is a constant temperature of fluid in which the body is submerged, T_0 is the body initial temperature, r_0 is the characteristic dimension of the body (a half-width of the plate or the cylinder/sphere radius), h is a heat transfer coefficient, λ is the solid thermal conductivity, and a is the solid thermal diffusivity. Employing the method of variable separation, a general solution of the one-dimensional transient conduction problem is found as,

$$\Theta(\xi,\tau) = \sum_{k=1}^{\infty} \frac{2\text{Bi}}{\mu_k^2 + \text{Bi}^2 + 2\nu\text{Bi}} \frac{\xi^\nu J_{-\nu}(\mu_k\xi)}{J_{-\nu}(\mu_k)} e^{-\mu_k^2\tau},$$ (9.69)

where $\nu = (1 - n)/2$ and thus $\nu = 1/2$ for a plate, $\nu = 0$ for a cylinder, and $\nu = -1/2$ for a sphere, and μ_k are eigenvalues found from the following transcendental equations,

$$\mu_k J_{-(\nu-1)}(\mu_k) = \text{Bi}J_{-\nu}(\mu_k).$$ (9.70)

We should note that the involved fractional order Bessel functions are as follows,

$$J_{-1/2}(x) = \left(\frac{2}{\pi x}\right)^{1/2}\cos x,$$ (9.71)

$$J_{1/2}(x) = \left(\frac{2}{\pi x}\right)^{1/2}\sin x,$$ (9.72)

$$J_{3/2}(x) = \left(\frac{2}{\pi x}\right)^{1/2}\left(\frac{\sin x}{x} - \cos x\right).$$ (9.73)

9.3 CONVECTION

Heat transfer effected by fluid motion is called *convection heat transfer*, or simply, *convection*. *Forced convection* takes place when fluid motion is resulting from an external pressure gradient, created by pumps, blowers or gravity head. In *natural convection*, only fluid density gradients are creating a driving force for fluid motion. A combined natural and forced convection is called the *mixed convection*. Additional classification takes into account the stability and character of fluid flow. Convection with laminar fluid flow is called the *laminar convection*, whereas when the flow is turbulent, it is called the *turbulent convection*.

One of the fundamental assumptions in the convective heat transfer theory is that the wall heat flux q'' is proportional to the temperature difference between the wall surface T_w, and the bulk or far-field fluid T_∞, with a coefficient of proportionality h, called a *heat transfer coefficient*,

$$q'' = h(T_w - T_\infty). \tag{9.74}$$

The whole complexity of the convective heat transfer is hidden in the way how the heat transfer coefficient is determined. This coefficient is frequently expressed in terms of a non-dimensional *Nusselt number* defined as,

$$\text{Nu} = \frac{hL}{\lambda}, \tag{9.75}$$

where L is the characteristic length and λ is the thermal conductivity of fluid.

9.3.1 FORCED CONVECTION

During forced convection heat transfer, the flow velocity and fluid material properties have the most significant influence on the heat transfer coefficient. Under such conditions, the Nusselt number is correlated to other non-dimensional numbers, such as the Reynolds number, Re, and the Prandtl number, Pr,

$$\text{Nu} = f(\text{Re}, \text{Pr}, ...), \tag{9.76}$$

where

$$\text{Re} = \frac{UL}{\nu}, \tag{9.77}$$

and

$$\text{Pr} = \frac{c_p \mu}{\lambda}. \tag{9.78}$$

Here ν—kinematic viscosity of fluid, U—characteristic velocity of fluid, L—characteristic length, c_p—specific heat of fluid, and λ—thermal conductivity of fluid.

Laminar Forced Convection

During laminar forced convection, the fluid motion is laminar and, in simple geometries, it can be described analytically. For example, when a fully-developed, laminar axisymmetric flow in a pipe is considered, the governing temperature change equation can be written as follows,

$$\frac{w}{a} \frac{\partial T}{\partial z} = \frac{1}{r} \frac{\partial}{\partial r} \left(r \frac{\partial T}{\partial r} \right) + \frac{\partial^2 T}{\partial^2 z}, \tag{9.79}$$

where $w(r)$ is the local flow velocity in the axial direction, a is the thermal diffusivity, T is the temperature, and r, z are radial and axial coordinates, respectively. The temperature change equation contains three terms: (i) a convective heat transfer on the left-hand side, (ii) a radial heat conduction, represented by the first term on the right-hand side, and (iii) an axial heat conduction, represented by the second term on the right-hand side. Not all three terms are equally important in various situations. For example, for fluids such as water or air, the axial conduction term is much less than the other two terms, and can be neglected. However, for highly conducting fluids like liquid metals, this term is not negligible and should be included in the analysis. In the present analysis, we assume that this term is small and can be dropped (thus the results we will obtain are not valid for liquid metals). We further assume a constant heat flux applied to the wall of the pipe and a constant heat transfer coefficient between the wall and the fluid. With these assumptions the fluid temperature in the pipe will increase in a uniform manner and the following relationships between axial temperature derivatives can be written,

$$\frac{\partial T}{\partial z} = \frac{dT_w}{dz} = \frac{dT_m}{dz}. \tag{9.80}$$

Here T is the fluid temperature at any radial location in the pipe, T_w is the wall surface temperature and T_m is the fluid cross-section mean temperature.

For fully developed laminar flow in a pipe, the velocity has the following distribution

$$w(r) = 2U \left[1 - \left(\frac{r}{R} \right)^2 \right], \tag{9.81}$$

where U is the cross-section mean fluid velocity and R is the pipe radius. Substituting this expression to Eq. (9.79), and employing the above-mentioned assumptions, we get,

$$\frac{2U}{a} \left[1 - \left(\frac{r}{R} \right)^2 \right] \frac{dT_m}{dz} = \frac{1}{r} \frac{d}{dr} \left(r \frac{dT}{dr} \right). \tag{9.82}$$

Double integration of the equation yields the following radial temperature distribution in the fluid,

$$T = T_w - \frac{2UR^2}{a} \left[\frac{3}{16} - \left(\frac{r}{2R} \right)^2 + \left(\frac{r}{2R} \right)^4 \right] \frac{dT_m}{dz}, \tag{9.83}$$

The found radial temperature distribution can be used to calculate the cross-section mean temperature of the fluid as follows,

$$T_m \equiv \frac{\int_A wT\,dA}{\int_A w\,dA} = \frac{\int_0^R 2\pi rwT\,dr}{\int_0^R 2\pi rw\,dr} = T_w - \frac{11}{96} \frac{2UR^2}{a} \frac{dT_m}{dz}. \tag{9.84}$$

Our goal is to find the convective heat transfer coefficient. To this end, we write Newton's equation of cooling and combine it with the above equation. As a result, we get the following expression for the wall heat flux,

$$q'' = h(T_w - T_m) = h\frac{11}{96} \frac{2UR^2}{a} \frac{dT_m}{dz}. \tag{9.85}$$

The same wall heat flux can be obtained from a boundary condition with a known fluid temperature at the wall surface. Thus,

$$q'' = -\lambda \left(\frac{dT}{dr} \right)_{r=R} = \frac{\lambda UR}{2a} \frac{dT_m}{dz}. \tag{9.86}$$

Combining the two expressions yields the following heat transfer coefficient and the corresponding Nusselt number,

$$h = \frac{48}{11} \frac{\lambda}{2R}, \quad \text{and thus} \quad \text{Nu} = \frac{h2R}{\lambda} = \frac{48}{11} \simeq 4.364. \tag{9.87}$$

This classical solution shows that, for fully-developed laminar flow in a pipe, the convective heat transfer coefficient, when the applied heat flux is uniform along the pipe axis, can be represented with a constant Nusselt number. Similar analyses can be performed for other conditions (such as for example a constant wall temperature) and channel shapes. Table 9.1 gives Nusselt numbers for several typical channel shapes.

The results presented in Table 9.1 are valid under an assumption that, between others, the fluid properties are constant and do not change with temperature. For example, velocity profile during laminar flow in a pipe has a parabolic shape only when the fluid viscosity does not change with the temperature. In reality, fluids have temperature-dependent viscosities. In addition, density variations with fluid temperature cause an onset of free convection, which during laminar flow can have a significant influence on the over-all heat transfer coefficient. These temperature influences on the flow

Table 9.1

Nusselt Number during Laminar Forced Convection in Channels of Different Shapes

Channel Geometry	Constant Wall Temperature	Constant Wall Heat Flux
Pipe	3.66	4.364
Equilateral triangle	2.35	3.00
Square	2.89	3.63
Rectangle $a/b = 0.713$	-	3.78
Rectangle $a/b = 0.5$	3.39	4.11
Rectangle $a/b = 0.25$	4.65	5.35
Parallel plates	7.54	8.235

Source: [7]

and heat transfer phenomena make the laminar convection analysis quite difficult, even using experimental methods. As a result, the theoretical and empirical expression, derived from measurements and the dimensional analysis, are frequently leading to quite significantly differing results.

A frequently used expression, in which the Nusselt number is based on a heat transfer coefficient derived from an arithmetic-average wall-fluid temperature difference at the inlet and outlet, was proposed by Sieder and Tate [79],

$$
\mathrm{Nu}_{ar} = 1.86 \left(\mathrm{Pe}\frac{D_h}{L} \right)^{1/3} \left(\frac{\mu_f}{\mu_w} \right)^{0.14}, \tag{9.88}
$$

where $\mathrm{Pe} = \mathrm{Re}\cdot\mathrm{Pr}$ is the Peclet number, μ_f is the dynamic viscosity of fluid calculated at the arithmetic average fluid temperature at inlet and outlet and μ_w is the same dynamic viscosity calculated at the wall temperature. The term $(\mu_f/\mu_w)^{0.14}$ represents the effect of heating or cooling on the heat transfer coefficient. All other fluid properties should be calculated at temperature T_f.

Turbulent Forced Convection

With an increasing Reynolds number, a laminar forced convection transits into a turbulent flow convection heat transfer. As a consequence, a heat transfer coefficient significantly increases. This behavior can be explained by realizing that, unlike in laminar convection, heat is now transported by turbulent eddies in the direction perpendicular to the wall surface.

Turbulent forced convection problems cannot be solved analytically, and only empirical methods can be used. Colburn [17] analyzed empirical data and observed that the heat transfer coefficient for turbulent duct flow can be described by the following correlation,

$$
\mathrm{St}\,\mathrm{Pr}^{2/3} \cong \frac{C_f}{2}, \tag{9.89}
$$

where now the non-dimensional heat transfer coefficient is expressed in terms of the *Stanton number* defined as,

$$
\mathrm{St} = \frac{h}{\rho c_p U}. \tag{9.90}
$$

Here h is the heat transfer coefficient, ρ is the fluid density, c_p is the specific heat at constant pressure and U is the averaged velocity. Equation (9.89) is called the *Colburn analogy*, which states that there exists an analogy between the convective heat transfer and the momentum transfer in turbulent flows.

The equation states a proportionality between the heat transfer coefficient, expressed in terms of the Stanton number, and the Fanning friction factor, C_f.

The Colburn analogy holds for $Pr \gtrsim 0.5$ and is to be used together with a correlation for the friction factor (e.g. Coolbrook or Haaland formula). For example, applying the Blasius correlation valid for smooth walls, and noting that $St = Nu/(Re\, Pr)$, we get,

$$Nu = 0.023\, Re^{0.8} Pr^{1/3}. \tag{9.91}$$

The validity range of the correlation is determined by the validity of the Blasius correlation and is $2 \times 10^4 < Re < 10^6$.

Sieder and Tate [79] modified Eq. (9.91) for applications in which temperature influence on properties is significant, and recommended,

$$Nu = 0.027\, Re^{0.8} Pr^{1/3} \left(\frac{\mu}{\mu_w}\right)^{0.14}, \tag{9.92}$$

where all fluid properties are evaluated at bulk fluid temperature and $\mu_w = \mu(T_w)$ is the fluid dynamic viscosity evaluated at the wall temperature. The correlation is valid for $0.7 < Pr < 16700$ and $Re > 10^4$.

The most popular formula is a correlation due to Dittus and Boelter [25],

$$Nu = 0.023\, Re^{0.8} Pr^n, \tag{9.93}$$

which is valid for fully developed turbulent flow ($L/D_h > 60$), $0.7 \leq Pr \leq 120$, and $2500 \leq Re \leq 1.24 \times 10^5$. The Prandtl number exponent is $n = 0.4$ when the fluid is heated and $n = 0.3$ when the fluid is cooled. All fluid properties are to be evaluated at the fluid bulk temperature.

One of the most accurate correlations based on the Colburn analogy was proposed by Gnielinski [34],

$$Nu = \frac{\left(Re - 10^3\right) Pr\, C_f/2}{1 + 12.7 \left(C_f/2\right)^{1/2} \left(Pr^{2/3} - 1\right)}. \tag{9.94}$$

The expression is to be used together with a correlation for the Fanning friction factor. Its validity range is $0.5 < Pr < 10^6$ and $2300 < Re < 5 \times 10^6$.

9.3.2 NATURAL CONVECTION

Natural (or *free*) *convection* occurs when a fluid flow is driven by buoyancy forces that are resulting from fluid density gradients. As a typical example, natural convection along a vertical wall can be investigated. Considering a vertical wall with height H in x-direction and infinite in z-direction, the steady-state, constant-property, conservation equations for mass, momentum and energy can be written in a two-dimensional $x - y$ plane as follows,

$$\frac{\partial(u)}{\partial x} + \frac{\partial(v)}{\partial y} = 0, \tag{9.95}$$

$$\rho \left(u\frac{\partial u}{\partial x} + v\frac{\partial u}{\partial y}\right) = -\frac{\partial p}{\partial x} + \mu \left(\frac{\partial^2 u}{\partial x^2} + \frac{\partial^2 u}{\partial y^2}\right) - \rho g, \tag{9.96}$$

$$\rho \left(u\frac{\partial v}{\partial x} + v\frac{\partial v}{\partial y}\right) = -\frac{\partial p}{\partial y} + \mu \left(\frac{\partial^2 v}{\partial x^2} + \frac{\partial^2 v}{\partial y^2}\right), \tag{9.97}$$

$$u\frac{\partial T}{\partial x} + v\frac{\partial T}{\partial y} = a \left(\frac{\partial^2 T}{\partial x^2} + \frac{\partial^2 T}{\partial y^2}\right). \tag{9.98}$$

Since y-axis is oriented perpendicular to the wall, and is thus horizontal, there is no gravity term in the y-momentum equation. We can further simplify the equations assuming that far from the wall (when $y \to \infty$), the fluid is stagnant ($u = v = 0$), isothermal with temperature T_∞, and constant density ρ_∞. The equations are then describing fluid motion and heat convection in a relatively narrow layer close to the wall, with dominant flow in the x-direction. Thus since $u \gg v$, the momentum equation in the y-direction can be ignored and all $\partial^2/\partial y^2$ terms can be dropped. As a result, the so-called *boundary layer equations* for momentum and energy are obtained,

$$\rho\left(u\frac{\partial u}{\partial x} + v\frac{\partial u}{\partial y}\right) = -\frac{dp_\infty}{dx} + \mu\frac{\partial^2 u}{\partial y^2} - \rho g, \tag{9.99}$$

$$u\frac{\partial T}{\partial x} + v\frac{\partial T}{\partial y} = a\frac{\partial^2 T}{\partial y^2}, \tag{9.100}$$

where in the momentum equation, the term $\partial p/\partial x$ is replaced by dp_∞/dx, since in the boundary layer, the pressure is a function of the vertical position only. Noting further that the pressure gradient is determined by the far-field hydrostatic pressure in fluid of density ρ_∞, we have $dp_\infty/dx = -\rho_\infty g$, and the momentum equation becomes,

$$\rho\left(u\frac{\partial u}{\partial x} + v\frac{\partial u}{\partial y}\right) = \mu\frac{\partial^2 u}{\partial y^2} + (\rho_\infty - \rho)g. \tag{9.101}$$

What we obtained is a simplified momentum equation in the boundary layer, consisting of three terms: (i) an inertia term on the left-hand side of the equation, (ii) a friction term $\mu\partial^2 u/\partial y^2$, and (iii) a buoyancy force term $g(\rho_\infty - \rho)$ on the right-hand side of the equation. It is desirable to express the buoyancy force term as a function of temperature. To this end we use a definition of a *coefficient of thermal expansion* for the fluid as,

$$\beta = -\frac{1}{\rho}\left(\frac{\partial \rho}{\partial T}\right)_p = \frac{1}{v}\left(\frac{\partial v}{\partial T}\right)_p, \tag{9.102}$$

where $v = 1/\rho$ is the fluid specific volume. Since

$$v \simeq v_\infty + \left(\frac{\partial v}{\partial T}\right)_p (T - T_\infty) = v_\infty\left[1 + \beta(T - T_\infty)\right], \tag{9.103}$$

we have,

$$g(\rho_\infty - \rho) = g\rho\left(\frac{\rho_\infty}{\rho} - 1\right) \simeq g\rho\beta(T - T_\infty) = g\rho\beta\Delta T. \tag{9.104}$$

Therefore, the mass, momentum, and energy conservation equations for the boundary layer can be written as,

$$\frac{\partial u}{\partial x} + \frac{\partial v}{\partial y} = 0, \tag{9.105}$$

$$u\frac{\partial u}{\partial x} + v\frac{\partial u}{\partial y} = v\frac{\partial^2 u}{\partial y^2} + g\beta\Delta T, \tag{9.106}$$

$$u\frac{\partial T}{\partial x} + v\frac{\partial T}{\partial y} = a\frac{\partial^2 T}{\partial y^2}, \tag{9.107}$$

where g, β, $v = \mu/\rho$, and $a = \lambda/(\rho_\infty c_p)$ are constants. The boundary conditions are as follows,

1. $u = v = 0$ and $T = T_w$ at $y = 0$,

2. $u \rightarrow 0$ and $T \rightarrow T_\infty$ as $y \rightarrow \infty$,
3. $u = 0$ at $x = 0$.

We introduce the following non-dimensional variables and coordinates,

$$\theta = \frac{T - T_\infty}{T_w - T_\infty}, \quad \xi = \frac{x}{H}, \quad \eta = \frac{y}{H}, \quad \hat{u} = \frac{u}{U}, \quad \hat{v} = \frac{v}{U}, \tag{9.108}$$

where U is a certain, not specified yet, velocity scale. Substituting the non-dimensional parameters into the momentum and energy equations, we get

$$\frac{U^2}{g\beta\Delta T H} \left(\hat{u}\frac{\partial \hat{u}}{\partial \xi} + \hat{v}\frac{\partial \hat{u}}{\partial \eta} \right) = \frac{\nu U}{g\beta\Delta T H^2} \frac{\partial^2 \hat{u}}{\partial \eta^2} + \theta, \tag{9.109}$$

$$\hat{u}\frac{\partial \theta}{\partial \xi} + \hat{v}\frac{\partial \theta}{\partial \eta} = \frac{a}{UH} \frac{\partial^2 \theta}{\partial \eta^2}. \tag{9.110}$$

We note that taking $U = a/H$ will simplify the energy equation to a parameter-free form as,

$$\hat{u}\frac{\partial \theta}{\partial \xi} + \hat{v}\frac{\partial \theta}{\partial \eta} = \frac{\partial^2 \theta}{\partial \eta^2}. \tag{9.111}$$

Substituting the newly defined velocity scale $U = a/H$ into the momentum equation yields,

$$\frac{a^2}{g\beta\Delta T H^3} \left(\hat{u}\frac{\partial \hat{u}}{\partial \xi} + \hat{v}\frac{\partial \hat{u}}{\partial \eta} \right) = \frac{\nu a}{g\beta\Delta T H^3} \frac{\partial^2 \hat{u}}{\partial \eta^2} + \theta. \tag{9.112}$$

Here we can identify the *Rayleigh number*, defined as follows,

$$\mathrm{Ra} = \frac{g\beta\Delta T H^3}{\nu a}. \tag{9.113}$$

Thus, the momentum equation can be written in terms of the Rayleigh number as,

$$\frac{1}{\mathrm{Ra}\,\mathrm{Pr}} \left(\hat{u}\frac{\partial \hat{u}}{\partial \xi} + \hat{v}\frac{\partial \hat{u}}{\partial \eta} \right) = \frac{1}{\mathrm{Ra}} \frac{\partial^2 \hat{u}}{\partial \eta^2} + \theta. \tag{9.114}$$

It is interesting to note the importance of the Prandtl number, which is purely a liquid-property quantity, on the balance of terms contained in the above equation. First, θ, which is always in the range between 0 and 1, represents the buoyancy force magnitude. When $\mathrm{Pr} \ll 1$, the first term on the right-hand-side (friction force) will become relatively less important than the term on the left-hand side (inertia force), which means that the flow will be determined by the balance between the inertia and buoyancy forces. On the contrary, when $\mathrm{Pr} \gg 1$ the two competing mechanisms will be the friction and buoyancy forces.

For low-Prandtl number fluids, since $\mathrm{Pr} = \nu/a$, we have $\nu \ll a$ and as a consequence, a heat transfer by thermal diffusion is more intensive than a momentum transfer. This means that the velocity boundary layer is located entirely within the thermal boundary layer.

The situation looks quite differently for the high-Prandtl number fluids, where the thermal boundary layer occupies only a fraction of the velocity boundary layer. In that case the velocity increases in the whole range of the thermal boundary layer due to a driving buoyancy force, and then decreases in the remaining, unheated part of the velocity boundary layer. This fundamentally different behavior of the low- and high-Prandtl number fluids is the reason that different models and correlations are needed for the two groups of fluids.

9.4 BOILING

Boiling is a process in which evaporation occurs due to addition of heat to the liquid-vapor mixture. In the **homogeneous boiling**, the vapor is generated in the whole volume of the mixture, whereas in the **heterogeneous boiling**, the vapor is generated at a heated solid wall only. In both types of boiling, the evaporation process takes place at the interface between the liquid phase and the vapor phase.

In this section we discuss heterogeneous boiling heat transfer in a pure substance, that is in a single chemical substance, such as water or liquid sodium. Water mixed with air or an alloy of two metals (e.g., a lead-bismuth alloy) are not pure substances. Such mixtures have thermodynamic properties that may substantially differ from those of the pure substances of which they are composed. In the continuation, the heterogeneous boiling heat transfer in pure substances will be simply referred to as the boiling heat transfer, and any departure from this assumption will be explicitly stated.

As already mentioned, during boiling heat transfer, a phase transition from the liquid phase into the vapor phase takes place at the liquid-vapor interface. For a given local pressure p, the phase transition occurs at a unique temperature, called a **saturation temperature** T_{sat}. The boiling curve $T_{sat}(p)$ for water is shown in Fig. 9.5.

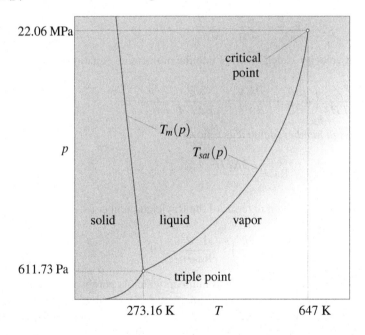

Figure 9.5 Phase plot for water in the pT plane. $T_m(p)$—melting curve, $T_{sat}(p)$—boiling curve.

Along the line that separates the liquid and vapor phases, both phases can co-exist at equilibrium. The relative proportion of the phases depends on the amount of heat added to the system. The more heat is added, the higher vapor fraction is present. The entire heat required for a transition of a unit mass of liquid into vapor is given by the **Classius-Clapeyron equation**,

$$i_{fg} \equiv i_g - i_f = \left(v_g - v_f\right) T \frac{dp}{dT}, \tag{9.115}$$

where i_{fg} is the **specific enthalpy of vaporization**, also called the **latent heat**, i_f is the specific enthalpy of saturated liquid phase, i_g is the specific enthalpy of saturated vapor phase, v_f is the specific volume of the liquid phase, v_g is the specific volume of the vapor phase, and T, p are the phase transition temperature and pressure, respectively. Since $i_{fg} > 0$ and $v_g > v_f$, Eq. (9.115) indicates that, with increasing pressure, the phase transition temperature increases, as confirmed by the shape of the boiling curve in Fig. 9.5.

9.4.1 NUCLEATION AND EBULLITION CYCLE

The boiling heat transfer is associated with a liquid-vapor phase change and appearance of vapor bubbles on the heater surface. The initial process of a bubble grow from the surface is called **nucleation** and the point on the surface from which the bubble grows, a **nucleation site**. Usually the nucleation sites are located at surface imperfections or cavities, where clusters of vapor can be easily entrapped. After nucleation, the bubble grows and finally departs from the surface, as depicted in Fig. 9.6.

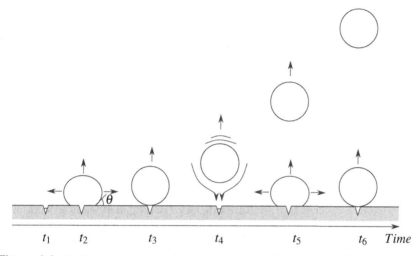

Figure 9.6 Nucleation from a cavity and an ebullition cycle: t_1—waiting time, t_2—bubble growth, t_3—bubble departure, t_4—waiting time, t_5—bubble growth, t_6—bubble departure, θ—contact angle.

Nucleation can only occur when the liquid around the nucleation site is superheated. There is also a certain cavity size range for which the bubble embryo can grow. If the conditions are satisfied, the bubble will grow as depicted at time instants t_2 and t_5 in Fig. 9.6. After achieving a certain diameter, the bubble will depart from the surface, as indicated by time instants t_3 and t_6. The bubble growth rate and the departure diameter depend to a large extend on a *contact angle* θ. The contact angle is measured in the liquid phase and is an angle of the interface with the solid surface. It appears to be a very important quantity, which is a constant for any combination of solid and fluid in a boiling system. For a perfectly hydrophobic system, the liquid does not wet the solid surface and $\theta = \pi$. On the contrary, for a perfectly hydrophilic system $\theta = 0$.

9.4.2 POOL BOILING

When a heat flux q'' is applied through a solid wall enclosing a pool of stagnant fluid at a temperature $T < T_{sat}$, the wall surface temperature on the wet side T_w will start increasing and, eventually, it will exceed the saturation temperature T_{sat}. The temperature difference $T_w - T_{sat}$ is called the **wall superheat**, T_{sup}. Figure 9.7 shows a curve $q'' = f(\Delta T_{sup})$ (frequently referred to as the **boiling curve**) for a system in which either the heat flux or the wall temperature is controlled. When the heat flux is controlled and increases between points A and B on the curve, the wall superheat increases as well and the rate of increase is governed by a natural convection heat transfer. In this region the bulk pool temperature may still be below the saturation temperature, but the liquid layer adjacent to the wall surface is already slightly superheated. However, the wall superheat is still too low to support the nucleate boiling.

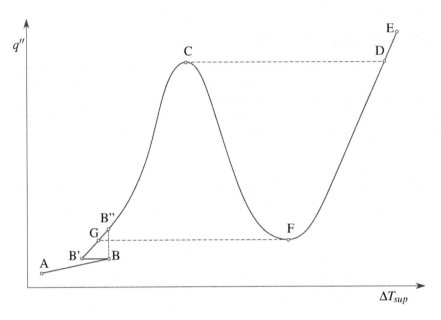

Figure 9.7 A boiling curve in pool boiling.

Only when the wall superheat exceeds B, a **nucleate subcooled boiling heat transfer** starts and due to an increased heat transfer intensity, the wall superheat may actually drop to point B'. For a temperature-controlled heat transfer, the resulting heat flux will suddenly increase from B to B", instead. Point B plays an important role in boiling heat transfer modeling and is known under such names as the **boiling incipient** point or the **onset of nucleate boiling** (ONB) point.

Curve B'C corresponds to boiling heat transfer in which, with a steadily increasing heat flux, the wall superheat increases as well. At point C a sudden deterioration of the heat transfer rate occurs and due to that the wall superheat suddenly jumps to point D. This jump is associated with a sudden change of the boiling heat transfer regime from a liquid-dominated to a vapor-dominated regime. The boiling regime transition CD is known under various names such as a **boiling crisis**, **burnout**, **departure from nucleate boiling** or **dryout**.

A literature devoted to boiling heat transfer is very rich and for understandable reasons it is not discussed in this text. An interested reader should consult, e.g., Carey [14], Rohsenow [73], Tong and Tang [84], Hsu and Graham [44], or Collier and Thome [19].

Many authors have proposed correlations based on experimental data obtained for various boiling systems. One of the earliest correlations of nucleate boiling heat transfer, applicable to a wide range of liquid-surface combinations, was proposed by Rohsenow [72]. The correlation has the following form,

$$\frac{q''}{\mu_g i_{fg}} \left[\frac{\sigma}{g \left(\rho_f - \rho_g \right)} \right]^{1/2} = C_{fs}^{-1/r} \mathrm{Pr}_f^{-s/r} \left\{ \frac{c_{pf} \left[T_w - T_{sat} \left(p_f \right) \right]}{i_{fg}} \right\}^{1/r} \tag{9.116}$$

where $r = 0.33$ and $s = 1.0$ for water and 1.7 for other liquids. Coefficient C_{fs} has been obtained from experimental data obtained with various liquid-surface combinations. A sample of available coefficient values is given in Table 9.2.

Table 9.2

Values of C_{fs} and s in Eq. (9.116) with $r = 0.33$

Liquid and Heating Surface Pair	C_{fs}	s
Water and mechanically polished stainless steel	0.0132	1.0
Water and chemically etched stainless steel	0.0133	1.0
Water and copper	0.013	1.0
Water and platinum	0.013	1.0
Ethyl alcohol and chromium	0.0027	1.7
Isopropyl alcohol and copper	0.0025	1.7
Benzen and chromium	0.010	1.7

Example 9.1:

A pool of water at atmospheric pressure ($p_f = 101325$ Pa) is heated from below through a mechanically polished stainless steel wall with heat flux $q'' = 100$ kW/m^2. Calculate the heated surface temperature and the wall superheat.

Solution

Eq. (9.116) yields the following expression for the wall temperature:

$$T_w = T_{sat}(p_f) + \frac{C_{fs} i_{fg} \mathrm{Pr}^s}{c_{pf}} \left\{ \frac{q''}{\mu_g i_{fg}} \left[\frac{\sigma}{g(\rho_f - \rho_g)} \right]^{1/2} \right\}^r .$$

For water and mechanically polished stainless steel we get from Table 9.2 $C_{fs} = 0.0132$, $s = 1.0$ and $r = 0.33$. For saturated water and vapor at atmospheric pressure, the properties are as follows, $i_{fg} = 2.257$ MJ/kg, $\mu_f = 2.819 \cdot 10^{-4}$ Pa·s, $\mu_g = 1.227 \cdot 10^{-5}$ Pa·s, $\sigma = 0.05892$ N/m, $\rho_f = 958.4$ kg/m^3, $\rho_g = 0.5976$ kg/m^3, $c_{pf} = 4.217$ kJ/kg·K, $\lambda_f = 0.6777$ W/m·K, $\mathrm{Pr}_f = \mu_f c_{pf}/\lambda_f = 1.754$ and $T_{sat} = 373.12$ K. Substituting the data we get $T_w = 398.74$ K. Thus the wall superheat is $\Delta T_{sup} = 25.62$ K.

9.4.3 FLOW BOILING

In **flow boiling**, liquid is not stagnant and experiences a macroscopic motion due to a forced or natural circulation. When a forced circulation is present, the macroscopic liquid motion is controlled by pumps or a gravity head, and the corresponding heat transfer regime is called **forced convective boiling**. In a similar way **natural convective boiling** is a flow boiling heat transfer regime when the macroscopic liquid motion is caused by buoyancy effects. This distinction is in analogy to the natural and forced convection heat transfer discussed in §9.3.

Flow boiling is one of the most efficient heat transfer regimes. On the one hand, the macroscopic liquid motion enhances heat transfer in a similar way as it is observed during the single-phase forced-convective heat transfer. On the other hand, the nucleation process creates additional microscopic liquid motion in the vicinity of the heated wall, further enhancing heat transfer rates. As a result, the flow boiling heat transfer can be considered as a superposition of microscopic and macroscopic effects.

The physics of flow boiling is very complex, leading to major difficulties as far as prediction of heat transfer rates is concerned. The difficulties stem from the fact that two-phase flow patterns and the corresponding heat transfer regimes are location-dependent. Typical distributions of important parameters in a heated channel in which flow boiling takes place are shown in Fig. 9.8. The figure is valid for a boiling channel where dryout occurs. At that point a sudden and significant increase of the wall temperature (and thus, a sudden and significant decrease of the heat transfer coefficient) occurs. An occurrence and location of dryout in a boiling channel depend on both the local conditions and the upstream flow history. Thus both these effects must be known to precisely predict the dryout occurrence. Selected methods to predict a boiling crisis and other flow boiling phenomena are presented in the following sections.

9.4.4 ONSET OF NUCLEATE BOILING

A transition from single-phase convective heat transfer to nucleate boiling heat transfer is called the **onset of nucleate boiling** (ONB). A location of the ONB in a boiling channel is important, since it determines both flow and heat transfer conditions. Upstream of the ONB point single-phase flow and heat transfer conditions prevail. Thus, correlations and models valid for single-phase flows can be used. However, downstream of the ONB point correlations and models valid for two-phase flows have to be employed.

9.4.5 SUBCOOLED BOILING

In a heated channel the bulk temperature of liquid increases with an increasing distance from the inlet. Assuming that the inlet liquid temperature is T_{in}, the bulk liquid temperature $T_b(z)$ at distance z from the inlet can be obtained from the energy balance as,

$$W c_p \left(T_b(z) - T_{in} \right) = q(z) = q'' P_H z, \tag{9.117}$$

where W is liquid mass flow rate, c_p—liquid specific heat, $q(z)$—total heat applied to the channel from the inlet to location z, P_H—heated perimeter, and q''—heat flux. In the above equation it is assumed that heat flux is uniformly distributed along the channel. Thus, the distance-dependent bulk liquid temperature is obtained as,

$$T_b(z) = T_{in} + \frac{q'' P_H z}{W c_p}. \tag{9.118}$$

As long as single-phase flow prevails in the channel, the wall temperature can be found from Newton's equation of cooling,

$$T_w = T_b + q''/h, \tag{9.119}$$

where h is a single-phase convective heat transfer coefficient. Thus the wall temperature can be now expressed as,

$$T_w = T_{in} + q'' \left(\frac{P_H z}{W c_p} + \frac{1}{h} \right). \tag{9.120}$$

Let us introduce wall superheat defined as,

$$\Delta T_{sup} \equiv T_w - T_{sat}, \tag{9.121}$$

and inlet subcooling defined as,

$$\Delta T_{subi} \equiv T_{sat} - T_{in}. \tag{9.122}$$

Note that the definitions are such that inlet subcooling is positive when liquid is subcooled at the inlet, and wall superheat is positive when the wall temperature is greater than the saturation temperature. Using the above definitions, the wall superheat can be expressed in terms of z as,

$$\Delta T_{sup}(z) = -\Delta T_{subi} + q'' \left(\frac{P_H z}{W c_p} + \frac{1}{h} \right). \tag{9.123}$$

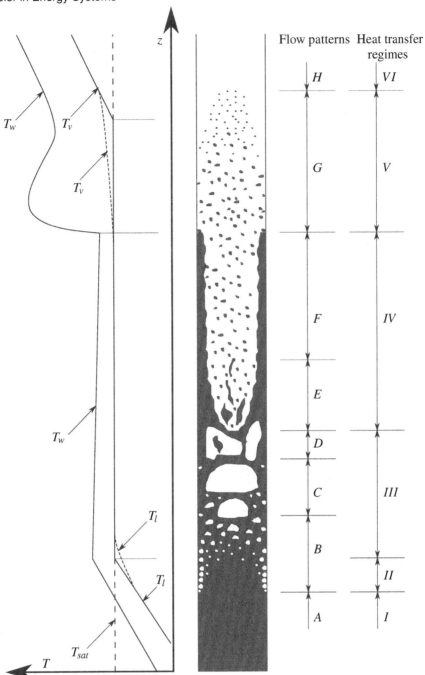

Figure 9.8 Wall temperature, fluid temperature, two-phase flow patterns and heat transfer regimes in a boiling channel with dryout. *A*—single-phase liquid flow, *B*—bubbly flow, *C*—slug flow, *D*—churn-turbulent flow, *E*—wispy-annular flow, *F*—annular flow, *G*—drop flow, *H*—single-phase vapor flow, *I*—convective heat transfer to liquid, *II*—subcooled nucleate boiling, *III*—saturated nucleate boiling, *IV*—evaporating liquid film, *V*—post-dryout heat transfer, *VI*—convective heat transfer to vapor, T_w—wall temperature, T_{sat}—saturation temperature, T_l—liquid bulk temperature, T_v—vapor bulk temperature (based on [19]).

Clearly, boiling in a channel is only possible at places where the wall superheat is positive, that is $\Delta T_{sup} > 0$. Experiments indicate that a certain minimum, positive wall superheat is needed for boiling to occur.

One of the earliest models to predict the wall superheat at the onset of nucleate boiling, $\Delta T_{sup,ONB}$, was proposed by Hsu [43], who postulated that nucleate boiling is possible in the thermal boundary layer only when the wall superheat is high enough to allow grow of bubbles at the wall. He provided the following equation,

$$\Delta T_{sup,ONB} = -\Delta T_{sub} + \frac{\theta + \sqrt{4\Delta T_{sub}\theta + \theta^2}}{2}, \quad \theta = \frac{12.8\sigma T_{sat}}{\rho_g i_{fg}\delta_t}, \quad \delta_t \cong \frac{\lambda_f}{h_{1\phi}}. \tag{9.124}$$

Here ΔT_{sub}—local subcooling, T_{sat}—saturation temperature, σ—surface tension, ρ_g—vapor saturated density, i_{fg}—latent heat, λ_f—liquid saturated thermal conductivity, and $h_{1\phi}$—single-phase heat transfer coefficient.

Using similar assumptions as Hsu, Sato, and Matsumura arrived at the following equation that should be satisfied at ONB,

$$q_w'' = \frac{\lambda_f i_{fg}\rho_g \Delta T_{sup,ONB}^2}{8\sigma T_{sat}}, \tag{9.125}$$

where q_w'' is the wall heat flux and other symbols have the usual meaning.

Davis and Anderson extended the model by Sato and Matsamura to include the effect of the contact angle as follows [22],

$$q_w'' = \frac{\lambda_f i_{fg}\rho_g \Delta T_{sup,ONB}^2}{8(1 + \cos\varphi)\sigma T_{sat}}, \tag{9.126}$$

where φ is the contact angle.

Basu et al. assumed that not all cavities will remain active in subcooled boiling and some of them will be flooded, especially when the wall surface is hydrophilic. Their equation is as follows [6],

$$q_w'' = \frac{F^2 \lambda_f i_{fg}\rho_g \Delta T_{sup,ONB}^2}{8\sigma T_{sat}}, \quad F = 1 - e^{-\varphi_{rad}^3 - 0.5\varphi_{rad}}, \tag{9.127}$$

where φ_{rad} is the contact angle expressed in radians.

9.4.6 SATURATED BOILING

When liquid bulk temperature in a boiling channel reaches the local saturation temperature, the heat transfer regime transits from subcooled boiling to saturated boiling. In this region the nucleate boiling is fully developed and is the dominating heat transfer mode. Due to increasing volume fraction of the vapor phase, the corresponding two-phase flow pattern gradually changes from bubbly flow, through slug and churn turbulent to wispy-annular flow (regions B through E in Fig. 9.8). The latter two-phase flow regime fundamentally differs from the former ones, since a continuous gas core is formed at the channel center, whereas a continuous liquid phase is limited to a layer covering channel walls. This layer, called a liquid film, is initially thick enough to accommodate vapor bubbles, allowing thus for nucleate boiling heat transfer. With thinning of the liquid film along the channel, a direct evaporation of the film on the film-core interface is becoming increasingly important. For very thin liquid films, the forced-convection vaporization is the dominant heat transfer mode.

Various approaches have been developed to predict heat transfer coefficient during saturated nucleate flow boiling. An extensive overview of the main methods and approaches, with a detailed description of the underlying theories have been given by Tong and Tang [84]. One of the most successful approaches was developed by Chen [16], who proposed to partition the heat transfer coefficient into two parts, representing the contributions due to nucleate boiling and force convection:

$$h_{tp} = h_{nb} + h_{fc}. \tag{9.128}$$

Chen suggested a Dittus-Boelter type of equation to calculate the heat transfer coefficient due to forced convection,

$$h_{fc} = 0.023 \mathrm{Re}_{tp}^{0.8} \mathrm{Pr}_{tp}^{0.4} \frac{\lambda_{tp}}{D} \tag{9.129}$$

where index tp indicates that a proper two-phase effective value of the corresponding quantity should be taken. Chen argued that λ_{tp} and Pr_{tp} can be replaced with their liquid-phase values, since the heat is mainly transferred through a liquid film in annular flow. The two-phase Reynolds number is expressed in terms of a liquid-only Reynolds number $\mathrm{Re}_{lo} = G(1-x)D/\mu_f$ through an introduction of the following parameter,

$$F \equiv \left(\frac{\mathrm{Re}_{tp}}{\mathrm{Re}_{lo}} \right)^{0.8} = \left[\frac{\mathrm{Re}_{tp} \mu_f}{G(1-x)D} \right]^{0.8}, \tag{9.130}$$

and the forced-convective heat transfer coefficient becomes,

$$h_{fc} = 0.023 F \left[\frac{G(1-x)D}{\mu_f} \right]^{0.8} \left(\frac{\mu c_p}{\lambda} \right)_f^{0.4} \frac{\lambda_f}{D}. \tag{9.131}$$

Here subscript f indicates the saturated liquid conditions, and parameter F is given by the following equation, derived from experimental data,

$$F = \begin{cases} 1 & X_{tt}^{-1} \leq 0.1 \\ 2.35 \left(0.213 + \frac{1}{X_{tt}} \right)^{0.736} & X_{tt}^{-1} > 0.1 \end{cases}, \tag{9.132}$$

where the so-called **Martinelli parameter** X_{tt} is given as,

$$X_{tt} = \left(\frac{1-x}{x} \right)^{0.9} \left(\frac{\rho_g}{\rho_f} \right)^{0.5} \left(\frac{\mu_f}{\mu_g} \right)^{0.1}. \tag{9.133}$$

The nucleate boiling contribution to the overall heat transfer coefficient is derived from the Forster and Zuber correlation valid for the nucleate pool boiling [30]. Since the mean superheat of fluid in pool boiling and in the convective boiling can significantly differ from each other, a correction parameter S is introduce into the original Forster and Zuber correlation, and the resulting heat transfer coefficient is as follows,

$$h_{nb} = 0.00122 S \left[\frac{\lambda_f^{0.79} c_{pf}^{0.45} \rho_f^{0.49}}{\sigma^{0.5} \mu_f^{0.29} i_{fg}^{0.24} \rho_g^{0.24}} \right] \Delta T_{\mathrm{sup}}^{0.24} \left(p_s(T_w) - p_f \right)^{0.75}. \tag{9.134}$$

Here parameter S, also known as a suppression factor, is derived from experimental data and is given as,

$$S = \left(1 + 2.56 \cdot 10^{-6} F^{1.463} \cdot \mathrm{Re}_{lo}^{1.17} \right)^{-1}. \tag{9.135}$$

The suppression factor ranges from zero to unity, as flow varies from high velocity to zero velocity.

9.5 BOILING CRISIS

A boiling crisis is a boiling heat transfer condition at which the heat transfer coefficient suddenly drops from a high to low value. This sudden reduction of the heat transfer coefficient is causing a sudden wall temperature increase (during heat-flux-controlled boiling) or a sudden decrease of the heat transfer rate (during temperature-controlled boiling). Similarly as for boiling heat transfer we distinguish pool boiling crisis and flow boiling crisis. This distinction is necessary, since the governing phenomena, and the corresponding correlations and models are different in both cases.

9.5.1 POOL BOILING CRISIS

Using dimensional analysis, Kutateladze derived the following correlation for a critical heat flux in pool boiling [57],

$$q''_{cr} = K^{1/2} i_{fg} \rho_g^{1/2} \left[\sigma \left(\rho_f - \rho_g \right) g \right]^{1/4}. \tag{9.136}$$

Various experimental data obtained by Kutateladze suggested that the average value of $K^{1/2}$ is 0.16, and it varies in a range from 0.13 to 0.19.

Similar equation for the critical heat flux in pool boiling was derived by Zuber, who used a hydrodynamic instability theory for transition boiling and obtained $K^{1/2} = \pi/24 \approx 0.13$ [90].

9.5.2 FLOW BOILING CRISIS

Depending on the corresponding two-phase flow regime, the flow boiling crisis is called either departure from nucleate boiling (DNB) or dryout. The DNB-type of boiling crisis occurs in bubbly flow, slug flow, and even wispy-annular flow, when nucleate boiling prevails. The dryout-type of boiling crisis occurs at high volume fractions of the gas phase, when the annular two-phase flow pattern exists and liquid film vaporization prevails.

Departure from Nucleate Boiling

Departure from nucleate boiling (DNB) occurs in channels with subcooled and low-quality nucleate boiling, at relatively high heat fluxes. The name of this type of boiling crises refers to the way it manifests itself. At boiling crisis, the local void spreads as a vapor blanket on the heating surface, and the nucleate boiling is replaced with a film boiling.

DNB has been intensively investigated both theoretically and experimentally, and the corresponding literature is very rich. For many practical application a correlation proposed by Levitan and Lantsman can be used [60],

$$q''_{cr} = \left[10.3 - 7.8 \frac{p}{98} + 1.6 \left(\frac{p}{98} \right)^2 \right] \left(\frac{G}{1000} \right)^{1.2\{[0.25(p-98)/98] - x_e\}} e^{-1.5 x_e}. \tag{9.137}$$

where q''_{cr}—critical heat flux [MW m^{-2}], p—pressure [bar], G—mass flux [kg m^{-2} s^{-1}], x_e—thermodynamic equilibrium quality. The correlation is applicable to DNB in round tubes with 8 mm inner diameter, with $29.4 < p < 196$ bar, and $500 < G < 5000$ kg m^{-2} s^{-1}. The root mean square deviation of the correlation against the experimental database used by Levitan and Lantsman is 15%. The correlation is applicable for other tube diameters by using the following correction factor,

$$q''_{cr,D} = q''_{cr} \left(\frac{8}{D} \right)^{0.5}, \tag{9.138}$$

where D is the tube diameter expressed in millimeters, and q''_{cr} is given by Eq. (9.137).

Dryout

Dryout occurs in channels with high void content, corresponding to annular flow, and at relatively low heat fluxes. Thus, it is customary to express a dryout condition in terms of a critical quality, rather than a critical heat flux. Similarly as for DNB, the dryout literature is very rich, and multiple correlations and models are available. Levitan and Lantsman [60] developed a simple dryout correlation given as,

$$x_{cr} = \left[0.39 + 1.57\frac{p}{98} - 2.04\left(\frac{p}{98}\right)^2 + 0.68\left(\frac{p}{98}\right)^3\right]\left(\frac{G}{1000}\right)^{-0.5}, \tag{9.139}$$

where x_{cr} is the critical quality, and other notation is the same as in Eq. (9.137). The correlation is valid for dryout in 8-mm tubes, with the following ranges of parameters: $9.8 < p < 166.6$ bar, and $750 < G < 3000$ kg m^{-2} s^{-1}. The predicted critical quality agrees with the experimental data used by the authors within ± 0.05 with the confidence level of 95%. The correlation can be used for other tube diameters by correcting the calculated values as follows,

$$x_{cr,D} = x_{cr}\left(\frac{8}{D}\right)^{0.15}. \tag{9.140}$$

Here D is the tube diameter in millimeters, and x_{cr} is given by Eq. (9.139)

9.6 POST-BOILING-CRISIS HEAT TRANSFER

A boiling heat transfer coefficient significantly deteriorates when boiling crisis occurs. Immediately after boiling crisis is reached, the boiling process is very unstable, and the surface is no longer wetted by liquid. The heat transfer rates under such conditions are very much dependent on flow conditions and geometry of the system under consideration.

A stable film boiling is developed in pools with stagnant liquid, when the wall temperature exceeds the **Leidenfrost point**. For a given pressure and liquid-solid system, such point is defined as the minimum wall temperature at which liquid will not wet the heater surface. Heat transfer in stable film boiling is accomplished by conduction and radiation across the vapor layer. An analytical model for the minimum heat flux during film boiling on horizontal plates was proposed by Berenson, who obtained the following equation for the heat transfer coefficient [9],

$$h = 0.425\left\{\frac{\lambda_g^3 i'_{fg}\rho_g g\left(\rho_f - \rho_g\right)}{\mu_g\left(T_w - T_{sat}\right)\left[\sigma\left(\rho_f - \rho_g\right)\right]}\right\}^{1/4} \tag{9.141}$$

where i'_{fg} is "effective" latent heat defined as,

$$i'_{fg} = i_{fg}\left[1 + 0.4\frac{c_p\left(T_w - T_{sat}\right)}{i_{fg}}\right]^2. \tag{9.142}$$

Post-dryout heat transfer mode is developed in channels with relatively high volume fraction of vapor, when the liquid film dries out on the heated surface. The primary mode of heat transfer at the wall is forced convection to the vapor phase, which contains dispersed liquid droplets. The heat transfer coefficient during post-dryout heat transfer is thus found in analogy to single-phase forced convection. Using best fit to experimental data, Groeneveld proposed the following correlation for heat transfer coefficient in the post-dryout heat transfer regime [36],

$$\mathrm{Nu}_g = \frac{hD}{\lambda_g} = a\left[\left(\frac{GD}{\mu_g}\right)\left(x + \frac{\rho_g}{\rho_f}(1-x)\right)\right]^b \mathrm{Pr}_{g,w}^c Y^d \tag{9.143}$$

where,

$$Y = 1 - 0.1 \left(\frac{\rho_f}{\rho_g} - 1 \right)^{0.4} (1-x)^{0.4}. \tag{9.144}$$

It should be noted that $\mathrm{Pr}_{g,w} = (c_p \mu / \lambda)_{g,w}$ is the vapor Prandtl number evaluated at the wall surface temperature. The exponents a, b, c, and d appearing in Eq. (9.143), valid in specific channel types, are provided in Table 9.3.

Table 9.3
Coefficients in Eq. (9.143)

Geometry	a	b	c	d
Tubes	0.00109	0.989	1.41	−1.15
Annuli	0.0520	0.688	1.26	−1.06
Tubes and annuli	0.00327	0.901	1.32	−1.5

Source: [36]

9.7 RADIATION

Any material at temperature higher than absolute zero radiates energy in the form of electromagnetic waves. Thermal radiation follows the **Stefan-Boltzmann law**, which states that the radiant power q emitted by a hot object depends only on its temperature T and surface area A as follows,

$$q = \sigma T^4 A, \tag{9.145}$$

where $\sigma \cong 5.670 \times 10^{-8}$ W/m^2 K^4 is the **Stefan-Boltzmann constant**. Objects that follow the Stefan-Boltzmann law given by Eq. (9.145) are called **perfect emitters** or **black bodies**, since they absorb and re-emit all radiation that impinges on them. Their radiation is known as **blackbody radiation**.

For real objects the Stefan-Boltzmann law is modified as,

$$q = \varepsilon \sigma T^4 A, \tag{9.146}$$

where ε is the object's emissivity, which equals its absorptivity α, the fraction of incident radiation that it absorbs. Both these quantities are in general frequency-dependent, and satisfy **Kirchhoff's law of thermal radiation**,

$$\alpha_\lambda = \varepsilon_\lambda, \tag{9.147}$$

where α_λ and ε_λ denote absorptivity and emissivity at wave length λ, respectively.

When electromagnetic radiation falls on a surface, it can be absorbed, reflected, or transmitted. The conservation of energy requires that the following equation is satisfied,

$$\alpha_\lambda + \rho_\lambda + \tau_\lambda = 1, \tag{9.148}$$

where ρ_λ is the reflectivity, defined as the fraction of the radiation with wave length λ reflected from the surface, and τ_λ is transmittance, defined as the fraction of the radiation with wave length λ transmitted through the surface. An opaque surface is one that transmits no radiation. Thus, for such surface $\tau_\lambda = 0$ and $\alpha_\lambda + \rho_\lambda = 1$. If an object has an emissivity less than one but independent on frequency, it is termed a **gray body**.

For black bodies, the net radiative heat transfer from surface A_1 to surface A_2 is,

$$q_{1\to2} = A_1 E_{b1} F_{1\to2} - A_2 E_{b2} F_{2\to1}, \tag{9.149}$$

where E_{b1}, E_{b2} are energy fluxes emitted from surfaces 1 and 2, respectively, and $F_{1\to2}$, $F_{2\to1}$ are view factors from surface 1 to surface 2, and from surface 2 to surface 1, respectively. The view factor $F_{1\to2}$ represents the fraction of radiation leaving surface A_1 that directly strikes surface A_2. Applying the reciprocity for view factors, $A_1 F_{1\to2} = A_2 F_{2\to1}$ and the Stefan-Boltzmann law $E_b = \sigma T^4$, the net rate of energy transferred from surface 1 to surface 2 is found as,

$$q_{1\to2} = \sigma A_1 F_{1\to2} \left(T_1^4 - T_2^4 \right). \tag{9.150}$$

For two grey-body surfaces forming an enclosure, the net heat transfer rate from surface 1 to surface 2 is,

$$q_{1\to2} = \frac{\sigma \left(T_1^4 - T_2^4 \right)}{\frac{1-\varepsilon_1}{A_1 \varepsilon_1} + \frac{1}{A_1 F_{1\to2}} + \frac{1-\varepsilon_2}{A_2 \varepsilon_2}}, \tag{9.151}$$

where ε_1 and ε_2 are the emissivities of the surfaces. For example, for two long, gray, coaxial cylinders $F_{1\to2} = 1$, and the net heat transfer rate is,

$$q_{1\to2} = \frac{\sigma \left(T_1^4 - T_2^4 \right)}{\frac{1}{A_1 \varepsilon_1} + \frac{1-\varepsilon_2}{A_2 \varepsilon_2}}. \tag{9.152}$$

Similarly, for two parallel, gray, equal-area surfaces we have $A_1 = A_2 = A$ and the net heat transfer rate from surface 1 to surface 2 becomes,

$$q_{1\to2} = \frac{A\sigma \left(T_1^4 - T_2^4 \right)}{\frac{1}{\varepsilon_1} + \frac{1-\varepsilon_2}{\varepsilon_2}}. \tag{9.153}$$

PROBLEMS

PROBLEM 9.1

A pipe with outer diameter 30 mm, inner diameter 25 mm and length 20 m is made of steel with heat conductivity 44 W/m K. The inner surface temperature of the pipe is 473.15 K and the outer surface temperature is 523.15 K. Calculate the total heat conducted through the pipe wall.

PROBLEM 9.2

A plane wall with thickness 25 cm and area 15.6 m^2 is made of material with heat thermal conductivity 1.5 W/m K. The air on one side of the wall has temperature 293.15 K and the heat transfer coefficient between the air and the wall is 5 W/m^2 K. On the other side of the wall the air temperature is 253.15 K and heat transfer coefficient between the air and the wall 15 W/m^2 K. Calculate the rate of heat in Watts transferred through the wall.

PROBLEM 9.3

For the same conditions as in Problem 9.2, calculate the mean wall temperature.

PROBLEM 9.4

For the same conditions as in Problem 9.2, but with an insulation layer of Styrofoam on the colder side of the wall, calculate the rate of heat transferred through the wall-insulation composite. The thickness of the insulation layer is 10 cm, and its thermal conductivity 0.04 W/m K.

PROBLEM 9.5

For the same conditions as in Problem 9.4, calculate the mean wall temperature.

PROBLEM 9.6

A pipeline with outer wall diameter 273 mm, wall thickness 16 mm, and length 70 m is made of material with thermal conductivity 45 W/m K. The outer surface of the pipeline is covered with 10 cm thick insulation layer with thermal conductivity 0.08 W/m K. A fluid with mean temperature 673.15 K flows inside the pipeline, and heat transfer coefficient between the fluid and the wall is 500 W/m^2 K. The air temperature surrounding the pipeline is 293.15 K, and the heat transfer coefficient between the air and the wall is 10 W/m^2 K. Calculate the total thermal loss rate (in Watts) in the pipeline.

PROBLEM 9.7

A copper pipe with outer diameter 12 mm and wall thickness 1 mm contains flowing hot water at temperature 363 K. The heat transfer coefficient between water and pipe wall is 500 W/m^2 K. The pipeline is surrounded by air at temperature 293 K, and the heat transfer coefficient between the air and the pipeline (the same for insulated and uninsulated pipe) is 5 W/m^2 K. Calculate the change of heat loss from the pipe per unit length, when an uninsulated copper pipe is covered with an insulation layer that has the critical thickness. Copper and insulation material thermal conductivity is 390 and 0.05 W/m K, respectively.

PROBLEM 9.8

Two infinite parallel gray planes with emissivities ε_1, ε_2, and temperatures T_1, T_2 exchange radiant heat. Calculate the ratio of the radiant energy transfer reduction between the plates when a thin parallel gray sheet of very high thermal conductivity, and with emissivity ε_s, is placed between them.

Part II

Energy Transformation Systems

10 Efficiency of Energy Transformation

An energy transformation technology refers to any system that transforms primary energy into secondary energy. Usually many transformations are required until the final energy form, that is such energy form that is delivered to the end users, is obtained. The final energy is not always directly consumable, however. Usually it must be further transformed into useful energy, such as heat, motion, light, and sound, to be consumed.

The aim of this chapter is to give a general overview of technologies that are necessary to obtain the final energy. Since the most important final energy forms are electricity and heat, we will pay a particular attention to generation of these energy forms. Electricity is particularly useful since it can be transmitted over long distances with relatively minor losses, and it can be efficiently transformed into virtually all useful energy forms. In reviewing various paths to generate electricity, we will discuss such important topics as the energy efficiency, exergy (which is a certain measure of energy quality), and energy storage.

10.1 POWER GENERATION TECHNOLOGIES

As mentioned in §2.4, power generation takes place in three types of power plants: electricity-only power plants, combined heat and power (CHP) plants, and heat-only power plants. These power plants can use various forms of primary energy, such as fossil fuel, moving water, wind, solar, nuclear fuel, and other energy forms, to transform into electricity and heat. In this part of the book we discuss the various technologies, both well-established and still under development, that are applicable in the power generation sector.

Thermal power plants are using fuel to generate thermal energy, which is further transformed into mechanical energy in turbines. Both steam turbines, operating in Rankine cycle, and gas turbines, operating in Brayton cycle, are used. Waste heat from turbines can be used in co-generation for district heating or as process heat. Combined cycle power plants use gas and steam turbines working in tandem. When used on land to generate electricity, a combined cycle gas turbine (CCGT) is employed[1]. Thermal power plants can use various thermal energy resources such as provided from combustion of fossil fuels, biomass and wastes, from fission of nuclear fuel, and from geothermal wells or concentrated solar radiation.

Kinetic energy of moving water is harvested in hydro power plants. In such power plants the mechanical energy of water is transformed into a mechanical energy of a hydraulic turbine. This energy is used in a connected to the turbine electric generator to generate electricity.

Wind turbines are used to transform the kinetic energy of wind into mechanical energy that drives electric generator. The dominating wind turbine design has a rotor placed on a horizontal axis and is known as horizontal axis wind turbine (HAWT).

Photovoltaic modules are used to transform solar radiation energy into direct-current (DC) power. An inverter is used to transform DC power into alternating-current (AC) power.

Fusion and other energy transformation devices are still under development and have not achieved commercial maturity yet. However, these devices show great potential for future and that is why they attract tremendous research resources.

---

[1]The same principle is used for marine propulsion, where it is called a combined gas and steam (COGAS).

DOI: 10.1201/9781003036982-10

10.2 ENERGY EFFICIENCY

So far in §7.7 we introduced the notion of efficiency as a measure of effectiveness for thermo-dynamic processes to transform one energy form to another. Two types of efficiency have been mentioned: the *first-law efficiency*, also called the *energy efficiency*, and the *second-law efficiency* also called the *exergy efficiency*. We have also learned in §2.5.7 that energy efficiency can be defined as an amount of energy that is needed to provide products and services. These two aspects of energy efficiency indicate that the quest for greater energy efficiency is both a scientific and engineering challenge.

The first-law efficiency was employed so far to determine the effectiveness of energy transformation in various thermodynamic cycles. A definition of the first-law efficiency is strict and useful from the thermodynamic point of view; however, it does not give a clear indication of how well a device performs compared to the best performance allowed by the laws of thermodynamics. As an example, let us consider a certain household furnace that is described as being 75% efficient. This means that the heat effectively delivered to the house is equal to 75% of the heat of combustion of fuel burned. Such definition suggests that a "perfect" furnace would deliver 100% of the heat of combustion, and that this would correspond to the best energy efficiency. However, taking into account various technologies currently available, one could do better. For instance, by employing a heat pump powered by a fuel cell, one could provide more heat to the house than just by direct burning of the fuel.

In §10.2.2 we further develop a concept of the second-law efficiency, which provides better indication how well a device or system performs. The efficiency will be derived and provided for several simple thermo-mechanical processes. As we will learn, the maximum possible value of the second-law efficiency, similarly to the first-law efficiency, is unity. However, unlike the first-law efficiency, it better describes the process from the environmental and energy-saving point of view.

10.2.1 FIRST-LAW EFFICIENCY

The **first-law efficiency** of a system is defined as a ratio of the energy of a desired form extracted from the system to the energy added to the system. When the desired energy form is work, and the added energy form is heat, the first-law efficiency is limited by the second law of thermodynamics.

The obvious issue with this definition is that the efficiency can be greater than unity for certain system, such as heat pumps. For such systems another name is used for the defined quantity. In case of heat pumps, the definition coincides with a parameter called the *coefficient of performance* (CoP). The first-law efficiencies for selected systems and devices are discussed below.

Heat Engine

A heat engine uses energy added in the form of heat (Q_a) to do work (L) and then exhausts the heat (Q_e) which cannot be used to do work. The first law of thermodynamics gives $L = Q_a - Q_e$, and the efficiency can be found as,

$$\eta = \frac{L}{Q_a} = \frac{Q_a - Q_e}{Q_a}. \tag{10.1}$$

As discussed in section §7.5.1, for Carnot cycle this efficiency can be expressed in terms of temperatures at which the heat is added (T_a) and exhausted (T_e),

$$\eta = \frac{T_a - T_e}{T_a} = 1 - \frac{T_e}{T_a}. \tag{10.2}$$

Expressions for first-law efficiencies of several heat engines are provided in section §7.5.

Electric Heat Pump

A heat pump is a heat extraction device that employs work to move heat from low to high temperature. The primary goal of a heat pump is to increase or maintain the temperature of the hotter region, which is usually a living space, using energy from the colder outside environment. If the heat exhausted to the hotter region is Q_e, and the work used to move heat is L, the coefficient of performance for heat pump is,

$$\text{CoP}_{\text{hp}} = \frac{Q_e}{|L|}. \tag{10.3}$$

According to the first law of thermodynamics, the work L can be expressed in terms of the heat added from colder region Q_a, and the heat exhausted Q_e to the hotter region as $|L| = Q_e - Q_a$. Thus, the coefficient of performance becomes,

$$\text{CoP}_{\text{hp}} = \frac{Q_e}{Q_e - Q_a}. \tag{10.4}$$

Assuming an ideal Carnot cycle, the coefficient can be expressed in terms of temperatures at which the heat is added (T_a), and at which the heat is exhausted (T_e):

$$\text{CoP}_{\text{hp}} = \frac{T_e}{T_e - T_a} = \frac{1}{1 - T_a/T_e}. \tag{10.5}$$

Since the heat is added from a colder region, we always have $T_a < T_e$ and $\text{CoP}_{\text{hp}} > 1$.

Air Conditioner or Refrigerator

Similarly as heat pumps, air conditioners and refrigerators are heat extraction devices that employ work to move heat from low to high temperature. However, unlike for heat pumps, the primary goal of them is to decrease or maintain the temperature of the colder region, which can be either a living space (air conditioning) or a refrigerator's compartment (refrigeration). This difference is reflected in the definition of the coefficient of performance for air conditioner or refrigerator, which in this case is given as,

$$\text{CoP}_{\text{r}} = \frac{Q_a}{|L|} = \frac{Q_a}{Q_e - Q_a}. \tag{10.6}$$

For an ideal Carnot cycle we have,

$$\text{CoP}_{\text{r}} = \frac{T_a}{T_e - T_a} = \frac{1}{T_e/T_a - 1}. \tag{10.7}$$

Here $T_e > T_a$, and CoP_{r} can be either greater or less than one.

Solar Collector

A solar collector is absorbing sunlight, which is approximately described by a black-body spectrum at the Sun's surface temperature $T_\odot$. If the absorbed heat is transformed into work, some heat must be expelled at the ambient temperature T_{amb}. Assuming an ideal Carnot cycle, the limit efficiency of the solar collector is,

$$\eta_{STH} = \frac{L}{Q_a} = 1 - \frac{T_{amb}}{T_\odot}. \tag{10.8}$$

Taking $T_\odot = 5778$ K and $T_{amb} = 288$ K, the maximum achievable efficiency of a solar collector is 95%. Due to technical and engineering limitations, the real efficiency is much less, however.

Solar Photovoltaic Devices

The theoretical maximum achievable efficiency of solar photovoltaic devices can be derived in a similar manner as for solar collectors. However, to accurately estimate the efficiency and power output of solar energy systems, it is necessary to have a more precise description of the solar spectrum than the simple black-body spectrum. In particular, the response of the photovoltaic system is highly dependent upon the frequency of incident radiation. A widely accepted standards are **air mass 0** (AM0) and **air mass 1.5** (AM1.5) reference spectra. Air mass 0 refers to the measured solar spectrum at the top of Earth's atmosphere, whereas air mass 1.5 represents terrestrial solar spectrum for the Sun at angle $\psi_z = 48.19°$ from the normal ($1/\cos 48.19° \approx 1.5$). The first law efficiency of a photovoltaic device may be defined as,

$$\eta_{PV} = \frac{L}{Q_{a,\text{AM1.5}}}, \tag{10.9}$$

where L is the work extracted from the photovoltaic device (in a form of an electric current), and $Q_{a,\text{AM1.5}}$ is the added solar energy at spectrum AM1.5.

More restrictive limits for specific solar transformation technologies are discussed later in this book. As an example, for a single-junction photovoltaic cells the *Shockley-Queisser bound* on the efficiency is valid, which for optimal conditions is approximately equal to 34%.

Mechanical and Electrical Energy Transformation

Transformations of electrical and mechanical energy are very efficient and incur very small losses. These losses result mainly from friction between moving parts, resistive drag between fluids and solid bodies, and from resistivity of electric conductors. By using various measures to reduce these losses, the first-law efficiency of devices that transform mechanical and electrical energy can approach one.

For some devices, such as wind turbines, ocean current turbines, and tidal stream turbines the *Betz limit* determines the maximum power that can be harvested by a single device from an unperturbed stream of fluid. This limit says that no more than 16/27 (59%) of the power that crosses an area occupied by the turbine's rotor can be converted into useful power, whereas the rest remains in the fluid. This remaining power can be harvested by another, downstream-located turbine, thus the Betz limit is not an absolute efficiency limit in the same manner as Carnot's limit on heat engines. Nevertheless, Betz's limit plays an important role in judging the effectiveness of wind and water-current turbines.

Transformation of Chemical Energy

Chemical energy can be transformed into heat by a chemical reaction (combustion) or into electrical energy by a battery or a fuel cell. The first-law efficiency of a system that transforms a chemical energy into another form of energy depends on which transformation path is followed. If a chemical reaction is used to supply space heating, the first-law efficiency can be close to one. If the thermal energy from the chemical reaction is use to derive work, the Carnot limit applies. For example, a flame temperature of burning methane (or natural gas) in the air at atmospheric pressure is about $T_b = 2223$ K. Assuming that the waste heat in the Carnot cycle i exhausted at temperature $T_a = 288$ K, the maximum possible efficiency is $(1 - T_a/T_b) \approx 0.87$. In a real cycle, the transformation efficiency is determined by a ratio of the extracted work divided by the absolute value of the *enthalpy of reaction* $|\Delta I_r|$.

If a battery or a fuel cell is used as a conversion device, the most electrical energy that can be harvested from a chemical reaction is given by an absolute value of the Gibbs free energy of reaction $|\Delta G_r|$. Thus, the first-law efficiency of batteries or fuel cells is limited by $|\Delta G_r/\Delta I_r|$.

10.2.2 SECOND-LAW EFFICIENCY

The **second-law efficiency** of a system is defined as a ratio of the useful energy extracted from the system to the maximum possible useful energy that could be extracted from the system using the same energy input and at the same ambient conditions. For example, if the work extracted from a heat engine is L, and the heat added to the system from a source at temperature T is Q_a, the first-law (or energy) efficiency is $\eta_E = L/Q_a$. The maximum possible useful energy that can be extracted from a heat engine is determined by the Carnot efficiency. Assuming that the heat sink is at the ambient temperature T_{amb}, this maximum possible work can be found as $L_{max} = \eta_{EC}Q_a = Q_a(1 - T_{amb}/T)$, where $\eta_{EC} = (T - T_{amb})/T$ is the first-law efficiency of a reversible Carnot engine operating with the same heat sources. Thus, the second-law efficiency η_B for the heat engine can be found as,

$$\eta_B = \frac{L}{L_{max}} = \frac{Q_a \eta_E}{Q_a \eta_{EC}} = \frac{\eta_E}{\eta_{EC}}. \tag{10.10}$$

This equation is valid for all kinds of heat engines, including an electric heat pump, air conditioner, and refrigerator. In all cases, η_E is the first-law efficiency of the device. However, for systems providing useful work and heat Eq. (10.10) is not valid and a more general definition of the second-law efficiency is needed.

An *exergy balance* of a system is the basis to calculate the second-law efficiency of the system, which explains why this quantity is also called exergy efficiency. It can be calculated as a ratio of the total useful exergy and work extracted from the system to the total work and exergy added to the system. For any process, the extracted useful exergy and work are as follows:

- useful exergy of products of the process, B_p, which are necessary to maintain the process,
- exergy of non-energetic raw materials, B_{rm}, which are necessary to maintain the process,
- useful work, L_u,
- useful electric energy, E_{uel},
- exergy increase of external heat reservoirs, whose cooling or heating is an objective of the process, ΔB_r,
- useful exergy increase of the system, ΔB_{us}.

The corresponding work and exergy added to the system are as follows:

- fuel exergy, B_f,
- added work, L_a,
- added electric energy, E_{ael},
- exergy drop of the external source of heat added to the system, ΔB_{as}.

Using the above quantities, the second-law efficiency can be calculated as,

$$\eta_B = \frac{B_p - B_{rm} + L_u + E_{uel} + \Delta B_r + \Delta B_{us}}{B_f + L_a + E_{ael} - \Delta B_{as}}. \tag{10.11}$$

Formula given by Eq. (10.11) can be applied for calculation of the second-law efficiency for many thermo-mechanical processes. Selected example applications are discussed below.

Mechanical or Electrical Energy

Useful mechanical or electrical energy can be involved in many processes. In an electric motor, a useful work L_u is extracted, whereas an electric energy E_{ael} is added. Since all other terms in Eq. (10.11) are equal to zero, the second-law efficiency of the electric motor is found as,

$$\eta_B = \frac{L_u}{E_{ael}}. \tag{10.12}$$

Mechanical energy can be transformed into electric energy in a generator, for which Eq. (10.11) yields,

$$\eta_B = \frac{E_{uel}}{L_a}. \tag{10.13}$$

When mechanical energy is extracted from a heat engine, to which heat is added from a heat reservoir at temperature T, the second-law efficiency is obtained from Eq. (10.11) as,

$$\eta_B = \frac{L_u}{-\Delta B_{as}}. \tag{10.14}$$

Since $-\Delta B_{as} = Q(1 - T_{amb}/T)$, the efficiency can be written as,

$$\eta_B = \frac{L_u}{Q(1 - T_{amb}/T)}, \tag{10.15}$$

where T_{amb} is the ambient temperature. In particular, this expression is valid for a geothermal power plant, which delivers useful work L_u and uses heat Q from a reservoir at temperature T.

Heating and Cooling

Processes with heating and cooling can be realized in different ways, depending on the energy form added to the process. In electrically driven heat pump, heat Q_e is exhausted from the system to the warm region at temperature T. Thus, the exergy increase of the warm region is $\Delta B_r = Q_e(1 - T_{amb}/T)$, and the second-law efficiency of the heat pump is,

$$\eta_B = \frac{\Delta B_r}{E_{ael}} = \frac{Q_e}{E_{ael}} \left(1 - \frac{T}{T_{amb}} \right), \tag{10.16}$$

where T_{amb} is ambient temperature.

If heating is provided from a thermal-solar system, the second law efficiency is given as,

$$\eta_B = \frac{\Delta B_r}{-\Delta B_{as}} = \frac{Q_e \left(1 - T_{amb}/T \right)}{Q_{ths} \left(1 - T_{amb}/T_{ths} \right)}, \tag{10.17}$$

where Q_e is the heat extracted from the system and supplied to the warm reservoir at temperature T, and Q_{ths} is the heat added from the solar hot water reservoir at temperature T_{ths} to the system.

Work-Heat System

For a system in which both useful work L_u and useful heat Q_u is delivered, and heat Q_a is added from a source at temperature T_a, the second law efficiency can be calculated as,

$$\eta_B = \frac{Q_u \left(1 - \frac{T_{amb}}{T} \right) + L_u}{Q_a \left(1 - \frac{T_{amb}}{T_a} \right)}. \tag{10.18}$$

Here T is the temperature of the heat reservoir to which heat Q_u is added. The above expression can be used to calculate the efficiency for *combined heat and power* (CHP) or *cogeneration* facilities.

10.3 ENERGY CONSERVATION AND STORAGE

Energy conservation has as a goal to reduce the consumption of energy by using the energy more efficiently. This goal can be achieved in a variety of ways, but usually it includes technical, economic, social, and policy choices. For example, using light-emitting diode light or compact fluorescent light bulb that requires less energy than an incandescent light bulb to produce the same amount of light can give substantial energy savings. Another example of energy conservation with a social dimension is turning the light off when leaving the room and recycling aluminium cans.

Energy storage is particularly important for integrating intermittent energy systems into the grid. The most important performance criteria for grid-scale energy storage are:

- Total energy storage capacity, which should be high enough to provide the grid stability.
- Round-trip storage efficiency, which is the fraction of energy that is recovered after storage.
- The rate at which energy can be removed from storage.

Transport applications require energy storage that in addition has high power density, low degradation rate, long cycle life, and high safety standards.

PROBLEMS

PROBLEM 10.1

A refrigerator operates with an ideal Carnot cycle between infinite heat reservoirs at temperatures 258 K and 298 K. Heat addition and extraction is taking place at temperature difference of 10 K. Calculate the second-law efficiency of the refrigerator.

PROBLEM 10.2

A heat pump is used in the Arctic climate to heat a building. The driving heat is taken from the sea water at temperature 276.5 K and the required temperature inside the building is 293 K. Calculate the second-law efficiency of the pump if the ambient temperature is 244 K and the ratio of the heat supplied to the building and the heat taken from the sea water is 0.5. The electric power provided to the pump can be neglected.

PROBLEM 10.3

A high-temperature heat pump is taking heat from a heat source with temperature 705 K and from the ambient air at temperature 288 K, and provides useful heat at temperature 355 K. Neglecting the electric power provided to the heat pump, calculate the second-law efficiency of the heat pump if the ratio of the useful heat to the heat taken from the high-temperature source is 1.65.

PROBLEM 10.4

A heat pump operates between two infinite heat reservoirs at temperatures 288 K and 375 K. The internal power used for driving the pump is 1.1 MW, and the thermal power provided by the pump to the high-temperature reservoir is 1.65 MW. Calculate the second-law efficiency of the pump.

PROBLEM 10.5

A solar thermal power will be used to concentrate solar rays to heat a molten salt which is stored at temperature 300 °C. The salt will be heated to 550 °C. How much energy can be stored in 6000 tons of molten salt, if its heat capacity is 1.5 kJ/kg K.

PROBLEM 10.6

Wind energy will be stored using a compressed air energy storage (CAES). Calculate the required volume to the air tank to be able to store the energy that the wind turbine can generate during 24 hours when operating at nameplate capacity of 8 MW. Assume an isothermal compression and decompression from ambient pressure $p_1 = 0.1$ MPa to pressure $p_2 = 7$ MPa. CAES operates as an open system.

11 Thermal Power

Thermal power transformations provide most of the electric power produced world-wide today. In this chapter we discuss the various technologies that are used to transform the thermal power into electricity and heat. Such aspects as the principles of operation and the main components of various thermal power plants are of particular interest. We analyze the energy and exergy efficiencies of the plants, since these parameters are fundamental from the point of view of efficient and economic use of primary energy resources. To be able to further improve thermal plant efficiencies, it is helpful to know the various method currently used. In particular, the advantages of using combined cycles and cogeneration are discussed.

11.1 INTRODUCTION

The development of thermal power plants became possible with the introduction of the first dynamo built for power generation by Werner von Siemens in 1866. The first central power station was built by Thomas Edison in New York in 1882, and two years later, Sir Charles Parson built the first steam-turbine generator. The energy efficiency of this first design was only 1.6%. Since then the development of thermal power plants has been very rapid, and already in early 1900s the power plants were rated in the range of 1 to 10 MW output per unit. At the same time the net plant efficiency increased to about 15%. In 2016 a record 62.22% efficiency was obtained in a combined-cycle power plant in Bouchain, France[1].

During this over 150 years long period of development, various configurations of thermal power plants have been implemented. Currently the following five basic types of thermal power plants are dominating: condensing power, non condensing power, stationary gas turbine power, combined cycle power, and poly-generation power.

Condensing power plants operate in the Rankine cycle that have condensing turbines, which receive steam from a boiler and exhaust it to a condenser. The exhausted steam is at a pressure well below atmospheric, to improve the turbine efficiency. The steam is condensed in a condenser using an external cooling water system. Condensing power plants are mainly used for generation of electricity.

Non condensing power plants operate in the Rankine cycle with back pressure, or non condensing turbines. A back pressure turbine is selected for process heat or cogeneration, when both electricity and process heat are required. Since enthalpy drop is less in a back-pressure turbine, more steam is required for the same operating load as in a condensing turbine. Back pressure turbines have several advantages, such as a simple configuration, low price as compared to condensing turbines, and no need for cooling water. The biggest disadvantage of this type of steam turbine is that it is highly inflexible. The output of this turbine cannot be regulated and it works best with the constant load.

Gas turbine engines are mainly used in transportation to power ships and aircrafts, but also in modern stationary power stations. They use fluid fuels such as petroleum and natural gas, to convert chemical energy into mechanical energy. The basic operation of the gas turbine is the Brayton cycle with air as working fluid. They range in size from portable mobile plants to large systems weighting more than a hundred tonnes. When a gas turbine is used for shaft power only, its energy efficiency is about 30%. This relatively low efficiency discourage to use it for electricity generation. However, much higher energy efficiency can be obtained when the waste heat from the gas turbine is recovered to power a steam turbine in a combined cycle configuration. Nevertheless, single-cycle gas turbines

---

[1]www.powermag.com/worlds-most-efficient-combined-cycle-plant-edf-bouchain, accessed on 2020-07-01.

DOI: 10.1201/9781003036982-11

are frequently used as energy backup system for intermittent electricity generating plants, such as wind turbine farms, since they can be turned on and off within minutes, covering unscheduled power supply losses.

Combined cycle power plants employ combined-cycle gas turbines (CCGT) to convert chemical energy of fuel into mechanical energy, as the intermediate stage, and to electricity, as the final product. Such plants have at least two heat engines working in tandem. After completing its cycle in the first engine, the working exhaust fluid is still hot enough to power the subsequent heat engine. Usually a heat exchanger is used to transfer heat between the engines so that they can use different working fluids. The overall efficiency of combined cycle power plants can be over 60%, with potential to reach 65% within nearest future.

Co- and poly-generation power plants have as a goal to produce, in addition to electricity, also other products, such as heat, synthetic gas, methanol, hydrogen, etc. Such systems offer various improvements, such as minimization of losses, improvement of energy and exergy efficiency, and reduction of negative environmental impacts. Cogeneration, also known as combined heat and power (CHP), is usually reserved to mean the simultaneous production of electricity and heat. Similarly, tri-generation produces electricity as well as provides hot water, and space heating and cooling.

A common feature of thermal power plants is that they use thermal energy to power the systems. The input thermal energy originates from a wide spectrum of resources, such as combustion of fuels, nuclear fission and fusion, but also from concentrated solar radiation, or from geothermal wells.

11.2 CONDENSING POWER

A condensing power plant is using a working fluid that undergoes phase change during the thermodynamic cycle. This is the most widespread technology used for generation of electricity. The main components of a condensing power plant are boilers, turbines, condensers, and pumps. The exhaust steam in a condensing power plant is at pressure well below atmospheric, and is in a partially condensed state. Typical quality of the exhaust steam is close to 90%. These parameters are chosen to maximize the plant over-all energy efficiency once generating electricity. When process steam is required, non-condensing or back pressure turbines are used. The exhaust pressure is controlled by a regulating valve to suit the needs of the process steam pressure.

11.2.1 SCHEMATIC OF A BASIC SYSTEM

The simplest condensing power cycle as a system of mechanical components is shown in Fig. 11.1. Figures 11.2 and 11.3 show the corresponding water-steam Rankine cycle in the TS- and is-planes. The heating and evaporation of water is represented by path $3-A-B-4$, where process $3-A$ represents heating of subcooled water entering the boiler until it becomes saturated. Water is evaporating at constant pressure, as shown by path $A-B$. As can be seen in Fig. 11.2, the temperature of the water-steam mixture is constant when following the path. However, the specific enthalpy of the boiling mixture increases along this path as shown in Fig. 11.3, and the enthalpy increase is equal to the latent heat of evaporation at the system pressure. Next, the saturated steam is superheated as indicated by point 4 in Figs. 11.2 and 11.3.

The superheated steam expands in the turbine along path $4-1$. In this process the thermal energy of superheated steam is converted into mechanical energy of rotation of the output shaft of the turbine. The process is irreversible and adiabatic. At point C the expanding steam becomes saturated, and along path $C-1$ the steam is wet. Path $4-1s$ shows the corresponding reversible process, in which there is no entropy increase between points 4 and 1s, that is $s_{1s} = s_4$. After exiting the turbine at point 1, the wet steam enters the condenser, in which it condenses following path $1-2$. At point 2, saturated condensate enters the pump, where it is pressurized to point 3. Path $2-3$ shows irreversible compression, whereas path $2-3s$ represents reversible compression, for which there are no losses,

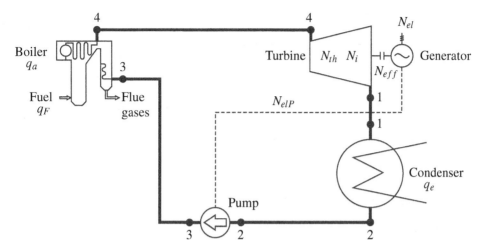

Figure 11.1 Schematic of a basic condensing power system containing: the boiler, turbine with generator, condenser, and pump.

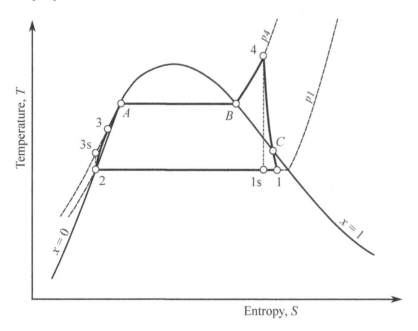

Figure 11.2 Water-steam Rankine cycle of a condensing power system shown in the TS-plane.

and the entropy is constant: $s_2 = s_{3s}$. At point 3 the compressed water enters the boiler and closes the cycle.

Boiler

Boilers are using heat to produce high pressure steam that can be expanded in steam turbines to generate mechanical energy. Fuel, such as coal, natural gas, or uranium, is used as the heat source. In a coal-fired steam power plant, coal is burned in a combustion chamber surrounded by an array of vertical pipes. The feedwater enters the pipes at the bottom and evaporates due to high-temperature heat transferred from combustion gases to pipe walls. The saturated steam is separated from water

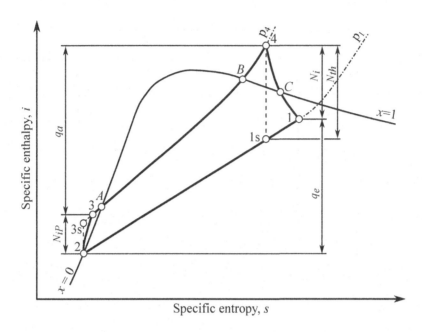

Figure 11.3 Water-steam Rankine cycle of a condensing power system shown in the *is*-plane.

and superheated before leaving to the turbine. Similar processes take place in boilers using other chemical fuels. In nuclear reactors the fission energy is used as the heat source. The fission energy is transferred from fuel rods to coolant that is passing through rod bundles. Usually saturated steam is generated directly in the reactor core (e.g. in boiling water reactors), or indirectly in a dedicated heat exchanger called steam generator (e.g. in pressurized water reactors).

Due to boiler heat losses, not all energy obtained from combustion or fission is used for rising the thermal energy of the working fluid. The boiler efficiency is defined as,

$$\eta_B = \frac{q_a}{q_F}, \tag{11.1}$$

where q_a is the actual thermal energy per unit time added to the working fluid passing through the boiler, and q_F is the chemical or fissile energy of the fuel consumed by the boiler per unit time.

The fuel thermal energy per unit time can be found from the known fuel consumption rate W_F and the fuel specific energy content c_F as,

$$q_F = W_F c_F. \tag{11.2}$$

Example 11.1:

A steam power plant produces $N_{el} = 500$ MWe at over-all energy efficiency of the plant cycle $\eta = N_{el}/q_a = 39\%$. Calculate the required amount of coal during one year, knowing that the coal heat content is 25 MJ/kg, and the boiler efficiency is 80%.

Solution

The plant generates $500/0.39 \approx 1282$ MW of thermal power. This corresponds to $1282 \cdot 365 \cdot 24 \cdot 3600 = 40.43 \cdot 10^9$ MJ of thermal energy during one year. The annual coal consumption is found as $40.43 \cdot 10^9/25/0.8 \approx 2.02 \cdot 10^9$ kg.

Turbine

The turbine uses the enthalpy of the working fluid to deliver a mechanical work.

The theoretical (or isentropic) power of the turbine is defined as,

$$N_{th} = W(i_4 - i_{1s}),\tag{11.3}$$

where i_4 is the specific enthalpy of the working fluid at the inlet to the turbine, i_{1s} is the specific enthalpy of the working fluid at the exit from the turbine assuming an isentropic expansion in the turbine (represented by path 4–1s in Figs. 11.2 and 11.3), and W is the mass flow rate of the working fluid.

The internal power of the turbine is defined as,

$$N_i = W(i_4 - i_1),\tag{11.4}$$

where i_1 is the actual specific enthalpy of the working fluid at the turbine exit.

The internal turbine efficiency is defined as,

$$\eta_i = \frac{N_i}{N_{th}} = \frac{i_4 - i_1}{i_4 - i_{1s}}.\tag{11.5}$$

It should be noted that the turbine internal efficiency is purely a thermodynamic quantity. It can be obtained by measuring the pressure and the temperature of the working fluid at the turbine inlet (p_4 and T_4, respectively) and outlet (p_1 and T_1). Assuming in addition that the superheated conditions at both ends of the turbine prevail, its internal efficiency can be obtained in the following steps:

- find the inlet specific enthalpy as $i_4 = i(p_4, T_4)$,
- find the inlet specific entropy as $s_4 = s(p_4, T_4)$,
- find the true specific enthalpy at the exit as $i_1 = i(p_1, T_1)$,
- find the isentropic specific enthalpy at the exit as $i_{1s} = i(p_1, s_4)$,
- find the internal turbine efficiency as $\eta_i = (i_4 - i_1)/(i_4 - i_{1s})$.

The turbine internal efficiency is a measure of the increase of entropy in the turbine. It can also be interpreted as a measure how far the expansion process departs from an ideal isentropic process.

The effective mechanical power of a turbine N_{eff} is always less than its internal power N_i. This is because there are additional mechanical losses in the turbine caused by, e.g., friction and vibrations. Thus, the turbine mechanical efficiency is introduced as follows,

$$\eta_m \equiv \frac{N_{eff}}{N_i}.\tag{11.6}$$

Condenser

The condenser extracts heat from the working fluid causing its transition from the vapor phase to the liquid phase. The energy conservation for the condenser yields,

$$q_e = W(i_1 - i_2),\tag{11.7}$$

where q_e is the thermal power extracted from the condenser, i_1 is the specific enthalpy of the working fluid at the inlet to the condenser, i_2 is the specific enthalpy of the working fluid at the exit from the condenser and W is the mass flow rate of the working fluid through the condenser.

Pump

The pump provides pressure increase in the system to allow boiling of the working fluid at a high pressure. We can write a generic energy conservation equation for a pump operating between points 2 and 3, shown in Fig. 11.3, as follows (see §7.2.2),

$$\frac{dE_T}{dt} = q - N_{iP} + W_2(i_2 + e_{P2} + e_{K2}) - W_3(i_3 + e_{P3} + e_{K3}) = 0. \tag{11.8}$$

Here we have to supply the pumping power, thus $N_{iP} < 0$ and $-N_{iP} = |N_{iP}|$. For steady-state conditions, with no heat added, and also neglecting the changes of the kinetic and potential energy of working fluid between pump inlet and outlet, the energy conservation equation can be written as,

$$|N_{iP}| = W(i_3 - i_2) = W\left(e_{I3} - e_{I2} + \frac{p_3 - p_2}{\rho_e}\right) = W\frac{p_3 - p_2}{\rho_e} + W\Delta e_I. \tag{11.9}$$

Here ρ_e is an equivalent fluid density for compression process 2–3. Since water has low compressibility, it can be assumed $\rho_e \approx \rho_2 \approx \rho_3$. The pumping power N_{iP} is called the internal pumping power, since this is the power that must be provided to the pump to increase the specific enthalpy of the working fluid from i_2 to i_3. Only a fraction of this power is used to increase the working fluid pressure, however. The last term on the right-hand side of Eq. (11.9), $W\Delta e_I$, represents the fraction of the pumping power that is used to increase the specific internal energy of the working fluid from e_{I2} to e_{I3}. Since the primary task of a pump is to increase the pressure of a working fluid, this term represents an internal loss of a pumping power $|N_{lossP}| = W\Delta e_I$. Accordingly, a useful pumping power is defined as $|N_{uP}| = W\Delta p/\rho_e$, where, in case of the cycle shown in Fig. 11.3, $\Delta p = p_3 - p_2$. Since $W = \rho_e Q$ and $\Delta p = \rho_e gH$, the useful pumping power can be written as,

$$|N_{uP}| = \rho_e gQH. \tag{11.10}$$

Here g is the gravity acceleration, Q is the volumetric flow rate, and H is the pumping head. In practice, the pump efficiency is determined experimentally by measuring the volumetric flow rate, the head, and the mechanical power input provided on the pump shaft. The pump efficiency is then found as,

$$\eta_P = \frac{\rho_e gQH}{|N_{mP}|} = \frac{|N_{uP}|}{|N_{iP}|}\frac{|N_{iP}|}{|N_{mP}|} = \eta_{iP}\eta_{mP}. \tag{11.11}$$

The mechanical pumping power N_{mP} appearing in Eq. (11.11) is equal to the internal pumping power, N_{iP}, minus the mechanical losses in the pump due to shaft sealing and in the shaft bearings. The mechanical efficiency of the pump is defined then as,

$$\eta_{mP} = \frac{|N_{iP}|}{|N_{mP}|}, \tag{11.12}$$

whereas the pump internal efficiency is given as,

$$\eta_{iP} = \frac{|N_{uP}|}{|N_{iP}|}. \tag{11.13}$$

Since the mechanical pumping power is usually provided by an electric motor, the needed electric power is as follows,

$$|N_{elP}| = \frac{|N_{mP}|}{\eta_{EM}}, \tag{11.14}$$

where η_{EM} is the efficiency of the electric motor. Combining Eqs. (11.9) through (11.14) yields,

$$|N_{elP}| = W\frac{p_3 - p_2}{\eta_{iP}\eta_{mP}\eta_{EM}\rho_e} = W\frac{p_3 - p_2}{\eta_P\eta_{EM}\rho_e}. \tag{11.15}$$

The above equation allows to find the needed electric power to provide a given pressure increase in a system with mass flow rate W, and with given efficiencies of the pump and the electric motor. The equation is useful to estimate the exergy losses in the system.

When applying energy balance for the system in Fig. 11.1, another form of Eq. (11.15) is useful, in which the enthalpy increase of the working fluid, when passing through the pump, is expressed in terms of the supplied electric power,

$$i_3 - i_2 = \frac{|N_{iP}|}{W} = \eta_{mP} \eta_{EM} \frac{|N_{elP}|}{W}. \tag{11.16}$$

Finaly, combining Eq. (11.15) with Eq. (11.16), the following relationship between the enthalpy and pressure increase is obtained,

$$i_3 - i_2 = (p_3 - p_2) \frac{\eta_{mP}}{\eta_P \rho_e} = \frac{p_3 - p_2}{\eta_{iP} \rho_e}. \tag{11.17}$$

11.2.2 BASIC SYSTEM EFFICIENCY

General efficiency definitions can be demonstrated for a simplified electric power station shown in Fig. 11.1. The key mechanical components in the system are boiler, turbine, generator, condenser, and pump.

Energy Efficiency of an Ideal Cycle

The energy efficiency of an ideal cycle is defined as,

$$\eta_{EIR} \equiv \frac{N_{th}}{q_a}, \tag{11.18}$$

where N_{th} is the theoretical turbine power and q_a is the added thermal power. Using the notation shown in Fig. 11.3, the energy efficiency of the ideal cycle is obtained as,

$$\eta_{EIR} = \frac{i_4 - i_{1s}}{i_4 - i_3}, \tag{11.19}$$

where i_{1s} is the specific enthalpy at the exit of the turbine assuming the isentropic expansion of the working fluid. It can be noted that the definition of η_{EIR} corresponds to the Carnot efficiency and represents the theoretical maximum mechanical power that can be achieved by the system. The efficiency η_{EIR} is sometime called the efficiency of the ideal Rankine cycle.

Energy Efficiency of a Power Plant

The energy efficiency of a power plant is defined as,

$$\eta_{EPP} \equiv \frac{N_{eff}}{q_F} = \eta_m \eta_i \eta_{EIR} \eta_B, \tag{11.20}$$

where η_m is the mechanical efficiency of the turbine, η_i is the internal energy of the turbine, η_{EIR} is the energy efficiency of an ideal cycle and η_B is the boiler efficiency. As can be seen, η_{EPP} describes how efficient the plant is to generate the turbine mechanical power N_{eff} from the provided fuel energy rate per unit time q_F.

Energy Efficiency of a Plant Cycle

The energy efficiency of the plant cycle is defined as,

$$\eta_{EPC} \equiv \frac{N_i - |N_{iP}|}{q_a}, \tag{11.21}$$

where N_i is the internal turbine power, N_{iP} is the internal pump power and q_a is the added thermal power. This efficiency determines how efficient the cycle is to convert the added thermal power q_a into the net internal power $N_i - |N_{iP}|$.

Energy Efficiency of an Electrical Power Plant

The energy efficiency of the electrical power plant is defined as,

$$\eta_{EEL} \equiv \frac{N_{el} - |N_{elP}|}{q_F}, \tag{11.22}$$

where N_{el} is the electrical power of the plant, N_{elP} is the electrical power needed by the pump and q_F is the energy rate of the fuel per unit time. This efficiency determines how efficient the cycle is to convert the added energy of fuel q_F into the net electrical power $N_{el} - |N_{elP}|$.

11.2.3 EFFICIENCY IMPROVEMENTS

Energy efficiency of an ideal Rankine cycle given by Eq. (11.18) is still, in most cases, much less than the theoretical limit determined by the Carnot efficiency, calculated from the high and low temperature values. At the same time, an improvement of even 0.5% in overall efficiency of a power cycle is a significant gain. Thus various methods are used to improve the energy efficiency, such as the intermediate reheating of the working fluid and the regeneration. The two methods are described in more detail below.

Intermediate Reheating of Working Fluid

In plants with the intermediate reheating, the working fluid, after passing through the high-pressure turbine, is returned to the boiler for reheating, and next it is directed to the low pressure turbine, as shown in Fig. 11.4.

Using the plant shown in Fig. 11.4 as a reference, the energy efficiency of the plant cycle is given as,

$$\eta_{EPC} \equiv \frac{N_{iH} + N_{iL} - |N_{iP1}| - |N_{iP2}|}{q_a}, \tag{11.23}$$

where $N_{iH} = W(i_1 - i_2)$ is the internal power of the high pressure turbine, $N_{iL} = W(i_3 - i_4)$ is the internal power of the low pressure turbine, $|N_{iP1}| = W(p_6 - p_5)/\rho_5/\eta_{iP1}$ is the internal power of the low-pressure (condensate) pump, $|N_{iP2}| = W(p_8 - p_7)/\rho_7/\eta_{iP2}$ is the internal power of the high-pressure (feedwater) pump and $q_a = q_{a1} + q_{a2} = W(i_1 - i_8) + W(i_3 - i_2)$ is the total added heat.

Regeneration

A second common method to improve efficiency of the basic Rankine cycle is to use regeneration. The main idea behind the regeneration is to split the turbine into a high-pressure part and a low-pressure part. The high-pressure steam expands first in the high-pressure turbine, and after leaving it, the steam splits into two streams. The main stream is directed to the low-pressure turbine for further expansion, whereas the remaining fraction is used in feedwater heaters to increase the temperature

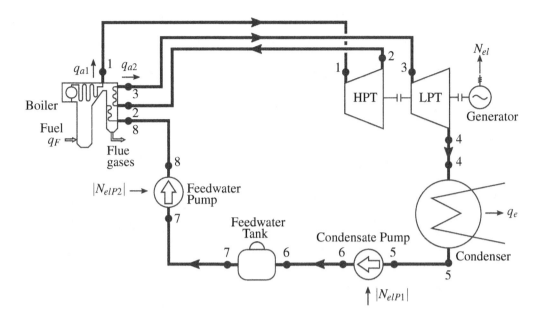

Figure 11.4 Schematic of a condensing power system with an intermediate reheating of steam between the high-pressure turbine (HPT) and the low-pressure turbine (LPT).

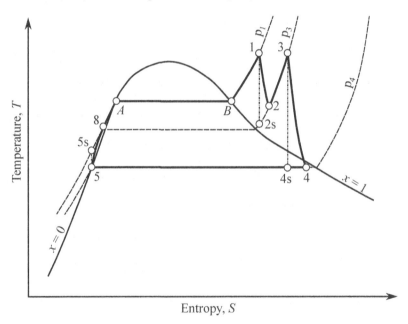

Figure 11.5 Condensing power cycle with intermediate reheating between high-pressure turbine and low-pressure turbine shown on TS-plane.

of the water flowing back to the boiler. A schematic of a simple Raknine cycle with regeneration is shown in Fig. 11.6.

Regeneration improves the cycle efficiency by using the latent heat of the steam for increasing the enthalpy of feedwater. In addition, a fraction of steam is bypassing the exit stages of a turbine and the condenser, allowing for a reduction of their size. The general rule is that these advantages

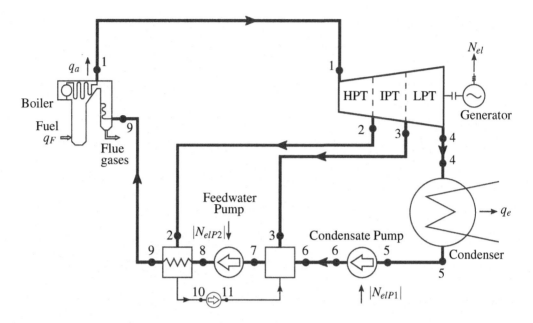

Figure 11.6 Schematic of a condensing power system with regeneration: HPT—high pressure turbine, IPT—intermediate pressure turbine, and LPT—low pressure turbine.

should outweigh the extra cost of the regeneration system. Several stages of regeneration may be used to improve the overall efficiency. In modern systems at least seven regeneration stages are employed.

The advantages of regeneration can be explained as follows. In a Ranking cycle with regeneration only a fraction of steam is passing through the condenser. This fraction follows the classic Ranking cycle, where certain amount of heat is extracted at low temperature and transferred to the environment. The remaining steam fraction expands in turbine stages and does a work as well, but does not dump any heat to the environment. In fact, neglecting the thermal losses, the steam passing through regenerators generates work with energy efficiency equal to one.

An efficiency of the Rankine cycle increases with an increasing number of feedwater heaters. As already mentioned, this number should be decided on the economic basis. Experience shows that the optimum number of feedwater heaters increases with increasing pressure and temperature of live steam, and with increasing total power output of turbines.

Since the feedwater in systems with regeneration has higher temperature, some modifications of the boiler are required, to avoid drop of the boiler efficiency. One such remedy is to use flue gases to heat up the intake air, rather than to heat up the feedwater. Once properly designed, the boiler size can be reduced, bringing about additional economic gains.

Two types of feedwater heaters are used:

- Open feedwater heater (OFWH), in which streams of steam and feedwater are directly mixed.
- Closed feedwater heater (CFWH), in which streams are separated, with feedwater flowing in tubes, and steam flowing in the shell surrounding the tubes.

The advantage of OFWH is that it allows to raise the feedwater temperature up to saturation temperature of the mixing steam. The disadvantage of such system is the required high number of pumps, which are used to equalize the feedwater pressure with the pressure of the mixing steam. In most systems CFWH is preferred, with only one OFWH, which is usually combined with a feedwater deaerator.

11.2.4 SYSTEM MODELING

The energy efficiency of a complex Ranking cycle system with the intermediate reheating and regeneration depends on many design and operational parameters. The calculations can be quite tedious, especially when the system optimization is desired. In 1960s the first software programs were created for monitoring power plants. Nowadays power plants are designed, optimized, and operated using dedicated computer codes.

In this section we will investigate a simple method to determine key parameters of a Ranking cycle system shown in Fig. 11.7. Our goal will be to find the plant efficiency for given plant data, such as:

- the electric power of generator,
- pressure and temperature of the live steam at the exit from boiler,
- steam pressure at the inlet to the intermediate reheater,
- steam pressure and temperature at the outlet from the intermediate reheater,
- internal efficiencies of turbines,
- the generator efficiency,
- the turbine mechanical efficiency.

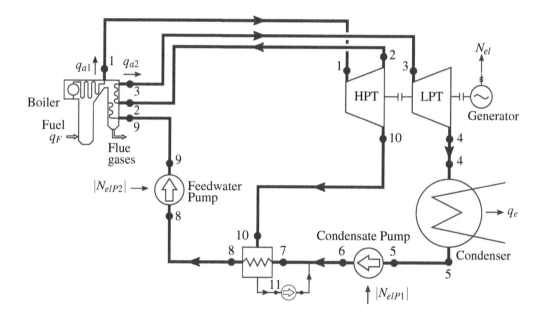

Figure 11.7 Schematic of a condensing power system.

We write first the mass and energy balance equations for main components. The mass balance equations for the high-pressure and low-pressure turbines are as follows, respectively,

$$W_1 - W_2 - W_{10} = 0, \tag{11.24}$$

$$W_3 - W_4 = 0. \tag{11.25}$$

The energy balance equation for turbines and generator reads,

$$W_1 i_1 - W_2 i_2 - W_{10} i_{10} + W_3 i_3 - W_4 i_4 = \frac{N_{el}}{\eta_m \eta_g}. \tag{11.26}$$

The specific enthalpy at the exit from the low pressure turbine with known internal efficiency η_{iLPT} and the high pressure turbine with internal efficiency η_{iHPT} can be found as,

$$i_4 = i_3 - (i_3 - i_{4s})\eta_{iLPT}, \quad i_2 = i_1 - (i_1 - i_{2s})\eta_{iHPT}, \tag{11.27}$$

where $i_{4s} = i(p_4, s_3)$ and $i_{2s} = i(p_2, s_1)$ can be found from the steam property tables.

For the feedwater heater we write the following mass and energy conservation equations

$$W_8 i_8 - W_7 i_7 - (W_{10} i_{10} - W_{11} i_{11})\eta_{FWH} = 0, \tag{11.28}$$

$$W_7 - W_8 = 0, \tag{11.29}$$

$$W_{10} - W_{11} = 0. \tag{11.30}$$

Here η_{FWH} is the overall heat transfer efficiency between streams 7–8 and 10–11 in the feedwater heater. When the design specification of the feedwater heater provides the temperature increase ΔT_{FWH} and the pressure drop Δp_{FWH} for the heated stream, we can find the exit temperature and pressure for the heated stream as,

$$T_8 = T_7 + \Delta T_{FWH}, \quad p_8 = p_7 - \Delta p_{FWH}. \tag{11.31}$$

Then the specific enthalpy of the heated stream is found from the water-property tables as $i_8 = i(p_8, T_8)$.

For the condensate pump with internal efficiency η_{iCP} and the feedwater pump with internal efficiency η_{iFP} we have, respectively,

$$i_6 - i_5 - \frac{p_6 - p_5}{\rho_5 \eta_{iCP}} = 0, \tag{11.32}$$

$$i_9 - i_8 - \frac{p_9 - p_8}{\rho_8 \eta_{iFP}} = 0. \tag{11.33}$$

The pressure of the regeneration stream p_{10} and the regeneration flow rate W_{10} should be chosen in such way that the overall plant efficiency is maximum. These optimum operating conditions can be found iteratively by assuming and modifying p_{10} and W_{10} until the optimum point is obtained. To establish pressure p_{10}, we usually assume that it should be equal to the saturation pressure at point 8 with some margin for the stagnation temperature: $p_{10} = p_{sat}(T_8 + \Delta T_{sFWH})$, where ΔT_{sFWH} is the stagnation temperature in the feedwater heater. It should be mentioned that in more realistic plant calculations pressure drops in all pipe connections should be taken into account.

The mass and energy conservation equations formulated for the whole plant can be combined in the following system of equations,

$$\mathbf{A} \cdot \mathbf{X} = \mathbf{Y}, \tag{11.34}$$

where $\mathbf{X} = [W_1, W_2, ..., W_n]^T$ is a vector of unknown mass flow rates in the system of equations, $\mathbf{Y}$ is a vector of known terms on the right-hand side of the system of equations and $\mathbf{A}$ is a matrix of known coefficients in the system of equations.

11.3 STATIONARY GAS TURBINES

Gas turbines have several advantages that make them popular for such applications as electric power generation, cogeneration, natural gas transmission, and various process applications. They are highly reliable and flexible. A power plant with gas turbines can be brought to full power operation in a relatively short time, measured in minutes. Gas turbines have long lifetime (of about 200000 hours), a compact and light structure, and occupy only half of the area in comparison to a steam turbine of the same power output. Availability of gas fuel, progress in material science, and technology improvements introduced in aviation and space exploration are other factors that promote the development of gas turbines and their growing application in the power sector.

The main elements in gas turbine engines include a compressor, a combustion chamber, and a turbine. These components can be assembled in an open, a closed, a partly closed, or a combined system. Example diagrams of simplest open and closed gas turbine systems are shown in Fig. 11.8.

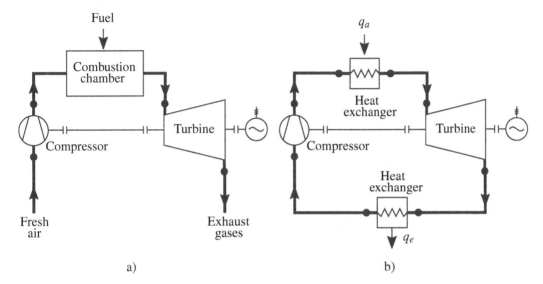

Figure 11.8 An open (a) and closed (b) gas turbine system.

In the open system the fresh air is taken from the environment and is adiabatically compressed in the compressor. The compressed air flows to the combustion chamber, in which the fuel is burned at constant pressure. The combustion produces a high temperature and high pressure gas stream that enters and expands in the turbine. The rotor of the turbine drives the compressor, to draw more air into the combustion chamber, and it powers the generator to produce electricity.

In the closed system a constant amount of working fluid is circulating and undergoing similar thermodynamic processes as in the open system. The working fluid is compressed in the compressor and absorbs heat in the heater (high-temperature heat exchanger). After expansion in the turbine, the working fluid is cooled in the cooler (low-temperature heat exchanger).

The closed system, unlike the open system, can operate with any working fluid, such as argon, helium, or supercritical water. It can be more compact thanks to a higher internal pressure. The power of the closed system can be regulated by changing the density of the working fluid. However, the system needs two heat exchangers, which increases its construction costs.

The energy efficiency of gas turbine power plants rapidly decreases with decreasing internal efficiencies of the turbine and the compressor. Initially this was a limiting factor to use such solutions to large-scale electricity generation. This barrier was removed only after development of high-efficiency components, with internal efficiency of compressors and turbines up to 88%.

11.4 COMBINED CYCLE POWER

The exergy efficiency of a gas turbine can be significantly improved if it operates together with a steam turbine. The low exergy efficiency of gas turbines is caused by irreversible combustion and mixing processes in the combustion chamber. The gas stream exhausted to the environment contains still high exergy that is lost. This loss can be reduced if the exhaust gases from a gas turbine are used as a heat source in a steam turbine operating in a Rankine cycle.

Example diagrams of combined cycle systems are shown in Fig. 11.9. Diagram a) shown in the

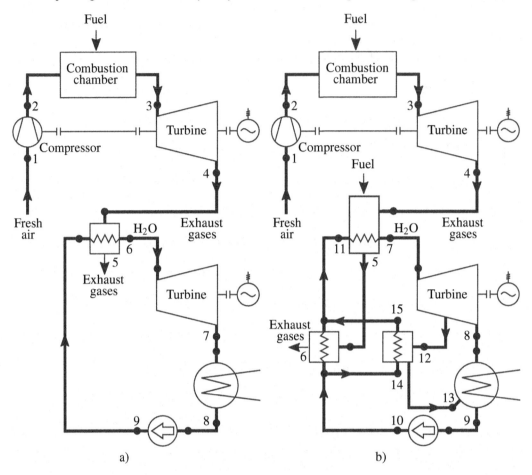

Figure 11.9 Combined cycle systems: a) 1–2–3–4–5 open Brayton cycle, 6–7–8–9 closed Rankine cycle; b) 1–2–3–4–5 open Brayton cycle, 7–8–9–10-11 closed Rankine cycle with regeneration.

figure represents a combined cycle system in which the exhaust gases are used in a single boiler to evaporate water and increase its specific enthalpy from i_9 to i_6. To achieve complete evaporation of water in the boiler, and to generate superheated steam at point 6, temperature T_4 of the exhaust gases must be high enough. This limitation can be removed in a system shown in diagram b), where additional fuel is burned in the boiler. For such systems, where additional regeneration is used, the over-all energy efficiency can be 42–44%.

11.5 COGENERATION AND TRIGENERATION

Cogeneration or combined heat and power (CHP) is a simultaneous generation of electricity and useful heat. Trigeneration or combined cooling, heat and power (CCHP) refers to the simultaneous generation of electricity and useful heating and cooling. Using a fuel to generate electricity and heat in a single unit is more efficient and cost effective than generating electricity and heat in two different units. The technology offers several advantages, such as, e.g., increased energy efficiency, reduced emissions, reduced energy costs, and enhanced energy system resilience.

Condensing steam turbines are designed to utilize steam exergy as efficiently as possible, which means that very little useful steam exits the turbines. Steam turbines for cogeneration require another design approach. They can be designed for extraction of some steam at lower pressures, after it has passed through a number of steam stages. The un-extracted steam passes through the entire turbine and condenses in the condenser. Another design approach is to exhaust steam from the turbine at back-pressure, without condensation. Such exhaust steam can be used for various purposes, such as process heating.

PROBLEMS

PROBLEM 11.1

The boiler in a power plant (see Fig. 11.1) generates superheated steam with pressure $p_4 = 3.6$ MPa and temperature $T_4 = 723$ K. The pressure in the condenser is $p_1 = p_2 = 5.5$ kPa. The steam mass flow rate in the system is $W = 18$ kg/s, and after passing the condenser, the condensate temperature is $T_2 = 307$ K. The feedwater pump has internal efficiency $\eta_{iP} = 0.80$, mechanical efficiency $\eta_m = 0.94$, and the efficiency of the electrical motor is $\eta_{EM} = 0.89$. The feedwater entering the boiler has pressure $p_3 = 3.9$ MPa. What is the energy efficiency of the plant if the generated electric power is $N_{el} = 16$ MW?

PROBLEM 11.2

The boiler in a power plant (see Fig. 11.1) generates superheated steam with pressure $p_4 = 4.2$ MPa and temperature $T_4 = 725$ K. The pressure in the condenser is $p_1 = p_2 = 7.5$ kPa. The ratio of steam mass flow rate in the system to the turbine internal power is $W/N_i = 1.13$ kg/MJ. The condensate leaving the condenser has temperature $T_2 = 312$ K. The energy efficiency of the boiler is $\eta_B = 0.82$ and the internal efficiency of the pump is $\eta_{iP} = 0.77$. The ratio of the chemical exergy to the lower heating value of the fuel is 1.05. The ambient temperature is 300 K. Calculate the relative exergy losses in the plant cycle due to process irreversibility. Calcualate the exergy efficiency of the cycle.

PROBLEM 11.3

A plant with intermediate reheating (Fig. 11.4) uses steam and water with the following parameters: $p_1 = 12.5$ MPa, $p_2 = 3.7$ MPa, $p_3 = 3.3$ MPa, $p_4 = p_5 = 5$ kPa, $p_6 = p_7 = 0.2$ MPa, $p_8 = 13.6$ MPa, $T_1 = 805$ K, $T_3 = 795$ K, $T_5 = T_6 = T_7 = T_8 = 309$ K. Calculate the plant energy efficiency if the internal efficiency of the high-pressure turbine is $\eta_{hp} = 0.76$, low-pressure turbine $\eta_{lp} = 0.81$, condensate pump $\eta_{icP} = 0.82$ and feedwater pump $\eta_{ifP} = 0.85$.

PROBLEM 11.4

A plant with intermediate reheating and regeneration (Fig. 11.7) uses steam and water with the following parameters: $p_1 = 13.5$ MPa, $p_4 = p_5 = 5$ kPa, $p_2 = p_6 = p_7 = p_8 = p_{10} = p_{11}$, $T_1 = T_3 = 810$ K, $T_5 = 309$ K, and the condensate leaving the reheater at point 11 has 2 K subcooling. The pressure drop in the intermediate reheater is $p_2 - p_3 = 0.4$ MPa, and in the boiler $p_9 - p_1 = 1$ MPa. The steam mass flow ratio $W_1/W_{10} = 0.25$. Find the optimum intermediate pressure $p_2 = p_{10}$ that gives maximum plant energy efficiency, if the internal efficiency of the high-pressure turbine is $\eta_{iHPT} = 0.78$, low-pressure turbine $\eta_{iLPT} = 0.83$ and the pumping power in the system is small and can be neglected.

PROBLEM 11.5

A plant with two feedwater reheaters (Fig. 11.6) uses steam and water with the following parameters: $p_1 = 3.5$ MPa, $p_2 = p_{10} = 0.9$ MPa, $p_3 = p_6 = p_7 = 0.12$ MPa, $p_4 = p_5 = 0.006$ MPa, $p_8 = 3.9$ MPa, $p_9 = 3.8$ MPa, $T_1 = 705$ K, $T_5 = 308$ K, $T_7 = 375$ K, $T_9 = 441$ K, $T_{10} = 445$ K. Find the plant energy efficiency, if the internal efficiency of the high-pressure turbine is $\eta_{hp} = 0.77$,

intermediate-pressure turbine $\eta_{ip} = 0.80$, low-pressure turbine $\eta_{lp} = 0.79$ and the internal efficiency of pumps is $\eta_{iP1} = \eta_{iP2} = 0.80$. Calculate the relative plant efficiency increase compared
to a system without regeneration, assuming in both cases the same operational parameters and the
same component efficiencies.

PROBLEM 11.6

For the combined plant systems shown in Fig. 11.9, derive expressions for the following efficiencies: a) energy efficiency of an ideal cycle, b) energy efficiency of a power plant, c) energy
efficiency of a plant cycle, d) energy efficiency of the electrical power plant. Assume that all
efficiencies of the system components are known.

12 Moving Water Power

Water in the Earth's hydrosphere is in a permanent motion, following a so-called water cycle. Water is found in solid, liquid, and gaseous form in rivers, lakes, oceans, and as vapor, in the atmosphere. The water cycle is a motion of the water from the ground and oceans to the atmosphere and back again. Many processes, such as evaporation, precipitation, runoff, and transpiration are involved in the water cycle. The driving forces for the water cycle are the solar energy and the gravity. The solar radiation is causing evaporation of water and its transport in the atmosphere. The gravity, together with heat transfer in the atmosphere, is essential for precipitation and runoff. Due to transpiration, water moves from soil through plants and evaporates to the atmosphere.

The gravitational pulls of the Moon and the Sun are causing the rise and fall in sea level called a tidal cycle. The ocean motions dissipate some of Earth's rotational energy through friction and also transfer some of that energy to the Moon. The tidal cycle can be observed as recurrence of high and low water on the coastline, which usually, but not always, reach a high and low level about six hours apart.

Gravity, wind, and variable water density are causing ocean currents, which are predictable, continuous, and directional movements of water. Ocean water moves in two directions: horizontally and vertically. Horizontal movements are referred to as currents, while vertical displacements are called upwellings or downwellings. Depending on the underwater topography and local conditions the flow velocity can be significantly enhanced, resulting in appreciable kinetic energy.

Wind passing over the surface of the sea generates waves. As long as the waves propagate slower than the wind speed, there is an energy transfer from the wind to the waves. The wave power is determined by wave height, wave speed, wavelength, and water density.

This chapter describes some of the major technologies that have been developed to extract energy from moving water in rivers, marine currents, surface waves and tides. Our focus is to determine the efficiency of the various technologies and to estimate the total power that can be harvested from the various water-related energy resources.

12.1 HYDROPOWER

The mechanical energy of water has been harnessed by humans since ancient times. Water wheels, used to convert water current energy into useful forms of power, were invented around 3rd Century BC. The Greeks used water wheels for grinding wheat into flour more than 2000 years ago. Waterwheels were gradually replaced by hydro turbines during the industrial revolution in the beginning of 19th Century. In 1849 James Francis developed the first modern water turbine, which remains the most widely-used water turbine in the world today. An impulse-type water turbine was invented by Lester Allan Pelton in 1870s. Viktor Kaplan developed a propeller-type water turbine with adjustable blades in 1913.

The world's first hydroelectric power plant began operation in the United States along the Fox River in Appleton, Wisconsin in 1882. At the present time, the hydropower is the main technology that provides clean and renewable electricity worldwide. Brasil and China became world leaders in hydropower with the Itaipu Dam in Brazil with capacity of 14 000 MW and the Three Gorges Dam in China with capacity of 22 500 MW.

Hydropower has several advantages, such as:

- The technology is mature, proven and available for more than a century.
- Hydropower plants can be easily upgraded and modernized utilizing recent technological advances.

DOI: 10.1201/9781003036982-12

- Hydropower does not require fuel and relies on renewable resources.
- In most cases hydropower is economically competitive.
- The storage hydropower facilities (dams, pumped storage) offer operational flexibility being able to easily ramp up or shut down.
- The creation of reservoirs can be useful for other purposes such as providing water for drinking or irrigation, and reducing human vulnerability to droughts.
- Reservoirs can provide flood protection, and can improve waterway transport capacity.

The main disadvantages of hydropower are as follows:

- High up-front investment costs compared to other technologies, such as thermal power.
- Reservoirs may have a negative impact on the inundated area, damage river flora and fauna, or disrupt river uses such as navigation.

12.1.1 HYDROPOWER POTENTIAL

As mentioned in §2.3.6 hydropower theoretical reserve is estimated to 39.1 PWh/y and roughly one-third of this reserve has a technical potential to generate electricity. However, taking into account the current economic constraints, the hydropower potential is 8.7 PWh/y. This corresponds to the amount of energy that is twice as large as the hydropower production in 2019.

12.1.2 TYPES OF WATER TURBINES

Impulse and reaction turbines are the two main types of hydro turbines. Schematic diagrams of typical hydraulic turbines are shown in Fig. 12.1. Impulse turbines use the velocity of the water to move the runner and discharge to atmospheric pressure. The water stream hits each bucket on the runner. This type of turbines is suitable for high head and low flow applications. A Pelton wheel is an example of the impulse turbine. Reaction turbines develop power from the combined action of pressure and moving water. The runner is placed directly in the water stream flowing over the blades rather than striking each individually. Such turbines are used for sites with lower heads and higher flows as compared with the impulse turbines. Kaplan and Francis turbines are two examples of the reaction turbine.

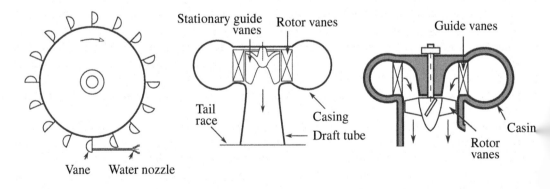

a) Impulse turbine (Pelton wheel)   b) Reaction turbine (Francis type)   c) Reaction turbine (Kaplan type)

Figure 12.1   Schematic diagrams of typical hydraulic turbines.

Pelton Wheel

The Pelton wheel is driven by one or more high-speed free water jets, which lie in the plane of the turbine runner. Each jet is accelerated in a nozzle external to the turbine wheel. If friction is neglected, neither the water pressure not its speed relative to runner change as it passes over the turbine buckets. The water jets strike each bucket in succession, are turned, and leave the bucket with relative velocity nearly opposite to that with which they entered the bucket. A schematic of the Pelton wheel installation is shown in Fig. 12.2.

Depending on water flow and design, Pelton wheels operate best with heads from 15 to 1 800 m. Power outputs can range from a few kilowatts up to tens of megawatts. The overall efficiency of the turbine can reach 90–95%.

Francis Turbine

In the reaction turbine of the Francis type, the incoming water flows circumferentially through the turbine casing. It enters the outside periphery of the stationary guide vanes and flows toward the runner. The flow pattern looks like a centrifugal pump in reverse. Water enters the runner nearly radially and is turned downward to leave nearly axially. Water leaving the runner flows through a diffuser known as draft tube before entering the tailrace. A schematic of the Francis turbine installation is shown in Fig. 12.3.

Well-designed Francis turbine can capture 90–95% of the water energy. The turbine demonstrates best performance with head heights between 100 and 300 m but it can be used within a wider range of heads from 3 to 600 m. The electric generators that use this turbine have a power output ranging from just a few kilowatts up to 700 MW[1].

Kaplan Turbine

The water entry in the Kaplan turbine is similar to the Francis turbine. However, it is turned to flow almost axially before encountering the turbine runner. Flow leaving the runner may pass through a draft tube. This turbine is an evolution of the Francis turbine to allow efficient power production in low-head applications. The head ranges from 10 to 70 m and the output ranges from 5 to 200 MW. The Kaplan turbine efficiencies are typically over 90%, but may be lower in very low head applications. A schematic of the Kaplan turbine installation is shown in Fig. 12.3.

12.1.3   TYPES OF HYDROPOWER PLANTS

There are three types of hydropower facilities: impoundment, run-of-river, and pumped storage. The most common type of hydroelectric power plant is an impoundement facility, which uses a dam to store river water in a reservoir. Water released from the reservoir flows through a turbine, which drives a generator to produce electricity. The water flow may be adjusted to either meet changing electricity needs or to maintain a constant reservoir level. A run-of-river facility channels a portion of a river through a canal or penstock. Often a dam is not required in this type of hydropower plant. A pump storage facility is used to store electricity generated by other power sources like solar, wind, or nuclear for use during high electricity demand. It works like a battery, which stores energy by pumping water uphill to a reservoir at higher elevation from a second reservoir at a lower elevation. This type of operation takes place when the demand for electricity is low. When the demand for electricity is high, the water is released back to the lower reservoir to generate electricity.

---

[1]The largest Francis turbines are installed at both the Itapu power plant on the Brazil-Paragway border and at the Three Gorges dam in China, and have generating capacity of 700 MW each.

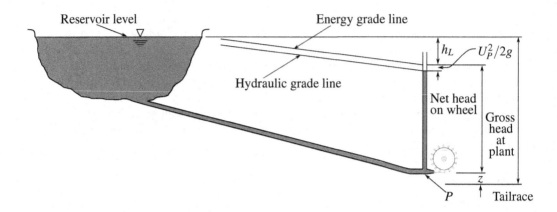

Figure 12.2 Schematic of impulse turbine installation, showing the definitions of the net head and the gross head on plant.

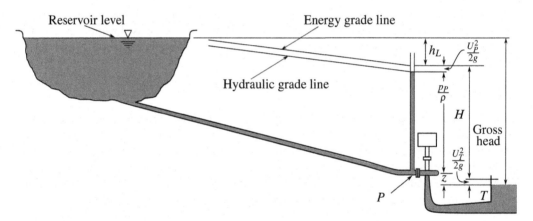

Figure 12.3 Schematic of reaction turbine installation, showing the definitions of the net head (H) and the gross head on plant.

Impoundment Hydropower

An impoundment facility uses a dam to store river water in a reservoir. Water released from the reservoir flows through a long conduit called penstock, next it passes through a turbine, spinning it, which in turn activates a generator to produce electricity.

When the available water head exceeds about 300 m, an impulse turbine is chosen. In such facilities the water is accelerated through a nozzle and discharges as a high-speed free jet at atmospheric pressure. The jet strikes deflecting buckets attached to the rim of the rotating wheel. When large amount of water is available, additional power can be obtained by connecting two wheels to a single shaft or by arranging two or more jets to strike a single wheel. Water discharged from the wheel at relatively low speed falls into the tailrace.

Two characteristic parameters for an impoundment facility are the available water flow rate and the water head. As shown in Figs. 12.2 and 12.3, the gross head available is the difference between the levels in the upper reservoir and the tailrace. For plant efficiency calculations the net head is used, which is taken as a difference between the total head (following the elevation of the energy grade line) at the nozzle entrance and the elevation of the nozzle centerline. In most practical situations at the rated power the net head is 85–95% of the gross head.

When lower heads are available, reaction turbines provide better efficiency than impulse turbines. Reaction turbines flow full of water, and they tend to be high flow, low head machines. Since their installation details differ from the impulse turbines, the definitions of the net head and the gross head are also different.

For reaction turbine installation, the gross head available is the difference between the upper reservoir head and the tailrace head. For the efficiency calculations the net head is used, which is defined as the difference between the elevation of the energy grade line just upstream of the turbine and that of the tailrace.

Run-of-river Hydropower

A run-of-river facility is a type of hydroelectric generation plant whereby little or no water storage is provided. A small dam may be used to ensure enough water passes through the penstock. A small water storage reservoir is sometime used to reduce the influence of seasonal water flows. This type of water reservoir is referred to as pondage. A plant without pondage will operate as an intermittent power source. To distinguish run-of-river facilities from impoundment facilities, more precise definitions are provided, even though they vary around the world. As an example, the European Network of Transmission System Operators for Electricity defines the run-of-river facility as such hydropower plant, which can hold enough water to allow generation for up to 24 hours. If the ratio of the reservoir capacity to the generating capacity is greater than 24 hours, the hydropower plant is defined as the impoundment facility[2].

The largest run-of-river plants have generating capacity comparable to typical capacities of the impoundment hydropower plants. For example, the Jirau hydroelectric power plant in Brazil is currently the largest run-of-river plant in the world and has the installed capacity of 3750 MW[3]. However, most of the run-of-river facilities are much smaller. This includes the so-called electricity buoys: small floating hydroelectric power plants, anchored to the ground. The moving water propels a power generator and thereby generates electricity.

Pumped Storage

Pumped-storage hydroelectricity allows excess energy to be saved for periods of higher demand. The excess electricity can be generated by solar and wind, when their electricity production is higher than the demand. It can also be generated by base load resources, such as coal and nuclear, during periods of low electric demands. The excess electricity is used to pump water from lower into the upper reservoir. When there is high demand, water is released back into the lower reservoir through a turbine to generate electricity. Usually a Francis turbine type is used for a combined pump and turbine operation. The latest large-scale units operate at variable speed for greater efficiency. They operate asynchronously (independent of the network frequency) during pumping, and synchronously with the network frequency when generating electricity.

As of 2021, the largest operational pumped-storage plant, with capacity of 3003 MW, is the Bath County Pumped Storage Station in the United States. About 70 pumped-storage plants are currently operating worldwide, and additional 37 are under construction, with planned capacity of over 1000 MW each.

---

[2]www.entsoe.eu

[3]www.esbr.com.br/a-usina, accessed 2020-09-02.

12.1.4 ANALYSIS OF WATER TURBINE EFFICIENCY

The **hydraulic power** of water flowing from a reservoir with the net head H is given as,

$$N_h = \rho g Q H, \tag{12.1}$$

where ρ—water density, Q—water volumetric flow rate, and g—gravitational acceleration. The **hydraulic turbine efficiency** is defined as,

$$\eta_t \equiv \frac{N_m}{N_h} = \frac{\omega T_s}{\rho Q g H}. \tag{12.2}$$

Here N_m is the turbine mechanical power measured on the turbine shaft, T_s—applied torque to the rotor, and ω—angular velocity of the rotor.

The **Euler turbomachine equation** can be derived from the angular momentum equation applied to a control volume enclosing the rotor, as shown in Fig. 12.4. The equation provides a relationship between the shaft torque and the fluid angular momentum as follows,

$$T_s = (r_2 U_{t2} - r_1 U_{t1}) W, \tag{12.3}$$

where r_1—fluid entrance radius, r_2—fluid exit radius, U_{t1}—fluid tangential velocity at inlet, U_{t2}—fluid tangential velocity at outlet, W—fluid mass flow rate.

The product of the rotor angular velocity and the applied torque is equal to the rate of work, or power, done on turbomachine rotor,

$$N_m = \omega T_s = \omega (r_2 U_{t2} - r_1 U_{t1}) W. \tag{12.4}$$

Since the tangential speed of the rotor at radius r can be found as $U_r = r\omega$, the above equation can be written as,

$$N_m = (U_{r2} U_{t2} - U_{r1} U_{t1}) W. \tag{12.5}$$

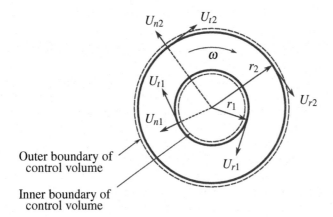

Figure 12.4 A turbomachinary rotor enclosed in a control volume. $U_{r1} = r_1 \omega$—rotor blade linear speed at $r = r_1$, $U_{r2} = r_2 \omega$—rotor blade linear speed at $r = r_2$, ω—angular speed of the rotor, r_1—rotor inner radius, r_2—rotor outer radius, U_{t1}—fluid tangential velocity at the inlet to control volume, U_{n1}—fluid normal velocity at the inlet to control volume, U_{t2}—fluid tangential velocity at the outlet from control volume, U_{n2}—fluid normal velocity at the outlet from control volume.

Optimum Speed of Impulse Turbine

We will now investigate the optimum speed of rotation of the impulse turbine to obtain the maximum power. A schematic of the turbine is shown in Fig. 12.5.

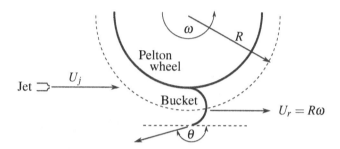

Figure 12.5 Optimum speed of impulse turbine. R—mean radius of rotor, U_r—bucket speed, U_j—jet speed, θ—jet turning angle, ω—angular speed of the Pelton wheel.

To simplify derivations, the following assumptions are adopted:

- The torque due to surface and body forces can be neglected.
- The mass of water on wheel can be neglected.
- All jet water acts on the bucket.
- There is no change in jet speed relative to the bucket.
- The steady state condition prevails.

The angular momentum principle gives,

$$T_s = (r_2 U_{t2} - r_1 U_{t1})W = [R(U_j - U_r)\cos\theta - R(U_j - U_r)]\rho U_j A =$$
$$R(U_j - U_r)\rho U_j A(\cos\theta - 1) = \rho QR(U_j - U_r)(\cos\theta - 1) \tag{12.6}$$

This equation gives a relationship for the torque T_s expressed in terms of main parameters, such as the volumetric flow rate Q, the wheel radius R, the jet speed U_j, the bucket speed U_r, and the jet turning angle θ. The output torque exerted by water on the wheel can now be found as,

$$T_{out} = -T_s = \rho QR(U_j - U_r)(1 - \cos\theta), \tag{12.7}$$

and the corresponding mechanical power of the turbine is,

$$N_m = \omega T_{out} = \rho QR\omega(U_j - U_r)(1 - \cos\theta) = \rho QU_r(U_j - U_r)(1 - \cos\theta) \tag{12.8}$$

The maximum power output as a function of the bucket speed U_r can be found from the following condition,

$$\frac{dN_m}{dU_r} = 0, \tag{12.9}$$

and the resulting optimum bucket speed is found as

$$U_{r,opt} = U_j/2. \tag{12.10}$$

The corresponding maximum power is found as

$$N_{m,max} = \frac{1}{4}\rho Q U_j^2 (1 - \cos\theta) \tag{12.11}$$

As can be seen, the absolute maximum would be obtained for $\theta = 180°$, for which $N_{m,max} = \rho Q U^2/2$. In practice, the θ angle can never reach $180°$ and it can be up to $165°$, for which the term $1 - \cos(165°) \approx 1.97$.

12.2 MARINE CURRENT POWER

The power of a marine current can be found from the following equation,

$$N_c = \rho Q_c \frac{U_c^2}{2}, \tag{12.12}$$

where N_c—marine current power, W, Q_c—water volume flux in the current, m^3/s, ρ—ocean water density, kg/m^3, U_c—average speed of the current, m/s.

Ocean currents carry tremendous amounts of water. It is estimated that the total worldwide power in ocean currents is about 5000 GW. Only Antarctic Circumpolar Current, which circulates around the South Pole, carries a water volume flux of about 1.35×10^8 m^3/s, and moves with an average speed of 1 m/s. Assuming the ocean water density of 1025 kg/m^3, the total power of the moving water is about $1.025 \times 135/2 \times 10^9$ W $\approx$ 69.2 GW. Similarly, the Gulf Stream in the Florida Straits carries 30×10^6 m^3/s of water with a speed of roughly 4 m/s, which gives the total power $1.025 \times 30 \times 4^2/2 \times 10^9$ W $\approx$ 246 GW. These values represent an absolute maximum to the power that could be extracted. Realistically, however, only a few percent of the power can be extracted by marine current turbines. In case of the Antarctic Circumpolar Current, which circulate around the South Pole at a latitude of approximately $60°$ S, harvesting this power would be extremely difficult, because the ocean is quite deep at that latitude. Emplacement of marine current turbines in the Florida Straights seems to be more realistic. However, removing significant fraction of energy from this current could have undesirable consequences for the global ocean systems and climate. Since many marine organisms use ocean currents for migration, harvesting current power may disturb their life pattern.

Flow patterns of marine currents are quite reliable, but they are significantly affected by tides. For this reason many ocean current energy extraction plants are placed in areas of high tidal flow rates.

The two dominating types of devices used for marine power generation are the axial-flow horizontal-axis propellers and the Darrieus rotors. Both these devices are derived from wind-power rotors and have a potential to achieve enough cost-effectiveness and reliability to be practical in marine-current future scenario.

12.3 WAVE POWER

Since waves are generated by wind, wave power at any location is as variable as wind power. In open oceans long and fast-moving waves can transport energy across thousands of kilometers. As for wind power, global wave power atlases are available. Such atlases provide the annular mean power density and its distribution along coastlines. Total wave power worldwide has been estimated at around 2000–3000 GW, whereas the average wave power density along the shore can range from a few kW/m to over 60 kW/m [37].

The deep-water surface wave has the following power density,

$$N' = \frac{\rho g^2}{64\pi} H_{m0}^2 T_e, \tag{12.13}$$

where N'—the wave energy flux per unit of wave-crest length, H_{m0}—significant wave height, T_e—the wave energy period, ρ—the water density, and g—the gravitational acceleration.

For moderate ocean conditions in deep water, a few kilometers off a coastline, the wave height can be about 3 m and a wave energy period of 8 s. Using Eq. (12.13) and assuming water density $\rho = 1025 \text{ kg/m}^3$, we get,

$$N' = \frac{1025 \times 9.81^2}{64\pi} \times 3^2 \times 8 = 35.3 \ \frac{\text{kW}}{\text{m}}. \tag{12.14}$$

This result indicates that the wave carries 35.3 kilowatts of power potential per meter of wave crest.

Various wave power devices are used to harvest the power potential of waves. The construction of the devices depend on their location, which can be either at the shoreline, near-shore, and offshore. They are typically using different methods to capture the energy from the waves such as hydraulic ram, elastometric hose pump, hydroelectric turbine, and linear electrical generator.

Wave-power generation is not widely established technology compared to wind power, hydropower, or solar power. However, due to its high power density, there have been attempts to use this energy resource at certain favorable locations. In 2000 the world's first commercial wave power device was installed on the coast of Islay in Scotland.

12.4 TIDAL POWER

Tidal power results from the gravitational interaction between Earth's oceans and the Moon. These interactions cause water motions in oceans, which dissipate some of Earth's rotational energy and also transfer some of that energy to the Moon.

Earth-Moon Energy Transfer

Precise measurements show that the average distance from Earth to the Moon increases about 38 mm each year. Due to that the energy and angular momentum of the Moon increases. Correspondingly, the angular momentum and rotational kinetic energy of Earth decreases. Direct measurements indicate that the length of a day is currently increasing by 1.7 ms/century. The increase from tidal effects alone would be in fact about 2.3 ms/century, but since Earth's shape is changing after polar compression in the last ice age, the angular velocity of Earth is slightly increasing, and partly offsets the slowing due to tidal effects.

Estimates show that the total rate of tidal energy dissipation is about [53],

$$N_{tidal} \cong 3.75 \text{ TW}. \tag{12.15}$$

This value sets a theoretical upper limit on the power that can be harvested from the tidal resources. In practice, however, only small fraction of the tidal power potential can be harnessed. The main reason for this is that most of this potential is lost in deep ocean. It is estimated that only 3% of the total rate of tidal energy dissipation can be practically recovered, corresponding to an average power of 60–120 GW [4].

The existing tidal power systems use the kinetic and potential energy of moving water. A tidal barrage uses the potential energy between the high and low tides. It is essentially a low-overhead dam that separates a tidal basin from the open ocean. The water that flows through the dam both during the incoming flood tide and during the outgoing ebb tide drives turbines.

Tidal stream turbines are used in places with significant tidal flow velocity. Similarly as for wind turbines, horizontal axis turbines are most common. However, other solutions, such as the vertical-axis rotors, oscillating devices, and shrouded turbines are investigated as well.

PROBLEMS

PROBLEM 12.1

A hydroelectric turbine rotates at speed $n = 35$ revolutions per minute (rpm). Express the turbine angular speed in radians per seconds.

PROBLEM 12.2

A Pelton wheel with a mean radius $R = 0.5$ m and angular speed $n = 120$ rpm is driven by a circular water jet, which has a cross-section radius $R_j = 145$ mm and which returns from a bucket at an angle $\theta = 175°$. Calculate the turbine power output if the volumetric water flow is $Q = 1$ m^3/s, and water density is 1000 kg/m^3.

PROBLEM 12.3

A Pelton wheel with a mean radius $R = 0.5$ m and angular speed $n = 150$ rpm is driven by a circular water jet, which has a cross-section radius R_j and which returns from a bucket at an angle $\theta = 178°$. Calculate the water jet radius R_j for which the turbine power output will be the highest possible. The water volumetric flow is $Q = 1.25$ m^3/s, and water density is 1000 kg/m^3.

PROBLEM 12.4

The following parameters have been measured for a hydroelectric turbine: rotor angular speed $n = 110$ rpm, torque measured on the shaft $T_s = 20.5$ kNm, volumetric water flow $Q = 2.25$ m^3/s, net head $H = 19.5$ m. Calculate the turbine efficiency, assuming that water density is $\rho = 1000$ kg/m^3.

PROBLEM 12.5

A hydroelectric turbine with known efficiency $\eta_t = 0.885$ is installed in a hydropower plant with a penstock cross-section area $A_p = 2.5$ m^2. The net head of the hydropower is $H = 20$ m and the mean velocity of water flowing in the penstock is $U_p = 1.35$ m/s. Calculate the mechanical power output of the turbine, assuming that water density is $\rho = 1000$ kg/m^3.

PROBLEM 12.6

Oceanic volume transport is typically measured in units of Sverdrups (abbreviated Sv), where 1 Sv = 10^6 m^3/s. The Gulf Stream in Florida Straits has the volumetric flux approximately 30 Sv, the speed of the flow 4 m/s, and the total power 246 GW. Estimate the total power of the stream near Scandinavia, where its speed drops to 2.5 m/s. Assume the same amount of water flowing in the stream and the same water density at both locations.

13 Wind Power

Since ancient time winds have been used for transportation on sea. One of the first recorded instances of using wind power is the windwheel of Hero of Alexandria (10 AD–70 AD). Wind-powered machines have been used to pump water and grind grain for more than two thousands years. The Netherlands polders were drained with wind-powered pumps already in 11th century. In remote places such as American mid-west or the Australian outback, wind pumps provided water for livestock.

For the first time electricity was generated by windmills designed by James Blyth in Scotland in 1887 and by Charles Brush in Cleveland, Ohio, in 1888. The rated power of Brush's windmill was 12 kW, and with 144 blades, it turned rather slowly. In 1957 Johannes Juul installed a wind turbine at Gedser in Denmark. The turbine had three blades with a horizontal axis of rotation, and can be seen as a precursor of the modern Horizontal Axis Wind Turbines (HAWT) that are dominating the development today.

In 1931 the Darrieus wind turbine was invented with a vertical axis of rotation. Such turbines can operate with wind from any direction and allow to place heavy equipment such as gearbox and generator on the ground.

Since 1974 large commercial wind turbines have been developed in the USA. The first multi-megawatt wind turbine was build in 1978 in Denmark. Nowadays wind power provides a significant contribution to the renewable power generation sector. In 2016 world wind electricity production was 958 TWh, corresponding to 4.6% of the total world production [47]. Efficiency and types of wind turbines, principles of their design and operation, and future perspectives of wind power are discussed in this chapter.

13.1 ENERGY OF MOVING AIR

Wind energy is the kinetic energy of naturally moving air masses in Earth's atmosphere. Similarly to almost all other renewable energy resources on Earth, the wind energy results from the nuclear fusion energy released by the Sun as a solar radiation. The solar radiation is warming air in Earth's atmosphere to different degrees in different locations causing air density and pressure gradients. These gradients produce motion of air masses on a global and a local scale, redistributing thermal energy and moisture contained in the air around the planet.

The kinetic energy of air with mass m, moving with average speed U is,

$$E = \frac{1}{2}mU^2. \tag{13.1}$$

If we assume that the air is moving with speed U, perpendicular to an area A, during time Δt, the total mass of the air flowing through the area is $m = V\rho = U\Delta tA\rho$, where ρ is the air density. Substituting the mass of air to Eq. (13.1) yields,

$$E = \frac{1}{2}\Delta tA\rho U^3. \tag{13.2}$$

Thus the power of wind flowing through area A is,

$$N \equiv \frac{E}{\Delta t} = \frac{1}{2}A\rho U^3. \tag{13.3}$$

The kinetic energy of winds can be converted in wind turbines into a rotational movement of blades attached to a slowly rotating shaft. The speed of rotation of the shaft is subsequently increased in gears, and electricity is generated in a generator attached to the high-speed shaft.

DOI: 10.1201/9781003036982-13

13.2 WIND POWER MACHINES

Design principles of wind power machines, previously called windmills, and now more properly wind turbines, have been changing over centuries of development. Currently, there are three primary wind turbine types, having either horizontal or vertical axis of rotation, as shown in Fig. 13.1.

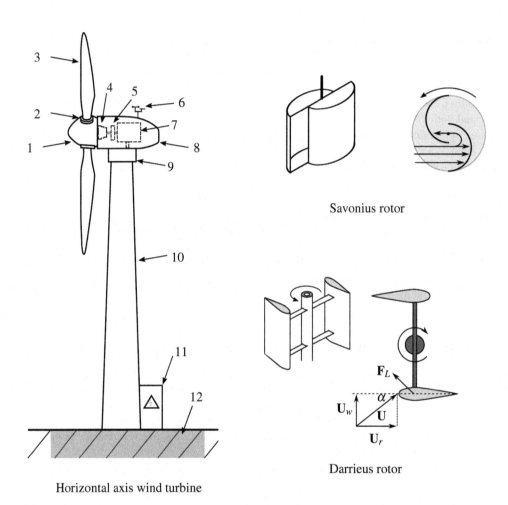

Savonius rotor

Darrieus rotor

Horizontal axis wind turbine

Figure 13.1 Wind turbine types: 1—rotor hub, 2—blade pitch control, 3—rotor blade, 4—gearbox, 5—break, 6—anemometer, 7—generator, 8—nacelle, 9—wind orientation control, 10—tower, 11—connection to the electric grid, 12—foundation, U_w—unperturbed wind velocity vector, U_r—relative air velocity vector due to rotation, U—resulting air velocity vector, F_L—lift force vector, α—angle of attack.

The *horizontal-axis wind turbine* (*HAWT*) harvests power from the wind by exploiting the lift force exerted on the airfoil-shaped blades as they circulate in the plane perpendicular to the wind direction. The rotor is mounted on a tower, allowing the turbine to reach stronger winds at greater heights. The *vertical-axis wind turbine* (*VAWT*) uses the lift force (when the *Darrieus rotor* is used) or the drag force (when the *Savonius rotor* is used) acting on the rotor to harvest power. Since the rotor has vertical axis, there is no need for *yaw control* to orient it toward the wind and it operates smoothly in winds with a rapidly changing direction. However, VAWTs are not self-starting as HAWTs are, since there is no net torque at wind speed approaching zero.

13.2.1 HORIZONTAL-AXIS WIND TURBINES

The **horizontal-axis wind turbine** is currently the dominating wind turbine type. Its design and performance have been extensively studied and significantly improved since 1990s. While the height and the rated power of wind turbines was in 1990s 50 m and 50 kW, respectively, the biggest wind turbines in July 2020 reached 190 m and 10 MW, with five biggest wind turbines listed below[1]:

1. SeaTitan[TM], developed by American energy technologies company AMSC, with 190 m rotor diameter, hub height of 125 m, and nameplate power 10 MW.
2. Sway Turbine ST10, developed by the Norwegian technology company Sway, with 164 m rotor diameter, and nameplate power 10 MW.
3. Areva 8MW, developed by the French energy company Areva, with 180 m rotor diameter, and nameplate power 8 MW.
4. Vestas V164-8.0MW, developed by Danish wind turbines manufacture Vestas, with 164 m rotor diameter, and nameplate power 8 MW.
5. Enercon E-126/7.5MW, developed by the German company Enercon, with 127 m rotor diameter, hub height of 135 m, and nameplate power 7.5 MW.

Design details of the horizontal-axis wind turbine are shown in Fig. 13.1. The **rotor**, consisting of two or three blades and a hub, is mounted on top of a tower, which is made from tubular steel, concrete, or steel lattice. The tower, build on a solid foundation, is the main supporting structure of the turbine. Because wind speed increases with height, taller towers enable turbines to capture more energy and generate more electricity.

Blades lift and rotate when wind is blowing over them, causing the rotor to spin. The **blade pitch system** enables blades to turn out of the wind to control the rotor speed, and to keep rotor from turning in winds that are too high or too low to produce electricity.

An anemometer and a wind vane are placed on top of wind turbine to measure the wind speed and direction. The wind speed and direction data are transmitted to the controller, which starts the turbine at wind speeds of about 3.5–7 m/s and shuts it off at about 25 m/s. The upper limit of the wind speed is chosen to protect the wind turbine from mechanical damages due to high vibrations and stresses caused by high winds.

The rotor is connected with a low-speed shaft to a gear box. The gear box increases the rotational speeds from about 30–60 rotations per minute (rpm) to about 1000–1800 rpm, which is the rotational speed required by most generators to produce electricity.

The gear box, low- and high-speed shafts, generator, controller and brake are placed in a nacelle, which is mounted atop the tower, through the wind orientation control system. This system contains yaw drive powered by yaw motor. The **yaw drive** orients upwind the turbine to keep it facing the wind. Downwind turbines, which face away the wind, do not require the yaw drive because the aerodynamic forces turn the rotor away from the wind.

13.2.2 DARRIEUS TURBINES

The **Darrieus turbine** is an example of a **vertical-axis wind turbine**. A rotor of the turbine has a number of curved aerofoil blades mounted on a rotating shaft or a framework. When the rotor is spinning, the aerofoils are moving forward and the wind velocity vector is added to the relative air flow vector. In this way the resultant air flow creates a small positive angle of attack to the blade. This generates a force pointing perpendicularly to the resultant air flow vector, giving a positive torque to the shaft. The design and operation principles of the Darrieus turbine with straight vertical airfoils are illustrated in Fig. 13.1. The analysis of these turbines is somewhat complicated, since the lift and drag forces change periodically as the angle of attack goes through 2π radians. However, related studies show that VAWTs can achieve comparable efficiencies to HAWTs.

---

[1] www.power-technology.com/features/featurethe-worlds-biggest-wind-turbines-4154395/, accessed in July 2020.

13.2.3 SAVONIUS TURBINES

The **Savonius turbine** is a drag-type vertical-axis wind turbine, with a simple, inexpensive and robust design. The rotor has a high drag coefficient and it generates non-zero lift when the rotor turns out of the wind, causing further increase of the power coefficient. A design and principle of operation of the turbine are illustrated in Fig. 13.1.

13.3 WIND POWER RESOURCES

It has been estimated that about 0.5–3% of the 120 000 TW of solar energy entering Earth's system is transformed into winds and dissipated in the atmosphere. Much of this energy, which is on the order of 1000 TW, is dissipated in the upper atmosphere and other inaccessible locations. As a result, world's *wind resource for electric power* is much less and is estimated in a wide range from as low as 1 TW [23] to 70 TW [3] or even higher. A more recent study shows that wind power production is limited to about 1 W m^{-2} at wind farm scales larger than about 100 km^2 [1]. The wind energy is significantly varying with a location and time. Winds are preferably developing over oceans, since they are not affected by surface non-uniformities (so-called roughness), which are present on land. Since the southern hemisphere is covered mostly by oceans, winds are stronger there.

Global winds are formed by several factors, including a solar heating, a spherical shape of the Earth, and Earth rotation. The global circulation of Earth's atmosphere is shown in Fig. 13.2.

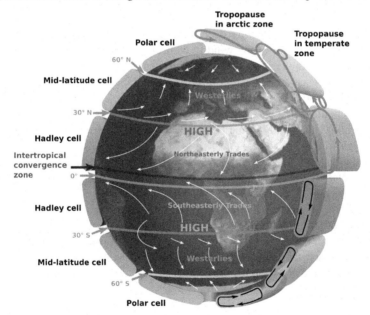

Figure 13.2 Global circulation in Earth's atmosphere. Source: Wikimedia Commons.

Hadley cells determine winds on both sides of the equator, producing rather gentle and reliable *trade winds* in the horse latitudes (latitudes from 0° to ±30°). The strongest westerly winds are produced in mid-latitude cells on both hemispheres, from 30° to 60°. The annual wind speed distribution, averaged over 10 years between July 1983 and June 1993, is shown in Fig. 13.3. The figure demonstrates a significant non-uniformity of averaged global winds in Earth's atmosphere. In-land winds are generally weaker than winds over oceans and vary between 1 to around 7 m/s. The strongest winds are on the southern hemisphere at around 60° latitude and locally exceed 12 m/s.

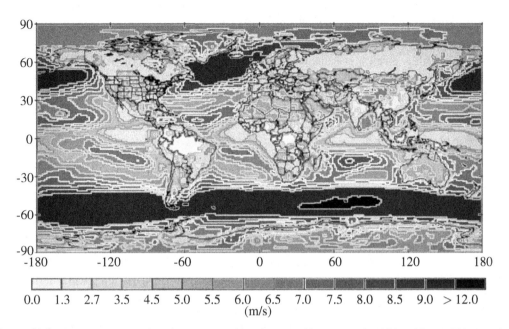

Figure 13.3 Annual wind speed 50 m over ground level averaged between July 1983 and June 1993. Based on NASA/SSE through Wikimedia Commons.

13.4 WIND CHARACTERISTICS

As the wind power is proportional to the cubic wind speed, it is crucial to have detailed knowledge of the site-specific wind characteristics, such as wind direction and speed, and their variation with time.

13.4.1 TEMPORAL VARIABILITY OF WIND

Wind speed varies significantly with time on many scales, starting from fraction of seconds and ending on many years. It is convenient to represent wind temporal variation on plots, where a fluctuation intensity is expressed in terms of a fluctuation frequency. Such plot, referred to as the **wind fluctuation spectrum**, has four notable peaks and a valley. The peaks correspond to wind variations on a few second, one day, four to five days, and one year scales. The valley represents less significant fluctuations at scales between one minute and several hours.

The high-frequency, short-term fluctuations in wind speed are referred to as the **wind turbulence**. These rapid fluctuations are experienced as gusts and lulls in the wind, and are mainly sources of stress on wind turbines. The peak at one day scale corresponds to the **diurnal wind variation**, which is caused by solar heating and nighttime cooling. These daily changing conditions are behind the land and sea breezes and mountain and valley winds. The peak around 4–5 days is caused by synoptic systems, such as cyclones and anticyclones. The annual peak with the frequency of one year corresponds to seasonal variations of insolation of southern and northern hemispheres. This causes strong winter winds in middle latitudes and the monsoons of the tropics.

The valley is also called the **spectral gap** and represents a noticeable low intensity of fluctuations with frequencies between one minute and several hours.

13.4.2 GLOBAL CIRCULATION IN ATMOSPHERE

The major driving force of atmospheric circulation is the uneven distribution of solar heating across the Earth, which is greatest near the equator and lesser at the poles. The atmospheric circulation transports energy poleward, thus reducing the resulting equator-to-pole temperature gradient. The mechanisms by which this is accomplished differ in tropical and extra-tropical latitudes.

Hadley cells exist on either side of the equator. Each cell encircles the globe latitudinally and acts to transport energy from the equator to about the 30th latitude. The circulation exhibits the following phenomena [40]:

- Warm, moist air converging near the equator causes heavy precipitation. This releases latent heat, driving strong rising motions.
- This air rises to the tropopause, about 10–15 kilometers above sea level, where the air is no longer buoyant.
- Unable to continue rising, this sub-stratospheric air is instead forced poleward by the continual rise of air below.
- As air moves poleward, it both cools and gains a strong eastward component due to the Coriolis effect and the conservation of angular momentum. The resulting winds form the subtropical jet streams.
- At about 30° latitude on either side of the equator, the jet streams become so much faster than the surface wind speed that baroclinic instability prevents the Hadley circulation from extending further poleward. This coincides with the beginning of the Ferrel cells.
- At this latitude, the now cool, dry, high altitude air begins to sink. As it sinks, it warms adiabatically, decreasing its relative humidity. Near the surface, a frictional return flow completes the loop, absorbing moisture along the way. The Coriolis effect gives this flow a westward component, creating the trade winds.

The Hadley circulation exhibits seasonal variation and the upward branch of the Hadley cell occurs not directly over the equator but rather in the summer hemisphere. In the annual mean, the upward branch is slightly offset into the northern hemisphere, making way for a stronger Hadley cell in the southern hemisphere. This evidences a small net energy transport from the northern to the southern hemisphere.

The Hadley system provides an example of a thermally direct circulation. The thermodynamic efficiency of the Hadley system, considered as a heat engine, has been relatively constant over the 1979–2010 period, averaging 2.6%. Over the same interval, the power generated by the Hadley regime has risen at an average rate of about 0.54 TW per year; this reflects an increase in energy input to the system consistent with the observed increasing of tropical sea surface temperatures.

Overall, mean meridional circulation cells such as the Hadley circulation are not particularly efficient at reducing the equator-to-pole temperature gradient due to cancellation between transports of different types of energy. In the Hadley cell, both sensible and latent heat are transported equatorward near the surface, while potential energy is transported above in the opposite direction, poleward. The resulting net poleward transport is only about 10% of this potential energy transport. This is partly a result of the strong constraints imposed on atmospheric motions by the conservation of angular momentum [40].

13.4.3 SYNOPTIC SCALE WINDS

In meteorology, **synoptic** refers to weather observations that are made simultaneously around the world, to give a snapshot of the weather at that time. Synoptic scale is considered to be weather features with horizontal diameters of about 700 to 4000 km. Data collected from weather stations are used to draw weather maps, which usually show regions of high pressure (highs or anticyclones), regions of low pressure (lows or cyclones), fronts, and winds.

Winds within high-pressure areas flow outward from their centers toward the low pressure areas at their peripheries. In low-pressure areas the directions of winds are reversed and they are flowing from the high-pressure peripheries to the low-pressure centers. Winds are usually stronger around low pressure areas. Their directions, as seeing from above, depend on the hemisphere. High-pressure systems on the northern hemisphere and low-pressure systems on the southern hemisphere rotate clockwise. Correspondingly, low-pressure systems on the northern hemisphere and high-pressure systems on the southern hemisphere rotate anticlockwise.

13.4.4 DIURNAL WIND CHANGES

In areas where the wind flow is light, sea breezes and land breezes are important factors in a location's prevailing winds. The sea is warmed by the solar radiation to a greater depth than the land and therefore has a greater capacity for absorbing heat than the land. As the temperature of the surface of the land rises, the land heats the air above it. The warm air is less dense and so it rises. This rising air over the land lowers the sea level pressure by about 0.2%. The cooler air above the sea, now with higher sea level pressure, flows toward the land into the lower pressure, creating a cooler breeze near the coast.

The strength of the sea breeze is directly proportional to the temperature difference between the land mass and the sea. If an offshore wind of 4 m/s or more exists, the sea breeze is not likely to develop.

At night, the land cools off more quickly than the ocean due to differences in their specific heat values, which forces the daytime sea breeze to dissipate. If the temperature onshore drops below the temperature offshore, the pressure over the water will be lower than that of the land, establishing a land breeze, as long as an onshore wind is not strong enough to oppose it.

Mountain and valley breezes form through a process similar to sea and land breezes. During the day, the sun heats up mountain air rapidly while valley remains relatively cooler. Convection causes it to rise, causing a valley breeze. At night, the process is reversed. During the night the slopes get cooled and the dense air descends into the valley as the mountain wind.

13.4.5 MODELING WIND SPEED VARIATION

Temporal wind speed distribution is generally modeled well with the **Weibull distribution**,

$$f(\upsilon, k, \lambda) = \frac{k}{\lambda} \left(\frac{\upsilon}{\lambda} \right)^{k-1} e^{-(\upsilon/\lambda)^k}, \tag{13.4}$$

where υ is a random variable representing the wind speed, which is allowed to vary in a range $0 \leq \upsilon < \infty$. The parameter λ characterizes the scale of typical wind speeds (a larger value of λ indicates a windier location) and k is a parameter to describe the variability of wind at the specific site, where smaller k corresponds to greater variability.

The function $f(\upsilon)$ is a wind frequency distribution for a given site. For each value of the speed υ, the function $f(\upsilon)$ corresponds to the fraction of the time that the wind speed lies between υ and $\upsilon + d\upsilon$. Thus, the probability $P(\upsilon)$ that the wind speed is less than υ can be calculated from the following integral,

$$P(\upsilon) = \int_0^\upsilon f(\upsilon')d\upsilon'. \tag{13.5}$$

This is known as the cumulative function of the probability distribution. Using the Weibull probability distribution, we get

$$P(\upsilon, k, \lambda) = \int_0^\upsilon f(\upsilon', k, \lambda)d\upsilon' = 1 - e^{-(\upsilon/\lambda)^k}. \tag{13.6}$$

Frequently we need to calculate moments of a wind speed distribution function. For example, to calculate the mean wind speed, the first moment is calculated. In general, the n-th moment of the Weibull distribution is given by,

$$\mu_n \equiv \int_0^\infty v^n f(v, k, \lambda) dv = \lambda^n \Gamma(1 + n/k), \tag{13.7}$$

where Γ is a special function known as the Gamma function[2] $\Gamma(x)$. The mean wind speed is now found as,

$$\langle v \rangle \equiv \mu_1 = \lambda \Gamma(1 + 1/k). \tag{13.8}$$

The relative power content of the wind is determined by the third moment, which is equivalent to the mean of the cube of the wind speed and can be found as,

$$\langle v^3 \rangle \equiv \mu_3 = \lambda^3 \Gamma(1 + 3/k). \tag{13.9}$$

Example 13.1:

Wind frequency distributions were measured at two sites and modeled with the Weibull distribution. The Weibull parameters were found to be $\lambda = 7.8$ m/s and $k = 1.75$ for site A and $\lambda = 5.8$ m/s and $k = 1.55$ for site B. Which of the two sites has higher mean wind speed and power potential, and by how much?

Solution

The mean wind speeds for both sites are found from Eq. (13.8) and are as follows:

$$\langle v \rangle_A = 7.8 \Gamma(1 + 1/1.75) = 7.8 \times 0.891 \cong 6.93 \text{ m/s},$$

$$\langle v \rangle_B = 5.8 \Gamma(1 + 1/1.55) = 5.8 \times 0.899 \cong 5.21 \text{ m/s}.$$

The mean speed ratio is $\langle v \rangle_A / \langle v \rangle_B = 6.93/5.21 \cong 1.33$ Thus, the mean wind speed at site A is 33% higher than at site B. The wind power potential is expressed with the mean of the cube of the wind speed at each site:

$$\langle v^3 \rangle_A = (7.8)^3 \Gamma(1 + 3/1.75) \cong 475 \times 1.55 \cong 736 \text{ (m/s)}^3,$$

$$\langle v^3 \rangle_B = (5.8)^3 \Gamma(1 + 3/1.55) \cong 195 \times 1.91 \cong 372 \text{ (m/s)}^3.$$

The ratio of these quantities is $\langle v^3 \rangle_A / \langle v^3 \rangle_B = 736/372 \cong 1.98$. This result indicates that, even though site A has only 33% higher mean wind speed, its wind power potential is 98% higher than at site B.

The Weibull distribution is a convenient and accurate representation of the wind speed variation, provided experimental data are available to fit λ and k parameters. Sometimes, however, when data are insufficient to fit both parameters, the parameter k is fixed to $k = 2$. In this case the Weibull distribution reduces to the **Rayleigh distribution**. The mean wind speed and the mean value of the cube of the wind speed can be expressed in terms of λ parameter as $\langle v \rangle = \lambda \Gamma(1.5) \cong 0.886\lambda$ and $\langle v^3 \rangle = \lambda^3 \Gamma(2.5) \cong 1.33\lambda^3$, respectively.

Wind speed increases with height through the surface layer of the atmosphere and can be approximated with the following logarithmic function,

$$U(z) = U_{ref} \frac{\ln(z/z_0)}{\ln(z_{ref}/z_0)}, \tag{13.10}$$

where z_0 is the **roughness length** to characterize the nature of the landscape, z_{ref} is a reference height, and U_{ref} is the wind speed at the reference height. Typical values of the roughness length for various landscape types are given in Table 13.1.

---

[2]See Appendix D.6.2

Table 13.1

Roughness Length for Various Terrains

Landscape Type	Roughness Length, m
very smooth (sand or ice)	10^{-5}
calm open sea	0.0002
blown sea	0.0005
snow surface	0.003
lawn surface	0.008
rough pasture	0.01
fallow field	0.03
crops	0.05
few trees	0.1
many trees, hedges, few buildings	0.25
forests and woodlands	0.5
suburbs	1.5
city centers, tall buildings	3.0

Source: [53]

13.4.6 WIND ROSE—WIND DIRECTION AND INTENSITY

A **wind rose** is a graphic tool used by meteorologists to give a succinct view of how wind speed and direction are typically distributed at a particular location. The compass is divided up into wedges to display the probability of wind from each direction. For each compass direction, the length of the wedge indicates the probability that the wind will be found in a given direction. The wedge is further divided in the radial direction into several bins to represent speed in a specific range.

Wind direction and intensity distributions are important in designing and siting wind farms. The v^3 weighted wind direction reveals the direction from which the most of the wind power will be harvested. Since wind turbines in a farm have significant downstream shadows, the farm design can be optimized to minimize the shadows for directions of the highest wind power potential.

Example 13.2:

At a certain site wind blows 75% of the time gently ($\lambda = 2.5$ m/s, $k = 1.75$) from the southwest, but 25% of the time strongly ($\lambda = 8.5$ m/s, $k = 1.75$) from the northwest. Calculate the wind power potential from each of the two directions and recommend how the wind farm should be aligned.

Solution

The wind power potential is expressed with the mean of the cube of the wind speed for each direction:

$$\langle v^3 \rangle_{SW} = (2.5)^3 \Gamma(1 + 3/1.75) \cong 15.6 \times 1.55 \cong 24.2 \ (\text{m/s})^3,$$

$$\langle v^3 \rangle_{NW} = (8.5)^3 \Gamma(1 + 3/1.75) \cong 614 \times 1.55 \cong 952 \ (\text{m/s})^3.$$

The relative wind power potential for southwest wind is thus $24.2 \times 0.75 \cong 18.2$ (m/s)^3, and for northwest wind $952 \times 0.25 \cong 238$ (m/s)^3. This result indicates that, despite the prevalence of southwest winds, the wind farm should be aligned to best capture the winds from the northwest.

13.5   WIND TURBINE AERODYNAMICS

For a turbine working element, which is moving with velocity $\mathbf{v}$ under action of a force $\mathbf{F}$, the power N that can be extracted is given as,

$$N = \mathbf{F} \cdot \mathbf{v}. \tag{13.11}$$

In realistic wind turbine design calculations[3], detailed distributions of the forces acting on turbine blades and of the corresponding velocities are required. For that purpose dedicated computational methods are used, such as the Blade Element Momentum Method (BEM), for fast, but still accurate predictions, or numerical models based on Euler and Navier-Stokes equations, suitable for yaw analysis and for prediction of interactions between wind turbines in parks [39].

A simple but important tool for understanding wind turbine aerodynamics is the **one-dimensional momentum theory**, from which the limit of obtained power can be derived. Consider an axial steady flow of air with unperturbed speed U_0 that passes through an idealized planar machine, called an **actuator disk**, as shown in Fig. 13.4. The speed of air far downstream the actuator disk, where the air has regained its undisturbed pressure, is U_1. We choose a control volume bounded by streamlines along its periphery called a **stream tube**. In this control volume we will apply mass, momentum, and energy conservation equations to derive an expression for the extracted power by the actuator disk.

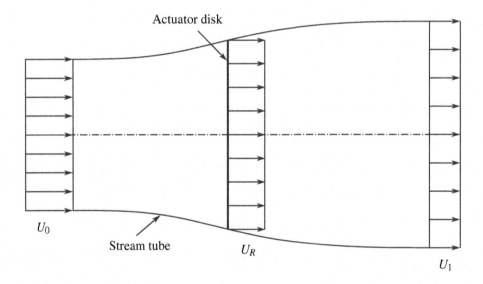

Figure 13.4   Stream tube used to analyze wind turbine performance.

The mas flow rate W of air flowing through the control volume is,

$$W = \rho U_0 A_0 = \rho U_R A_R = \rho U_1 A_1 \tag{13.12}$$

where U_R and A_R is the air velocity passing through the actuator disk (rotor) and the actuator disk surface area, respectively.

The axial momentum balance yields the following expression for the thrust T,

$$T = W(U_0 - U_1) = \rho U_R A_R (U_0 - U_1). \tag{13.13}$$

---

[3]IEC 61400-1:2019 specifies essential design requirements to ensure the structural integrity of wind turbines.

From the Bernoulli equation we find the pressure drop Δp as follows,

$$\Delta p = \frac{1}{2}\rho \left(U_0^2 - U_1^2 \right). \tag{13.14}$$

It is reasonable to assume that the pressure drop given by Eq. (13.14) occurs across the actuator disk and it is related to the thrust as,

$$T = A_R \Delta p. \tag{13.15}$$

Combining Eq. (13.13) through Eq. (13.15) yields the air velocity at the actuator disk:

$$U_R = \frac{1}{2} \left(U_1 + U_0 \right). \tag{13.16}$$

This result is consistent with the **Rankine propeller theory** which predicts that half the velocity change occurs upstream and half downstream from the propeller plane.

13.5.1 MAXIMUM POWER OF A WIND TURBINE

Let us introduce the **axial interference factor** defined as,

$$a = \frac{U_0 - U_R}{U_0}. \tag{13.17}$$

We can now express the air velocity at various locations in the stream tube in terms of unperturbed air velocity U_0 and the interference factor a as follows,

$$U_R = (1 - a)U_0, \tag{13.18}$$

$$U_1 = (1 - 2a)U_0. \tag{13.19}$$

The thrust and retrieved power can be written as,

$$T = 2\rho A_R U_0^2 a(1 - a), \tag{13.20}$$

$$N = U_R T = 2\rho A_R U_0^3 a(1 - a)^2. \tag{13.21}$$

It is customary to introduce the dimensionless thrust and power coefficients as follows,

$$C_T \equiv \frac{T}{\frac{1}{2}\rho A_R U_0^2} = 4a(1 - a), \tag{13.22}$$

$$C_P \equiv \frac{N}{\frac{1}{2}\rho A_R U_0^3} = 4a(1 - a)^2 \tag{13.23}$$

The maximum power that can be obtained for a given unperturbed air speed and a given rotor diameter depends on the axial interference factor only and corresponds to the maximum of the power coefficient. Differentiating the power coefficient with respect to the axial interference factor and equating to zero yields,

$$a_{max} = \frac{1}{3}, \quad C_{Pmax} = \frac{16}{27} \cong 0.593. \tag{13.24}$$

Similarly, the maximum thrust could be found to occur for $a = \frac{1}{2}$, for which the downstream velocity U_1 would be zero and the stream tube area A_1 would diverge. In a real situation the downstream wake will have a finite area. The discrepancy between the theory and the real behavior is due to the neglect of turbulence and vorticity. Nevertheless, this simple theory provides a valuable solution for ideal and lossless flow conditions. Thus, under such assumptions, the maximum power of a wind turbine is as follows,

$$N_{max} = \frac{16}{27} \cdot \frac{1}{2} \rho A_R U_0^3. \tag{13.25}$$

The condition for the maximum power of a wind turbine given by Eq. (13.24) is usually referred to as the **Betz limit**[4]. It states that the upper maximum energy, which can be extracted and converted into useful energy, is less than 59.3% of the kinetic energy contained in a stream tube having the same cross section area as the actuator disk.

13.5.2 WIND TURBINE EFFICIENCY

The wind turbine efficiency can be obtained by dividing the wind turbine power given by Eq. (13.21) with the power of unperturbed air flowing with speed U_0 through area A_0 as follows,

$$\eta = \frac{2\rho A_R U_0^3 a(1-a)^2}{\frac{1}{2}\rho A_0 U_0^3} = 4a(1-a)^2 \frac{A_R}{A_0} = 4a(1-a). \tag{13.26}$$

Substituting $a_{max} = 1/3$, the corresponding wind turbine efficiency at the maximum power is,

$$\eta_{max} = \frac{8}{9} \cong 0.889. \tag{13.27}$$

Interestingly, according to Eq. (13.26), the wind turbine efficiency could be further increased by increasing a toward $1/2$, but this would decrease the total power produced by the turbine.

13.6 ENVIRONMENTAL EFFECTS OF WIND POWER

Wind power has rather fewer effects on the environment than many other energy technologies. Wind turbines do not release emissions once generating electricity and they do not need water for cooling. However, modern wind turbines can be very large machines and they may visually affect the landscape. In particular, they can generate noise and cause shadow flickering. They can have negative impact on birds and other sensitive species. Most wind power projects on land require service roads that add to the physical effect on the environment. Producing the metals and other materials used to make wind turbine components has impacts on the environment. Since wind turbines have a shorter life span than most other technologies, and their rotor blades are huge and are made of a mix of resin and fiberglass, their decommissioning creates an additional environmental concern.

13.6.1 NOISE

Wind turbine noise is of great concern when the turbines are placed in rural environments. The noise is generated by machinery and aerodynamic effects. The machinery noise is generally not as important as the aerodynamic noise, since it can be relatively easy reduced by proper insulation of the nacelle. The aerodynamic noise is determined by the speed of a tip of wind turbine blade. As a rule of thumb, it increases with the fifth power of that speed.

The noise level is measured in decibels (dB), weighted by the sensitivity of a human ear. The **sound power** is the power of the sound at the source, whereas the **sound pressure** refers to the power of the sound at the receiver at a distance. The sound power can be interpreted as the total sound energy emitted by a source per unit time. This is a theoretical quantity that is not measurable

---

[4]Known also as the Lanchester-Betz-Joukowsky limit.

and is expressed in watts. Using a reference sound power $P_0 = 10^{-12}$ watts (the lowest sound persons with excellent hearing can discern), the **sound power level**, expressed in dB, is defined as,

$$L_W = 10 \log (P/P_0). \tag{13.28}$$

The **sound pressure level** quantifies in decibels the strength of a given sound source. It is defined as,

$$L_p = 20 \log (p/p_0), \tag{13.29}$$

where $p_0 = 2 \times 10^{-5}$ Pa = 20 μPa is the reference sound pressure.

Sound pressure vary substantially with distance from the source and can be found as,

$$p = \sqrt{\frac{Q \cdot \rho \cdot c \cdot P_{ac}}{4\pi \cdot r^2}}, \tag{13.30}$$

where p—sound pressure in Pa, ρ—air density in kg/m³, c—speed of sound in m/s, P_{ac}—sound power in W, r—distance from the sound source in m, and Q—sound directivity factor (equal to 1 for a spherical sound propagation, 2 for half-spherical sound propagation, 4 for quarter-sphere sound propagation, etc.). The formula for converting the sound power level L_W (in dB) to sound pressure level L_P (in dB) is as follows,

$$L_p = L_W - 10 \times \log \left(\frac{Q}{4\pi r^2} \right). \tag{13.31}$$

Models with various levels of approximation are used to calculate wind turbine noises. Some of the models are quite simplistic, and can be used as an initial estimation of the noises. There are also attempts to predict the noises using complex CFD solvers, which can be applied to realistic rotors and flow conditions. However, the most commonly used models are based on semi-empirical relationships, derived from experimental data and scaling laws. In general, mechanical noises are not considered and the following five mechanisms are studied:

- turbulent boundary layer trailing edge noise,
- separation-stall noise,
- laminar boundary layer vortex shedding noise,
- tip vortex formation noise,
- trailing edge bluntness vortex shedding noise.

The scaling laws for the different mechanisms are assumed to have the following form,

$$L_{p,i} = 10 \times \log \left(\frac{\delta_i^* M^{r_i} L Q}{r^2} \right) + F_i(\text{St}) + G_i(\text{Re}), \tag{13.32}$$

where δ_i^*—boundary layer displacement thickness, M—Mach number, r_i—exponent that depends on the particular noise mechanism i, L—airfoil section semi-span, Q—sound directivity factor, and r—distance to the observer. The additional terms F_i and G_i are functions of the Strouhal number St and the Reynolds number Re.

Based on analyses and calculations, wind farm noise emission maps are created that show the sound pressure levels at different points and distances from the farm. A modern wind turbine has a sound power level of 90–100 dB. In 350–1000 m distance, the sound pressure level is less than 45 dB. Generally accepted noise levels for wind turbines are 35 dB during night and 45 dB during the day in front of an open window. Wind turbines, especially those with downwind rotors, produce infra-sounds (< 20 Hz) below human perception. However, presently there are no specific recommendations as far as their allowable level is concerned.

In 2018 World Health Organisation (WHO) issued guidelines for the European region concerning environmental noises, their adverse effects, and their prevention and control[5]. According to the guidelines the noise levels produced by wind turbines should be reduced below 45 dB, as wind turbine noise above this level is associated with adverse health effects. No recommendation was made for average night noise exposure due to poor quality of evidence of night-time exposure to wind turbines.

13.6.2 SHADOW FLICKER

Wind turbines cast a shadow on their vicinity in direct sunlight. As the blades are turning they may cut through light beams, causing a flickering effect. Typical flickering frequency is below 2 Hz and can be annoying for affected population. The location of the turbine shadow varies by time of the day and season. Shadow flicker has been a concern in Northern Europe where the high latitude and low Sun angle intensify the effect. Medical research has shown that a flicker rate of three flashes per second or slower has a very low health risk. Guidelines for wind farm development follow this observation and recommend a flicker rate of less than three per second (3 Hz). On a typical three-bladed wind turbine, this would correspond to a rotation speed of one complete rotation per second (60 rpm).

Shadow flicker can be addressed in various ways, including landscaping to block the shadows or stopping the turbines during the sensitive times. However, there is no standard regulation on the limit flickering time. Many municipalities regulate the amount of wind turbine flicker through zoning ordinances that require wind turbine owners to analyze shadow flicker to determine where the shadows would fall and for how long over the course of one year. Typically a limit of 30 hours of shadow flicker on a occupied building per year is mandated.

13.6.3 VISUAL IMPACT

Due to their height, wind turbines are highly visible structures in any landscape. Such factors as the distance, the angle of the light, the contrast, and the movement of the blades affect the perceptibility of a wind farm. In general too many wind turbines can ruin a landscape. The number of wind turbines should be limited (and sometime one bigger wind turbine is a better choice than several small) to such a level, that their visual impact is still acceptable.

Unlike with noise and shadow flicker, there are no general rules to determine the visual impact of proposed wind farms. Methods using photo montages, 2D or 3D animations, or maps with zones of visual influence can be useful to determine the impact.

In a photo montage, pictures are taken from certain view points, which could be houses or roads, and the planned wind turbines are superimposed into them. The visual impact can be assessed by comparing the pictures with and without superimposed wind turbines.

Maps with zones of visual influence are created to show how many turbines would be visible from each point. It is generally accepted that the impact of wind farms that are further away than 30 km will be negligible, as the human eye will not register them in the field of view from that distance. However, at shorter distances, especially in the high impact zone (within a 4 km radius), a detailed analysis is necessary.

13.6.4 BIRD COLLISIONS

Wind turbines should be placed in location where a probability of bird collisions is as low as possible. Wind farms should be located far from bird conservation areas and breeding grounds of sensitive species. In addition to direct collisions, the huge pressure changes around wind turbines can cause

---

[5]http://www.euro.who.int

serious internal damages to some birds, and especially to bats. The risk of collisions with birds can be reduced by,

- increasing the visibility of rotor blades,
- siting wind farms away from bird migration corridors,
- using warning flashing lights (which should be white, rather than red).

13.6.5 SITE PLANNING

The site planning has to take into account effects of noise, shadow flickering, visual impact, disturbance to birds, and other factors that may influence the environment around the future wind farm.

Best practices for wind site development have emerged and are often enforced by the licensing authority. Typical rules are expressed as minimum distances from wind turbines to affected objects. Examples of the minimum distances are as follows,

- to major roads 50 m,
- to power lines 100 m,
- to parks and leisure zones 200 m,
- to detached houses 500 m,
- to urban settlements 1000 m.

PROBLEMS

PROBLEM 13.1

Calculate the specific energy (energy per unit air mass flowing through the area occupied by a single wind turbine rotor) for a wind turbine operating with the maximum power given by the Betz limit, when the unperturbed wind speed is $U = 8$ m/s. Assume the interference factor $a = 1/3$.

PROBLEM 13.2

Find the maximum power that can be generated by the Savonius turbine with radius $R = 2$ m and height $H = 3$ m when unperturbed wind speed is $U_0 = 15$ m/s and air density is $\rho = 1.25$ kg/m³.

PROBLEM 13.3

Wind speed, measured at a reference distance from the ground $z_{ref} = 25$ m, is equal to $U_{ref} = 15$ m/s. Find the area-mean wind speed for HAWT rotor with axis at 120 m above the ground and the rotor diameter 90 m. Assume the roughness length equal to 1 m.

PROBLEM 13.4

A cylindrical drum with height 1 m and diameter 0.75 m is cut along the height into two identical pieces, such that both pieces have identical drag coefficients equal to 2.3 when the open side is facing flow, and 1.2 when the open side is facing downstream. Construct a windmill rotor similar to the Savonius rotor by attaching the two parts reversed along the shorter or the longer edge. Estimate which design will give higher power and by how much.

PROBLEM 13.5

Derive an expression for the power coefficient C_P of a drag machine of the Savonius design in terms of the drag coefficient c_D and the **tip-speed ratio** $\lambda = \Omega R/U$, where Ω is the angular velocity of the rotor, R is the rotor radius, and U is the incoming axial wind speed. Find the maximum value of the power coefficient.

PROBLEM 13.6

A wind pump based on the HAWT design with blade length $R = 5$ m and tip-speed ratio $\lambda = \Omega R/U = 1$, where Ω is the angular velocity of the blades, and U is the incoming axial wind speed, is used for pumping water. Estimate what is the theoretical maximum amount of water that can be pumped during one hour from a depth of 30 m when the wind speed is 6 m/s.

14 Solar Power

Even though solar panels are a relatively new technology, dating back to the 1970s, humans have started harnessing Sun's power long time ago. Ancient civilizations from the 7th century B.C. used mirrors to light torches for religious rituals and even for military purposes. Homes and bath houses captured sunlight for heating and lighting purposes.

The first patents for solar collectors and solar water heaters started appearing in the eighteen century, and in 1839 Edmond Becquerel[1] discovered the *photovoltaic effect* by experimenting with electrolytic cells and observing that more electricity was produced when the cells were exposed to sunlight. In 1876 a group of researchers discovered that selenium could turn light into electricity without heat. This element was later used to create the first solar cell in 1883. However, a real breakthrough in the development of solar cells came in 1954, when the first silicon photovoltaic cell was invented[2]. Its solar efficiency reached only 6%, but, unlike its selenium predecessor, it managed to generate enough electricity to power electrical equipment. However, the first solar cells were still prohibitively expensive for the public to purchase.

Since the 1970s through the 1990s solar energy was increasing rather slowly, as solar manufactures continued making solar cells smaller and less expensive. Only after the 2000s the higher efficiency accompanied with lower prices of solar panels encouraged more widespread adoption of solar energy.

14.1 SOLAR RADIATION ON EARTH

Radiant energy of the Sun is produced in fusion reactions that were described in §6.2. The luminosity of the Sun, which is an absolute measure of Sun's radiant power, is $L_\odot = 3.828 \times 10^{26}$ W. Since the radius of the Sun is $R_\odot = 6.957 \times 10^5$ km, the average energy flux at the Sun surface can be found as $I_s = L_\odot/(4\pi R_\odot^2) \cong 63.11$ MW/m^2. Similarly, the average radiant energy flux on Earth's orbit, often referred to as the **solar constant**, is,

$$I_\odot = \frac{L_\odot}{4\pi R_{\text{orb}}^2} \cong 1366 \text{ W/m}^2, \tag{14.1}$$

where $R_{\text{orb}} = 149.4 \times 10^6$ km is the mean radius of Earth's orbit around the Sun.

The total energy incident on Earth's atmosphere can be computed as a product of the solar constant and the projected cross-section area of Earth as,

$$q_{\text{insol}} = \pi R_\oplus^2 I_\odot \cong 173\,000 \text{ TW}, \tag{14.2}$$

where $R_\oplus \approx 6\,378$ km is Earth's mean equatorial radius. Thus, about 173 000 TJ of energy from the Sun enters Earth's atmosphere every second as electromagnetic radiation. Since around 30% of this energy is reflected back into space by Earth's atmosphere and surface, 120 000 TJ is the net solar energy absorbed and transformed into other forms of energy on Earth. Even though this amount of energy is still enormous, the quantity of solar energy available at any particular location is highly dependent on latitude, time, and atmospheric conditions.

---

[1] French physicist (1820–1891) who studied the solar spectrum, magnetism, electricity, and optics.

[2] American scientists Daryl Chaplin, Calvin Fuller, and Gerald Pearson developed the silicon photovoltaic cell at Bell Labs in New Jersey.

DOI: 10.1201/9781003036982-14

14.1.1   ENERGY OF THE SUNLIGHT

The Sun radiates the electromagnetic energy because its surface temperature is high. The relationship between the radiated energy, q_{rad}, and the radiating body temperature, T, is described by the **Stefan-Boltzmann law**,

$$q_{rad} = \sigma \varepsilon A T^4, \tag{14.3}$$

where A is the emitting body's surface area, σ is the **Stefan-Boltzmann constant**, $\sigma = 5.670\,374\,419...\times 10^{-8}$ W·m^{-2}·K^{-4}, and ε is the emissivity of the material. Some materials, called black bodies, absorb and re-emit all radiations that impinges them, and their radiation is known as black-body radiation. The Sun radiates approximately like black body with emissivity $\varepsilon_{sun} \approx 1.0$. Using Eq. (14.3) and taking $q_{rad} = L_\odot$ we can calculate the effective temperature of the Sun at its surface as,

$$T_\odot = \left(\frac{L_\odot}{4\pi \sigma R_\odot^2} \right)^{1/4} \cong 5780 \text{ K}. \tag{14.4}$$

The spectral distribution of the thermal energy radiated by a blackbody, or the pattern of the intensity of the radiation over a range of wavelengths or frequencies, depends on its temperature only. The spectral energy density of the emission at each wavelength and at particular temperature is given by **Planck's law of blackbody radiation**,

$$B_V(v,T) = \frac{2hv^3}{c^2} \frac{1}{e^{\frac{hv}{k_B T}} - 1}, \tag{14.5}$$

where $B_V(v,T)$—spectral radiance, h—Planck constant, v—frequency of the electromagnetic radiation, c—speed of light in vacuum, k_B—Boltzmann constant, T—absolute temperature of the body.

As solar irradiation passes into Earth's atmosphere, molecules in the atmosphere absorb radiation of certain frequencies associated with their excitation spectra. Additional light is redistributed by Raleigh scattering, which is responsible for the atmosphere's blue color. Solar radiation spectra, which are different at different locations in the atmosphere, and a corresponding blackbody spectrum are shown in Fig. 14.1. The figure shows the solar radiation spectrum for direct light at both the top of the Earth's atmosphere (AM0) and at sea level (AM1.5). The Sun produces light with a distribution similar to what would be expected from a 5780 K blackbody. These curves[3] are standards adopted by the photovoltaics industry to ensure consistent test conditions. Such standards are necessary to accurately estimate the efficiency and power output of solar energy systems, in particular photovoltaic systems.

The **air mass m spectrum (AMm)** is defined as the spectrum of sunlight passing through a mass of air M_a such that $m = M_a / M_{a0}$, where M_{a0} is the mass of air passed by normally incident sunlight at sea level. With incidence angle ψ_z from the normal at sea level, this ratio, often refereed to as the **air mass number**, can be approximated as,

$$m = \frac{1}{\cos \psi_z}. \tag{14.6}$$

The above definition indicates that at the top of Earth's atmosphere $m = 0$, since $M_a = 0$, whereas at the sea level for normally incident sunlight $m = 1$, since $M_a = M_{a0}$. Using the approximation given by Eq. (14.6), we find that $m = 1.5 = 1/\cos \psi_z$, thus $\psi_z = 48.19°$.

When the air mass number needs to be determined with higher accuracy, the following correlation can be used [55],

$$m = \left[\cos \psi_z + \frac{0.50572}{(96.07995 - \psi_z)^{1.6364}} \right]^{-1}. \tag{14.7}$$

---

[3]National Renewable Energy Laboratory (NREL), Reference solar spectral irradiance: air mass 0 and air mass 1.5 standards, available online at https://www.nrel.gov/grid/solar-resource/spectra.html.

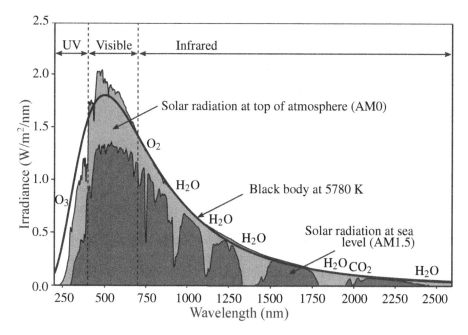

Figure 14.1 Solar radiation spectrum at the top of the atmosphere and at sea level, indicating sunlight absorption by ozon, oxygen, water vapor, and carbon dioxide.

Solar intensity at Earth's surface reduces with increasing airmass coefficient. This dependence is quite complex and is influenced by many factors, such as the presence of water vapor, aerosols, and air pollution. The last factor is particularly important toward the horizon where effects of the lower layers of atmosphere are dominating. A formula that can be used to capture the influence of the airmass on the solar intensity is as follows [66],

$$I(m) = 1.1 \times I_\odot \times 0.7^{\left(m^{0.678}\right)}, \tag{14.8}$$

where $I_\odot$ is the solar intensity external to the Earth's atmosphere equal to the solar constant.

14.1.2 SUN POSITION

Earth orbits the Sun at an average distance of 149.60 million km, and one complete orbit takes 365.256 days. Earth's orbit is an ellipse with **perihelion** (closest distance from Earth to the Sun, occurring around January 3) equal to 147.10 million km and **aphelion** (greatest distance from Earth to the Sun, occurring around July 4) equal to 152.10 million km.

Earth's axis of rotation is tilted to the orbit plane with angle of about 23.44°. The maximum tilt of the Earth's axis toward the Sun occurs near June 21 in the northern hemisphere and near December 21 in the southern hemisphere. Due to these variations, the actual position of the Sun in the sky is a function of the time of a day, Earth location in the orbit, and the geographic location on Earth's surface.

The position of the Sun can be found in three steps:

- calculate the Sun's position in the ecliptic coordinate system,
- convert to the equatorial coordinate system,
- at given time and geographical location, convert to the horizontal coordinate system.

These three different coordinate systems are celestial coordinate systems used for various applications. The ecliptic coordinate system is used for representing the apparent position and orbit of Earth and other Solar System objects. The equatorial coordinate system is used to specify the positions of celestial objects. It can be implemented in spherical or rectangular coordinates. The geocentric equatorial coordinate system has its origin at the center of Earth, a fundamental plane consisting of the projection of Earth's equator onto the celestial sphere, and a primary direction toward the vernal equinox. It follows the right-handed convention.

The horizontal coordinate system uses the observer's local horizon as the fundamental plane. Coordinates of an object in the sky are expressed in terms of two angles, corresponding to the altitude (or elevation) and azimuth.

In solar energy applications it is customary to use correlations to determine two coordinates: ψ—the altitude of the Sun, and α—the azimuth of the Sun. These coordinates are functions of the local longitude λ and latitude φ.

The **declination of the Sun** δ is the angle between the rays of the Sun reaching Earth and the plane of the Earth's equator. At solstices, the declination reaches its maxima, and at equinoxes its zero values. The simplest expression for the Sun declination is given by the following formula,

$$\delta = -23.44° \times \cos\left[\frac{360°}{365}(N+10)\right],\tag{14.9}$$

where N is the day of the year beginning with $N=0$ at midnight Universal Time as January 1 begins. This simple formula overestimates the declination near the September equinox by up to $\pm 1.5°$. For more accurate calculations, as required by solar energy applications, a more complex formula can be used,

$$\delta = -\arcsin\left\{\sin(-23.44°) \times \cos\left[\frac{360°}{365.24}(N+10) + \frac{360°}{\pi} \times 0.0167 \times \sin\left(\frac{360°}{365.24}(N-2)\right)\right]\right\},\tag{14.10}$$

where N is defined as in Eq. (14.9). Typical declination of the Sun as a function of the day of the year is shown in Fig. 14.2.

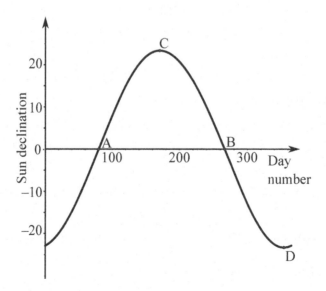

Figure 14.2 Sun declination: A—March equinox, B—September equinox, C—June solstice, D—December solstice.

In addition to the north-south solar oscillations caused by the declination, there is also east-west oscillation which causes a discrepancy between the apparent time based on the tracked Sun position and the local (wall-clock) time, with 24 hours apart. The reason of this oscillations is the tilt of the Earth's axis and the changes in the speed of its orbital motion around the Sun. The correction to remove this discrepancy is traditionally called the **equation of time** and can be found from various correlations, such as, e.g.,

$$ET = 0.0066 + 7.3525 \times \cos\left(N'_d + 85.9°\right) + 9.9359 \times \cos\left(2N'_d + 108.9°\right) \\ + 0.3387 \times \cos\left(3N'_d + 105.2°\right) \tag{14.11}$$

where ET is expressed in hours and $N'_d = 360°N_d/N_{dy}$, N_d—day number starting with $N_d = 1$ for January 1, $N_{dy} = 366$ for the leap year, otherwise $N_{dy} = 365$.

The **solar time** ST can be found from the local time LT and the equation of time ET as follows,

$$ST = LT - 4 \times (15° - \lambda)\frac{\text{min}}{°} + ET. \tag{14.12}$$

The **hour angle** is one of the coordinates used in the equatorial coordinate system to give the direction of a point on the celestial sphere. It is defined as the angle between two planes containing the Earth's axis: one containing the point and the other containing the zenith. For a given solar time ST, the hour angle of the Sun can be found as,

$$\omega = (12.00\text{h} - ST)\frac{15°}{\text{h}}. \tag{14.13}$$

The Sun coordinates in the horizontal coordinate system can now be found as follows,

$$\psi = \arcsin\left(\cos\omega\cos\varphi\cos\delta + \sin\varphi\sin\delta\right), \tag{14.14}$$

where ψ is the altitude of the Sun, ω is the hour angle, φ is the local latitude, and δ is the declination of the Sun. The azimuth α of the Sun is found as,

$$\alpha = \begin{cases} 180° - \arccos\left(\frac{\sin\psi\sin\varphi - \sin\delta}{\cos\psi\cos\varphi}\right) & \text{for } ST \leq 12.00\text{h} \\ 180° + \arccos\left(\frac{\sin\psi\sin\varphi - \sin\delta}{\cos\psi\cos\varphi}\right) & \text{for } ST > 12.00\text{h} \end{cases} \tag{14.15}$$

The Sun position in different solar energy applications is determined by dedicated solar tracking systems. Such systems play an important role not only in the power and efficiency gains compared to the fixed systems, but also in the economic analyses of the large-scale solar energy applications. There are two main types of the solar tracking systems: single-axis systems that are using a single pivot point to rotate, and dual-axis systems that are using two pivot points to rotate. The single-axis trackers are particularly useful for parabolic trough and linear Fresnel solar systems. The dual-axis trackers usually have both horizontal and vertical axes, and are applied for solar dish and solar tower systems.

The currently existing solar tracking systems are based on five different principles of operation: active, passive, semi-passive, chronological, and manual solar tracking. An active solar tracking system determines the position of the Sun in the sky during the day with sensors. These sensors trigger actuators to position the solar absorber during the day. A passive solar tracking system uses fluid thermal expansion or evaporation to move the fluid from one side of the solar absorber to the other side, and in this way to adjust the position of the absorber to maximize the power gain. The system does not use motors or actuators to control the tracking. A semi-passive solar tracking system requires a minimal mechanical effort to adjust the absorber's position by combining active and passive elements. A chronological solar tracking system moves and adjusts the absorber's position continuously during the day based on the calculated Sun position in the sky. A manual solar tracker is the simplest one, but it requires personnel action. For practical reasons, it is used only to adjust the absorber's position during different seasons of a year.

14.1.3 COMPONENTS OF SOLAR RADIATION

The solar radiation has three main components: the direct, the diffuse, and the reflected solar radiation. The direct radiation is intercepted in a direct line from the Sun. The diffuse radiation is scattered in the atmosphere by, e.g., clouds and dust particles. Reflected radiation is reflected from surface features.

14.1.4 SOLAR RADIATION ON INCLINED SURFACES

The total radiation intensity on a surface inclined with angle β to the horizontal plane is a sum of three terms,

$$I_\beta = I_{b\beta} + I_{d\beta} + I_{r\beta}, \tag{14.16}$$

where $I_{b\beta}$—direct beam radiation intensity, $I_{d\beta}$—diffuse radiation intensity, and $I_{r\beta}$—reflective radiation intensity.

For isotropic dispersed and reflected radiation, the radiation intensity on an inclined surface is given as,

$$I_\beta = I_b R_b + I_d \left(\frac{1+\cos\beta}{2} \right) + (I_b + I_d) \rho_g \left(\frac{1-\cos\beta}{2} \right), \tag{14.17}$$

where,

$$R_b = \frac{\cos(\varphi - \beta)\cos\delta\cos\omega + \sin(\varphi - \beta)\sin\delta}{\cos\varphi\cos\delta\cos\omega + \sin\varphi\sin\delta}. \tag{14.18}$$

Here φ—geographic longitude, δ—the Sun declination, ω—the hour angle, β—collector inclination angle, ρ_g—reflectivity of the ground.

14.2 SOLAR THERMAL POWER

Passive solar technology can be used for solar energy collection, storage, distribution, and heating of homes. In passive systems the solar energy is not transformed into other useful forms of energy. This technology has been used, between others, in the development of so called passive houses.

In most solar energy applications active solar systems are used. Such systems collect solar radiation and use mechanical and electrical equipment for the conversion of solar energy to heat or electric power. The active solar technologies can be broadly divided into two groups: the solar thermal technology and the photovoltaic technology. In this section we discuss the principles of design and operation of the solar thermal energy systems.

14.2.1 ABSORPTION OF RADIATION

Consider a blackbody system that exchange energy with the surroundings through radiation only. The system is in **radiative equilibrium** when it radiates energy at the same rate at which it absorbs energy. For such conditions, incoming normal radiation I_0 on flat surface with area A balances the energy radiated as blackbody radiation $A\sigma T_b^4$, where T_b is the blackbody temperature. Thus, in the radiative equilibrium the blackbody temperature is found as,

$$I_0 = \sigma T_b^4 \quad \Rightarrow \quad T_b = \left(\frac{I_0}{\sigma} \right)^{\frac{1}{4}}. \tag{14.19}$$

Consider now that incoming radiation I_0 is perfectly transmitted through a glass layer parallel to a flat absorber, but the glass layer absorbs all radiation transmitted by the absorber. The glass layer and the absorber are exchanging energy through radiation only. Assuming that the temperature of glass layer is T_g, and that it radiates heat $I_g = \sigma T_g^4$ in directions normal on each side, the heat balance for the absorber is,

$$\sigma T_b^4 = I_0 + \sigma T_g^4, \qquad (14.20)$$

and the corresponding energy balance for the glass layer is,

$$\sigma T_b^4 = 2\sigma T_g^4. \qquad (14.21)$$

Solving the above two equations for unknown T_b in terms of I_0 yields,

$$T_b = \left(\frac{2I_0}{\sigma}\right)^{\frac{1}{4}}. \qquad (14.22)$$

This result shows that the absorber temperature, when covered with a glass layer, increases with factor $\sqrt[4]{2} \cong 1.189...$ and becomes approximately 19% higher than the temperature of the bare absorber. Even though in reality other heat transfer modes such as convection and conduction are very important and cannot be neglected, the result demonstrates **heat trapping** phenomenon that exists in nature and is commonly referred to as the **greenhouse effect**.

The key condition for the greenhouse effect to occur is presence of a material that transmits one radiation but absorbs another. Such materials exist in Earth's atmosphere as **greenhouse gases**, such as water vapor and CO_2, which admit incoming solar radiation while blocking some of the outgoing infrared radiation from Earth. Nevertheless, heat trapping phenomenon can be exploited when constructing solar energy systems to increase the temperature of the solar collector.

14.2.2 COLLECTORS

Solar collectors are mainly used to collect incoming solar radiation and to transform it into heat for space heating, domestic hot water, or cooling with an absorber chiller. A **flat-plate collector** represents the simplest type of low-temperature (less than 60 °C) solar collector. It contains pipes running through a dark absorber plate. The pipes carry a fluid such as water, which evacuates heat from the collector. The fluid is actively pumped through the system to increase heat transfer efficiency. Transparent glazing over the absorber is used to trap heat and to suppress heat losses due to convection on pipe walls.

Energy balance of the flat-plate collector takes into account the solar radiation, the useful heat, and the heat losses. From the total radiation I_β incident on glass cover, a part $(\tau \cdot I_\beta)$ reaches the absorber plate where it is transformed into heat. Here τ is the transmittance of the glass cover, which, in addition, reflects in space radiation $(\rho \cdot I_\beta)$ and absorbs radiation $(\alpha \cdot I_\beta)$, where ρ and α are the glass cover reflectance and absorptance, respectively. The coefficients τ, ρ, and α depend on the incident angle and material used for glazing, and satisfy the condition,

$$\tau + \alpha + \rho = 1. \qquad (14.23)$$

The useful heat collected by the collector is,

$$q_u = Wc_p\left(T_{co} - T_{ci}\right), \qquad (14.24)$$

where W is mass flow rate of the working fluid, c_p is the specific heat of the working fluid, T_{co} and T_{ci} are the outlet and inlet temperatures of the working fluid of the collector, respectively. The instantaneous efficiency η_c of a collector operating under steady conditions is defined as,

$$\eta = \frac{q_u}{A_c I_\beta} = \frac{Wc_p\left(T_{co} - T_{ci}\right)}{A_c I_\beta}, \qquad (14.25)$$

where A_c is the collector surface area.

The useful heat of flat-plate collector can be expressed in terms of an overall heat loss coefficient U_L using the Hottel-Whillier-Bliss expression [27],

$$q_u = A_c F_R \left[S_a - U_L \left(T_{ci} - T_\infty \right) \right], \tag{14.26}$$

where T_{ci} is the collector inlet temperature, T_∞ is the ambient air temperature, F_R is the collector heat removal factor, and S_a is the radiation absorbed by the collector's absorber.

For a flat collector with the angle of inclination to the horizon β, the absorbed radiation can be found as

$$S_a = I_b R_b (\tau\alpha)_b + I_d (\tau\alpha)_d \left(\frac{1 + \cos\beta}{2} \right) + \rho_g (I_b + I_d)(\tau\alpha)_r \left(\frac{1 - \cos\beta}{2} \right). \tag{14.27}$$

Here $(\tau\alpha)_b$, $(\tau\alpha)_d$, and $(\tau\alpha)_r$ are the transmittance-absorptance products for the direct beam radiation, diffusive radiation and reflective radiation, respectively. Each $(\tau\alpha)_x$ term ($x = b, d, r$) takes into account multiple reflections of the each of the radiation component between the glazing and the absorber. It can be calculated for the radiation component x as,

$$(\tau\alpha)_x = \frac{\tau_x \alpha_x}{1 - (1 - \alpha_x)\rho_{1x}}, \tag{14.28}$$

where ρ_{1x} is the glazing reflectivity for the radiation component x reflected from the absorber.

The collector heat removal factor F_R depends on the collector geometry and design. For a flat collector with parallel pipes with outer diameter D and inner diameter D_i, attached to the absorber plate with thickness δ_p with distance L from each other, this factor is given as,

$$F_R = \frac{W c_p}{A_c U_L} \left[1 - \exp\left(-\frac{A_c U_L F'}{W c_p} \right) \right], \tag{14.29}$$

where F' is the collector efficiency factor found as,

$$F' = \frac{\frac{1}{U_L}}{L \left[\frac{1}{U_L (D + (L-D)F)} + \frac{\delta_w}{\lambda_w D} + \frac{1}{\pi D_i h_f} \right]}. \tag{14.30}$$

Here δ_w is the mean thickness of the weld between pipe and the collector plate, λ_w is the thermal conductivity of weld material, and h_f is heat transfer coefficient between the working fluid and the pipe wall. Parameter F is the standard fin efficiency given as,

$$F = \frac{\tanh[m(L-D)/2]}{m(L-D)/2}, \quad m = \sqrt{\frac{U_L}{\lambda_p \delta_p}}. \tag{14.31}$$

Most of roof-top collectors have a fixed angle. To maximize energy collected through the day, a flat-plate collector should be placed at an angle facing the Sun at noon (south in the Northern Hemisphere) toward the horizon. A typical choice of β should be latitude tilt, $\beta = \lambda$. Since heating needs are generally greater in the winter months, flat-plate collectors are mounted at even greater angles such as $\beta = \lambda + 15°$.

For practical applications, the instantaneous efficiency of flat collector is calculated from the following empirical expression,

$$\eta = F_R (\tau\alpha)_{e,n} - a \frac{T_{ci} - T_\infty}{I_\beta} - b \left(\frac{T_{ci} - T_\infty}{I_\beta} \right)^2, \tag{14.32}$$

where $(\tau\alpha)_{e,n}$ is an effective transmittance-absorptance product, determined when the direct beam radiation is normal to the absorber plate. The first term on the right-hand-side of Eq. (14.32) is often denoted η_o and is referred to as the **optical efficiency of collector**. Coefficients a and b are determined from fitting expression given by Eq. (14.32) to experimental data.

The efficiency of flat-plate thermal collectors increases with increasing absorption of the solar irradiance and transformation into heat. The efficiency decreases with increasing heat loss coefficient U_L. This coefficient depends on the collector temperature and varies from 1 W/(m^{2}K) to 10 W/(m^{2}K). However, for most flat-plate collectors this coefficient is between 3 and 6 W/(m^{2}K).

An **evacuated-tube collector** can be used to reduce the loss coefficient to around 1 W/(m^{2}K). In this collector a space around the absorber is evacuated, and the absorber is coated with a selectively absorbing material to reduce losses to infrared radiation. Collectors of this type operate in a temperature range from 70 to 180 °C and they have much higher efficiency than the basic flat-plate collectors. However, they are significantly more expensive and not generally used in domestic applications.

14.2.3 CONCENTRATORS

In order to achieve much higher temperatures that are desirable for conversion to mechanical or electrical energy, it is necessary to concentrate the incoming solar radiation by means of mirrors, or other optical devices. The simplest concentrator consists of two flat mirrors reflecting incident light onto a cylindrical absorber. Since the light reflected from each mirror adds to the direct light, in this configuration the absorber receives three times as much incident radiation as it would without the mirrors.

A further improvement of the concentration ratio can be achieved in a parabolic concentrator. In such concentrators all light incident along a line parallel to the symmetry axis of the parabolic concentrator is reflected to a common point, called the focal point of the parabola. **Parabolic trough** solar collectors and **parabolic dish** solar collectors are two main types of geometry that are used.

Parabolic trough systems are the most developed concentrated solar power systems. The main part of a parabolic trough is a linear parabolic reflector made of polished metal mirrors that concentrates light onto a receiver located along the reflector's focal line. The reflector follows the Sun during the daylight hours along a single axis in such a way that sunlight enters the mirror parallel to its plane of symmetry. A tube located along the focal line serves as a receiver. It carries a working fluid, such as molten salt, which is heated up to 150–350 °C and used as a heat source in a thermal power plant.

A parabolic dish concentrator tracks the Sun along two axis to concentrate light onto a receiver positioned at the reflector's focal point. The working fluid can be heated up to 250–700 °C, which provides high thermal efficiency of the order of magnitude 30–32% when combined with the *Stirling engine*.

The effectiveness of concentration is expressed in terms of parameter C referred to as the **effective concentration**, defined by

$$C = \frac{A_{\mathrm{apr}}}{A_{\mathrm{a}}},\tag{14.33}$$

where A_{a} is the absorber surface area and A_{apr} is the total area, normal to the direction of the incident radiation, across which all incoming light rays continue to the absorber. Once the effective concentration C is known, the analysis of collector performance can be performed in the usual manner. As an example, the radiative equilibrium condition for the absorber with surface area A_{a} and incident radiation I is as follows,

$$\sigma A_{\mathrm{a}} T_{\mathrm{a}}^4 = A_{\mathrm{apr}} I = C A_{\mathrm{a}} I, \quad \Rightarrow \quad T_{\mathrm{a}} = \left(\frac{CI}{\sigma}\right)^{\frac{1}{4}}.\tag{14.34}$$

Thus, the temperature of the absorber changes as $\sqrt[4]{C}$ with the effective concentration.

Unlike flat-plate collectors, the concentrated collectors are very sensitive to the changes of incident light direction. Typically concentrators accept light from a limited range of angles, expressed in terms of the **acceptance angle**, which is defined as the angle within which all light impinging on a cross-sectional area A_{apr} reaches the absorber.

14.3 PHOTOVOLTAIC SOLAR CELLS

A photon energy can be directly transformed into an electric current in a device called **photovoltaic solar cell** or **PV cell**. The PV cell is composed of layers of various materials, such as a front glass or plastic protection layer, a transparent electric contact layer, a semiconductor absorption layer, and back protection layers. When photons hit the PV cell, they may be reflected, absorbed in the semiconductor layer, or pass right through it. The absorbed photons can transfer their energy to electrons, allowing the electrons to flow through the material to the electric contact layers as electric current.

The first PV cells, developed in the 1950s, used **crystalline silicon** as the semiconductor material. The crystalline silicon exists either as a polycrystalline silicon, consisting of small crystals, or monocrystalline silicon, in which the crystal lattice of the entire sample is continuous and unbroken, and has no grain boundaries. Even today the crystalline silicon is the dominant semiconducting material used in the photovoltaic technology for the production of solar cells. During the 1990s, polycrystalline silicon cells became increasingly popular. These cells offer lower efficiency than their monocrystalline silicon counterparts, but their production is significantly cheaper.

14.3.1 THEORY

A solar cell is an electronic device which directly converts sunlight into electricity. This process involves such phenomena as absorption of photons in a semiconductor material, generation of electric charge carriers, collection of the electric charge carriers to generate a current, generation of a large voltage across the solar cell, and dissipation of power in the load and in parasitic resistances. The theory behind this process is shortly described in present section, but we first start with a characteristics of the sunlight.

Sunlight and Photons

Sunlight is a form of electromagnetic radiation and it constitutes only a fraction of total energy emitted by the Sun incident on the Earth. As shown in Fig. 14.1, the maximum of solar electromagnetic radiation is within the visible fraction of the spectrum, with the photon wavelength from 400 to 750 nm. The corresponding photon energy is between 1.65 and 3.11 eV.

Observations reveal both the wave nature and the particle nature of light. Wave nature of light gained acceptance already in early 1800s, when experiments showed interference effects in light beams. In 1900 Planck proposed that the energy stored in matter is made up of quanta of energy. Five years later, while examining the photoelectric effect[4], Einstein introduced the idea that light itself is made of discrete units of energy.

The light quantum, called a **photon**, is a massless and stable particle with no charge. In empty space, the photon moves with speed of light $c = 299\ 792\ 458$ m·s^{-1}. The energy E and the magnitude of the momentum vector p of a photon can be expressed in terms of its frequency ν and wavelength λ as follows,

$$E = \hbar\omega = h\nu = \frac{hc}{\lambda}, \quad p = \hbar k = \frac{h\nu}{c} = \frac{h}{\lambda}, \tag{14.35}$$

---

[4]The emission of electrons from a metal plate caused by photons.

where $\hbar = h/2\pi$ is the **reduced Planck constant**, $k = 2\pi/\lambda$ is the **wave number**, and $\omega = 2\pi\nu$ is the **angular frequency**.

The process of the generation of voltage and electric current in a material upon exposure to light is called the **photovoltaic effect**. This effect is closely related to the photoelectric effect. In both cases, light is absorbed, causing excitation of an electron to higher energy state. The photovoltaic effect occurs when the excited electron is still contained in the material, rather than ejected out of the material. As a result, an electric potential is produced by the separation of charges due to diffusion. The incident photons need to have a sufficient energy to overcome the potential barrier for excitation.

Fermi-Dirac Distribution

As in an atom, the number of electrons matches the number of protons in a neutral solid. The electrons fill successively higher energy levels until the electron charge cancels the proton charge making the solid electrically neutral. The energy of the highest filled state for a material at zero temperature is called the **Fermi energy**, E_F^0. With increasing temperature, some electrons are thermally excited above this level. The probability that electrons will have a given energy E is a function of the absolute temperature T and is expressed by the **Fermi-Dirac distribution**,

$$f(E) = \frac{1}{1 + e^{(E-E_F)/k_B T}}, \tag{14.36}$$

where k_B is the Boltzmann constant. This function goes from 1 for very low energy values to 0 for very high energy values. The **Fermi level** E_F is defined as energy where the probability of finding the level occupied by an electron passes $1/2$.

Collection Efficiency

Physical properties of a solid are determined by the position of the Fermi level relative to the band structure. When the Fermi energy is in the middle of a band, the material is a good conductor, and the band is called the **conduction band**. However, when the Fermi energy is between two bands with a large energy separation, the lower band, known as a **valance band**, is filled with electrons, whereas the upper conduction band is empty. Such material does not easily conduct electrons, and it acts as an insulator. A similar situation exists in a **semiconductor**, but in that case the **band gap**, defined as the difference in energy between the highest state in the valence band and the lowest state in the conduction band, is small.

When the band gap is not too large, the thermal fluctuations can excite an electron up to the conduction band. At the same time a vacancy is created in the valance band called a **hole**. The electron and the hole can propagate in the conduction band and the valance band as negative and positive charge carriers, respectively. In a semiconductor at room temperature, both electrons and holes conduct electric current.

When a photon is absorbed in the semiconductor layer of a solar cell, its energy is given to an electron, usually located in the valence band in the crystal lattice. With high-enough photon energy, the excited electron can be promoted to the conduction band where it is free to move around within the semiconductor. A photon only needs to have energy greater than that of the band gap in order to excite an electron from the valence band to the conduction band.

The total incident power of photons with the Planck radiation spectrum can be found as,

$$N_{tot} = KA \int_0^\infty \frac{E^3}{e^{E/k_B T} - 1} dE, \tag{14.37}$$

where K is a proportionality constant, $E = \hbar\omega$ is the energy of a single photon of angular frequency ω, A is the area exposed to the solar radiation. Assuming a semiconductor with band gap E_g, only photons with energy $E = \hbar\omega > E_g$ can excite an electron from the valance band to the conduction band. Any excess photon energy $E - E_g$ is rapidly lost to thermal energy of the material. Even though research efforts are directed to finding a way to capture this excess energy, this fundamental physical constraint limits the over all efficiency of commercially available solar cells. Assuming that we get E_g of energy from each photon with $E > E_g$, the maximum power that is collected by the solar cell is,

$$N_{max} = KAE_g \int_{E_g}^{\infty} \frac{E^2}{e^{E/k_BT} - 1} dE. \tag{14.38}$$

The solar cell **collection efficiency** can be now found as,

$$\eta_{coll,max} = \frac{N_{max}}{N_{tot}} = \frac{E_g}{k_BT} \frac{\int_{E_g}^{\infty} \frac{E^2}{e^{E/k_BT}-1} dE}{\int_0^{\infty} \frac{E^3}{e^{E/k_BT}-1} dE}. \tag{14.39}$$

Thus, a solar cell can extract at most only a fraction of the radiation energy. The value of this fraction depends on E_g and the radiation spectrum.

However, it is still not clear how the radiation energy can be transformed into an electric current. In general, an excited electron will rapidly drop back down to the valance band and recombine with a hole. In addition, an electric current will exist if the symmetry of the system is broken so that electrons move in a preferred direction. These technical challenges have to be resolved by proper construction of photovoltaic cells.

The symmetry of a semiconductor can be broken by non-homogeneous doping the semiconductor material using impurities to produce a *p-n* junction. The most common types of doping are **n-type doping** and **p-type doping**. In the case of *n*-type doping, a semiconductor atom is replaced by an atom that has one more electron in its outer orbital than the semiconductor atom. In case of crystalline silicon, *n*-type doping is achieved if a silicon atom is replaced with a phosphorus or arsenic atom. The extra electron is only weakly bound to the nucleus of the atom and can be easily promoted into the conduction band. At room temperature there is enough thermal energy available to promote most of these electrons into the conduction band, and, as the consequence, the Fermi level E_F moves up. Similarly, introducing atoms with one fewer electron than the semiconductor atom (for silicon it can be atoms of aluminium or gallium) has the opposite effect on the Fermi level, which is now moved down toward the top of the valance band.

Semiconductors that have been doped by adding *n*-type and *p*-type impurities are known as *n*-type and *p*-type materials, respectively. In a *n*-type material electrons become mobile charge carriers in the conduction band, while in a *p*-type material, holes, which are mobile positive charge carriers in the valance band, play a similar role.

A **p-n junction** is created when the *n*-type material and the *p*-type material are brought into contact, as shown in Fig. 14.3. Random thermal motions lead neutrons to diffuse from the *n*-material into the *p*-material, while holes diffuse in the opposite direction. The system comes into a dynamic equilibrium when the **diffusion current**, I_{diff}, is balanced by the **field current**, I_{field}, resulting from the local electric field and the potential difference across the junction. When the dynamic balance is achieved, the Fermi levels of the two pieces of the *p-n* junction become equal. In a *p-n* junction photocell, an incoming photon excites an electron into the conduction band leaving behind a hole. When this occurs in the *p*-doped region, the electron is pushed across the junction by the potential difference, whereas the hole moves into the opposite direction. These charge motions contribute to the **photo-current**, I_{ph}, that can drive an external connected circuit. The process is depicted in Fig. 14.4(a).

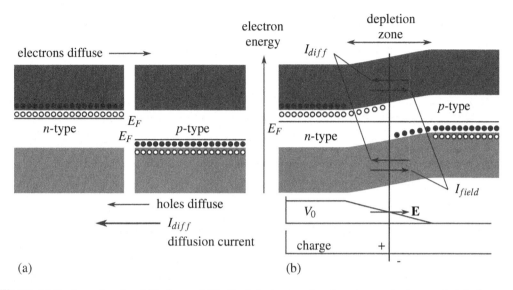

Figure 14.3 A p-n junction (a) before and (b) after bringing together the n-type and p-type materials. I_{diff}—diffusion current, I_{field}—field current, V_0—potential difference across the junction, E_F—Fermi level, **E**—electric field.

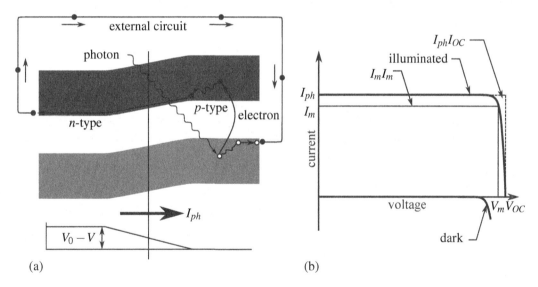

Figure 14.4 A p-n junction photocell: (a) Photo-current I_{ph} generation in an illuminated junction. (b) The current-voltage characteristics under dark and illuminated conditions.

Current-Voltage Characteristics

With no external voltage applied to a p-n junction, the diffusion current and the field current are in balance, thus,

$$|I_{diff}| = |I_{field}| = I_0, \tag{14.40}$$

where I_0 is the equilibrium current. In particular, $I_{field}(V = 0) = I_0$, where I_0 is referred to as the **reverse saturation current**. In presence of an external potential V, the total current across the p-n junction is given by the **Shockley diode equation**,

$$I(V) = I_{diff}(V) - I_{field}(V) = I_0 \left(e^{eV/k_BT} - 1 \right), \tag{14.41}$$

where T is the junction temperature, e is the elementary charge, and k_B is the Boltzmann constant.

The p-n junction acts as a **photodiode** when it is illuminated with an electromagnetic radiation. When a photodiode formed by a p-n junction is connected to an external voltage V, the total current passing through the external circuit is equal to a sum of the photo-current, directed from the n-side to p-side of the junction, the field current, with the same direction as the photo-current, and the diffusion current, with an opposite direction to the two previous currents. Thus, we have,

$$I = I_{ph} + I_0 - I_{diff} = I_{ph} - I_0 \left(e^{eV/k_BT} - 1 \right). \tag{14.42}$$

The above equation indicates that the maximum current occurs at zero external voltage $V = 0$. Therefore this maximum current is known as the **short-circuit current** $I_{SC} = I_{ph}$. The maximum voltage occurs when no current is drawn and is known therefore as the **open-circuit voltage** V_{OC}.

Setting $I = 0$ in Eq. (14.42) gives a relation between the open-circuit voltage and the short-circuit current,

$$V_{OC} = \frac{k_BT}{e} \ln \left(\frac{I_{ph}}{I_0} + 1 \right). \tag{14.43}$$

A detailed balance limit of efficiency of p-n junction solar cell was derived by Shockley and Queisser [78]. Using thermodynamic principles, they provided the following expression for the open-circuit voltage,

$$V_{OC} \leq \frac{k_BT}{e} \ln \left(\frac{\omega_s}{\pi} \frac{\int_{E_g}^{\infty} \frac{E^2 dE}{e^{E/k_BT_\odot} - 1}}{\int_{E_g}^{\infty} \frac{E^2 dE}{e^{E/k_BT} - 1}} \right), \tag{14.44}$$

where T is the solar cell temperature (≈ 300 K), $T_\odot$ is the characteristic temperature of blackbody for the sunlight spectrum (≈ 6000 K), E_g is the band gap, and ω_s is the solid angle subtended by the Sun, $\omega_s = \pi R_\odot^2/d^2$, with $R_\odot$—Sun's radius, and d—Earth's mean orbital distance from the Sun.

Fill Factor

An overall behavior of a solar cell is expressed with the **fill factor**, denoted ff, defined as a ratio of the maximum total electric power of the solar cell N_{max} to the product of the open circuit voltage V_{OC} and the short circuit current I_{SC},

$$\text{ff} = \frac{N_{max}}{V_{OC} I_{SC}}. \tag{14.45}$$

For single-junction terrestrial silicon $V_{OC} \approx 0.492 - 0.738$ V and the fill factor ranges approximately from 0.721 to 0.849 [35].

The Shockley-Queisser Ultimate Efficiency

An overall efficiency limit of a single-junction photovoltaic cell taking into account the minimum photon energy, the electron-hole recombination, and the fill factor, is called by Shockley and Queisser the **ultimate efficiency**. This efficiency limit is given as,

$$\eta_{max} = \text{ff} \times \frac{eV_{OC}}{E_g} \times \eta_{coll,max}. \tag{14.46}$$

Here ff is the fill factor, e is the elementary charge, V_{OC} is the open-circuit voltage, E_g is the band gap, and $\eta_{coll,max}$ is the collection efficiency.

The ultimate efficiency reflects the fact that photons with an energy below the band gap of the junction material cannot generate an electron-hole pair. Instead, such photons only generate heat when absorbed. For photons with an energy above the band gap energy, the excess energy above the band gap is converted into kinetic energy of the carriers that are slowing down to equilibrium while generating heat. Traditional single-junction cells have the maximum theoretical efficiency limit of 33.16% [74]. Solar cells with multiple band gap absorber materials improve efficiency by dividing the solar spectrum into smaller bins where the efficiency limit is higher for each bin.

Solar Cell Efficiency

The efficiency of a solar cell is determined as the ratio of the maximum electric power output of the cell, N_{max}, and the total incident solar power, N_{in},

$$\eta = \frac{N_{max}}{N_{in}} = \frac{V_{OC} \cdot I_{SC} \cdot \text{ff}}{N_{in}}, \tag{14.47}$$

where V_{OC}—open-circuit voltage, I_{SC}—short-circuit current, and ff—filling factor. The performance of a solar cell is measured by its efficiency. Since the incident solar power is changing from place to place, the conditions under which the efficiency is measured must be carefully controlled. Terrestrial solar cells are measured under solar spectrum equivalent to AM1.5, radiation intensity equal to 1 kW/m^2, and temperature of 25 °C. Progress in efficiency of solar cells is reported on a regular basis every six months [35].

14.3.2   SILICON SOLAR CELLS

Silicon is the main semiconductor material that is used to produce solar cells. From the point of view of the silicon structure, we distinguish monocrystal cells and polycrystal cells. Thin-film cells use amorphous silicon or other materials. Silicon dioxide (SiO_2) is the most abundant mineral in the Earth's crust. It is used to manufacture the pure silicon for photovoltaics. The most common raw material for electronic grade is high purity quartz rock, with low concentrations of iron, aluminium and other metals. In the first stage the oxygen is removed to produce metallurgical grade silicon. It is done through a reaction with carbon in the form of coal (charcoal) and heating to 1500–2000 °C in an electrode arc furnace. The obtained silicon is 98% pure and is extensively used in the metallurgical industry. This metallurgical grade silicon is produced at the rate of millions of tons per year, at a cost of few USD/kg, and an energy cost of 14–16 kWh/kg. A semiconductor grade silicon is obtained through further purification processes that require a lot of energy. Solar cells silicon is obtained in a crystallization process.

The silicon atom has 14 electrons, with four valence electrons. The atoms bond with each other by means of their valence electrons to form a crystal. The crystal has a three-dimensional diamond cubic structure known as the silicon lattice.

Almost all of the silicon crystals grown for solar cells are produced by the **Czochralski process**. The process begins with melting raw silicon in a crucible. Small precise amounts of impurity atoms such as boron or phosphorus are added to the molten silicon, changing it into *n*-type or *p*-type silicon. When the silicon is fully melted, a small precisely oriented seed crystal mounted on the end of a rotating shaft is slowly lowered until it just dips below the surface of the molten silicon. The rotating rod is then drawn upward, allowing an ingot to be formed. A large, single crystal, cylindrical ingot is extracted from the melt by precisely controlling the temperature gradient, rate of pulling, and speed of rotation.

The ingot can be one or two meters in length and up to 450 mm in diameter, depending on the amount of silicon in the crucible. This process is normally performed in an inert atmosphere, such as argon, and in inert chamber. The ingot is then sliced with a wafer saw and polished to form wafers.

The size of wafers for photovoltaics is 100–200 mm^2 and the thickness is 100–500 μm. Wafers are cleaned with weak acids and textured to create a rough surface to increase the solar cell efficiency.

The silicon used for solar cells can have various crystalline forms such as,

- single crystalline (sc-Si) silicon, with grain size above 10 cm,
- multicrystalline (mc-Si) silicon, with grain size in a range 1 mm to 10 cm,
- polycrystalline (pc-Si) silicon, with grain size in a range 1 μm to 1 mm,
- microcrystalline (μc-Si) silicon, with grain size below 1 μm.

The preferred types of silicon used for solar cells are the single crystalline silicon and the multicrystalline silicon. The single crystalline silicon have better material properties, however, they are also more expensive. Single crystalline Czochralski wafers are the most commonly used type of silicon wafer by both the solar cell and integrated circuit industry.

A typical crystalline silicon solar cell consists of a glass or plastic layer, an anti-reflective layer, a front contact to a circuit, a semiconductor layer, and a back contact to close the circuit. Since the amount of power produced by a single solar cell is relatively small, the cells are grouped into panels, or modules, to produces higher voltages and currents. The higher voltage can be achieved by connecting cells in series, where the positive terminal of one cell is connected to the negative terminal of the next cell. The higher current is obtained by connecting cells in row, where the positive terminals of cells are connected together, and the same is done with the negative terminals.

14.3.3 ADVANCED SOLAR CELLS

Solar photovoltaic technology is still under intensive development and can be so far divided into three categories: first generation, including mainly crystalline silicon solar cells; second generation, including thin-film photovoltaic cells that employ semiconductors with a direct band gap[5]; and third generation, which contains various technologies that are still in the early stage of their development, e.g., multi-junction cells and organic cells.

Thin film cells are much cheaper to make because they use only 1–2% of the expensive semiconductor material of a crystalline silicon cell. However, they are not as efficient as silicon cells. Thin film cells are made from thin films of semiconductors, just a few microns thick, chemically vapor deposited or sputtered on glass or metal. Since large surfaces can be covered, panels are using much large cells with dimensions of about 0.5 m by 1 m. The panels are fairly robust and can operate in a wide temperature range, and in a variety of light conditions, including dawn and dusk. Since the semiconductor material in a thin film cell is only a few microns thick, the incoming light with wavelength longer than the thickness of the cell most likely will not be absorbed. This means that wavelength in the infrared part of the spectrum will be weakly absorbed. For example, monocrystalline or polycrystalline silicon must be on the order of 100 μm thick to absorb a substantial fraction of incoming solar radiation. However, a material with a direct band gap can absorb most solar radiation over a distance of order 1 μm. Most promising materials for thin film cells are cadmium telluride (CdTe), copper indium/gallium (di)selenide (CuIn$_x$Ga$_{1-x}$Se$_2$), also known as CIGS, and hydrogenated amorphous silicon (a-Si:H). However, all of the thin-film materials used so far in a commercial scale contain chemical elements that are far less abundant than silicon. The quest for more abundant thin-film materials is an active research area in the field of photovoltaic solar power.

---

[5]The band gap is called "direct", when the momentum of electrons and holes is the same in both the conduction band and the valance band of the crystal.

14.3.4 PHOTOVOLTAIC MODULES

Individual solar cells are encapsulated and connected together into **photovoltaic modules**. The encapsulation protects the solar cells from the environment and the user from electric shock. The connection of solar cells increase their output power.

Many different types of photovoltaic modules exists, depending on their construction material and application. Amorphous silicon solar cells are often encapsulated into a flexible array, while crystalline silicon solar cells are usually rigid with glass front surface. The front surface of a module must have a high transmission in the wavelengths that can be used by solar cells. Typical wavelength range for the silicon solar cells is from 350 to 1200 nm. To reduce the light reflection from the front surface, it is made rough. The front surface can be made of various materials such as acrylic polymers and glass. Tempered, low iron-content glass is frequently used thanks to its low cost, strength, stability, high transparency, water resistance, and good self-cleaning properties. The solar cells are encapsulated in an adhesive, optically transparent and thermally conductive material such as EVA (ethyl vinyl acetate), and placed between the front surface and the rear surface. The rear surface must have a high thermal conductivity, and at the same time, it must prevent the ingress of water or water vapor. These requirements are satisfied by the polyvinyl fluoride (PVF)—$(CH_2CHF)_n$—a polymer material with excellent resistance to weathering and staining.

The voltage from a photovoltaic module is determined by the number of solar cells. In a typical module, 36 cells connected in series are required to charge a 12 V battery. A single silicon solar cell under AM1.5 illumination and temperature 25 °C has a voltage at the maximum power point around 0.5 V. Taking into account expected various voltage drops, the 36-cell module has an operating voltage of about 17 to 18 V.

The current from a photovoltaic module depends mainly on the size of the solar cells. At standard conditions such as AM1.5 and an optimum tilt, the current density from a commercial solar cell is approximately in a range of 30–36 mA/cm^2. Since the single crystal solar cells are typically $15.6 \times 15.6 \cong 243$ cm^2, one module can give a total current of almost 9–10 A.

A photovoltaic module consisting of identical solar cells and operating at identical conditions will have the same current-voltage characteristic as that of the individual cells, except that the voltage and the current are changed as follows,

$$I_C = N_p \cdot I_{SC} - N_p \cdot I_0 \left(e^{\frac{eV_C/N_s}{mk_BT}} - 1 \right), \tag{14.48}$$

where N_s—the number of cells in series, N_p—the number of cells in parallel, I_C—the total current from the circuit, V_C—the total voltage from the circuit, I_0—the saturation current from a single solar cell, I_{SC}—the short-circuit current from a single solar cell, m—the ideality factor of a single solar cell, e—the elementary charge, k_B—the Boltzmann constant, and T—the system temperature.

PROBLEMS

PROBLEM 14.1

Calculate the Sun's surface temperature knowing that, due to fusion, it looses mass 4.3×10^9 kg per second. Assume a black-body radiation from the Sun's surface.

PROBLEM 14.2

Find the Sun's coordinates in the horizontal coordinate system as seen from a location with coordinates (59.3293° N, 18.0686° E) on 21 June at 13:00 (CET).

PROBLEM 14.3

Calculate the declination of the Sun on June 21. Assume that the year has 365 days.

PROBLEM 14.4

Find equation of time of the Sun on June 21. Assume that the year has 365 days.

PROBLEM 14.5

Calculate hour angle of the Sun on June 21 at 12:00 (CET) at a location with coordinates (59.3536° N, 18.0578° E) Assume that the year has 365 days.

PROBLEM 14.6

The band gap of silicon is $E_g = 1.107$ eV. Estimate the range of wavelength of the sunlight to cause the photovoltaic effect.

PROBLEM 14.7

Estimate the maximum possible collection efficiency for a silicon solar cell, with band gap $E_g = 1.107$ eV.

15 Nuclear Power

Even though almost all energy available on Earth has its origin in the nuclear energy, a dedicated human transformation of the nuclear energy into other useful forms of energy has a relatively short history.

Nuclear fission was discovered in December 1938 by physicists Lise Meitner and Otto Robert Frisch and chemists Otto Hahn and Fritz Strassmann. Four years later, in December 1942, the first human-made self-sustaining nuclear chain reaction was initiated in Chicago Pile-1 (CP-1) reactor, during an experiment led by Enrico Fermi. On September 3, 1948, a nuclear reactor X-10 in Oak Ridge, Tennessee, US, was used to generate electricity to power a light bulb, and on June 27, 1954 the first nuclear power station generated electricity for a power grid in Obninsk, Soviet Union.

Nuclear fusion was suggested to be the primary source of stellar energy by Arthur Eddington already in 1920. However, research into fusion began in the early 1940s as part of the Manhattan Project for military purposes. As a result, human-made fusion was accomplished in 1951 with the Greenhouse Item nuclear test, and on a large scale, in an explosion carried out on 1 November 1952 in the Ivy Mike hydrogen bomb test. Research into developing controlled fusion inside fusion reactors has been ongoing since the 1940s, but the technology is still in its development phase.

15.1 INTRODUCTION

Nuclear energy is energy released during nuclear reactions or nuclear disintegrations. To make this energy useful, it is necessary to control nuclear reactions as a self-sustained chain-reaction process, from which the energy can be retrieved in a safe and reliable manner. The processes and devices which are necessary to achieve this goal are described in this chapter. Sections §15.1 through §15.7 are devoted to nuclear fission reactors, whereas fusion reactors and other technologies a shortly discussed in §15.8.

15.1.1 NEUTRON REACTIONS

Nuclear reactions in which neutrons are involved belong to the most important ones in nuclear reactor physics. This is because neutrons are necessary to sustain a chain reaction and their properties, such as kinetic energy, flight direction and distribution in space, are very important as far as the nuclear reactor operation is concerned.

Two main reaction categories are absorption and scattering. The absorption reaction can be either the radiative capture, fission, (n,2n), (n,p) or (n,α), whereas the scattering can be either elastic or inelastic. The most important reactions are discussed below in a more detail.

Neutron absorption reactions can lead to various outcomes, such as:

- radiative capture or (n,γ) reaction,
- fission or (n,f) reaction,
- (n,2n) reaction,
- (n,α) reaction.

The probability of a particular outcome depends on the incident neutron speed and on the type of nuclide. These probabilities are expressed in terms of cross sections and are discussed later in this section.

Neutron scattering reactions can be either elastic or inelastic. During an inelastic scattering the product nucleus is in an excited state. To bring the nucleus back to stable or ground state, one

DOI: 10.1201/9781003036982-15

or more photons of gamma radiation are emitted. Elastic scattering can occur for neutrons of any energy, whereas inelastic scattering requires neutrons to have large enough energy to cause the excited state of the nucleus.

In thermal-spectrum reactors neutron slowing down occurs by a series of scattering events. Such events play important role in the reactor behavior and thus they must be examined in detail. In this section we discussed the elastic scattering events which are relatively simple. First we should note that a neutron-nucleus collision is elastic when kinetic energy is conserved. We distinguish:

- Potential scattering corresponding to a single diffusion of the wave associated with the neutron by the potential field of the nucleus. This type of reaction can occur for all nuclei with neutrons of any energy.
- Resonant scattering in which the incident neutron is first absorbed, a compound nucleus is created and then a neutron is re-ejected. After the reaction the target nucleus remains at the ground state (is not excited).

Elastic scattering has no threshold and it can occur with neutrons of any energy. It plays the most important role in reactors with a moderator, that is with a reactor material whose purpose is to slow down neutrons. In this section we make no further distinction between the potential and the resonant elastic scattering, since the presented analysis is equally valid for both of them.

A neutron collision with a nucleus is inelastic when kinetic energy is not conserved and some of the energy is used to modify the internal state of the target nucleus. Thus, during such collision the target nucleus, initially in a ground energy state, reaches an excited state after the reaction. The nucleus is later decaying by gamma emission. Unlike the elastic scattering, inelastic scattering has a reaction threshold of a few MeV for light nuclei, and a few tens of keV for heavy nuclei. This means that in reactors, where neutrons have energies below a few MeV, inelastic scattering will occur mainly in the fuel materials, especially in ^{238}U.

Neutrons are electrically neutral subatomic particles with the rest mass $M_n \cong 1.675 \times 10^{-27}$ kg (939.566 MeV/c^2), spin 1/2, and non-zero magnetic moment. Free neutrons (outside a nucleus) are beta-decaying and their half-life is 611 s. They exist as a small natural neutron background on Earth caused by muons that are produced in the atmosphere by high energy cosmic rays. Neutrons are very important in nuclear applications, since they can cause fission when absorbed in heavy nuclei of such nuclides as $^{233}_{92}$U, $^{235}_{92}$U, $^{239}_{94}$Pu, and $^{241}_{94}$Pu. Since in fission reactions new neutrons are released, self-sustained *chain fission reactions* can be achieved.

The term **neutron reaction** is used very broadly to describe any of a wide array of interactions involving neutrons and nuclei. Many types of neutron reactions can occur in the laboratory or in nuclear reactors, but in considering energy from nuclear fission, our interest is limited almost entirely to the following four reactions: elastic scattering, inelastic scattering, radiative capture, and fission.

Experiments show that most of the neutrons striking a thin target just pass through the target without interacting with any of the nuclei. The number of interactions Υ per unit time and unit area of the target is found to be proportional to the number of neutrons I striking the target per unit time and area, to the atom density N of the target, and to the target thickness Z. This observations can be expressed as,

$$\Upsilon = \sigma I N Z, \tag{15.1}$$

where σ is the proportionality constant called the **microscopic cross section**. Since the product NZ is the number of nuclei per unit area, $\sigma I = \Upsilon / NZ$ can be interpreted as the number of interaction per unit time and per single nucleus. Consequently, the microscopic cross section is equal to the number of interactions per single incident neutron per one nucleus and is given as follows,

$$\sigma = \frac{\Upsilon}{NZI} \frac{\frac{\text{reaction}}{\text{m}^2\text{s}}}{\frac{\text{nucleus}}{\text{m}^3} \cdot \text{m} \frac{\text{neutron}}{\text{m}^2\text{s}}} = \frac{\Upsilon}{NZI} \, \text{m}^2 \frac{\text{reaction}}{\text{nucleus} \cdot \text{neutron}}. \tag{15.2}$$

The standard unit for measuring the microscopic cross section is the **barn**, abbreviated by the letter b, which is equal to 10^{-28} m^2 or 10^{-24} cm^2.

Equation (15.2) is useful to determine the microscopic cross section from experimental data, when the reaction rate is measured, and characteristic data of the target and the neutron beam are known. Experiments showed that σ depends on the type of nucleus and on the neutron speed v. To determine the function $\sigma(v)$, a beam of monoenergetic neutrons is used. If there are n neutrons per unit volume (a quantity called the **neutron density**) in the beam of neutrons moving in the direction perpendicular to the target, the number of neutrons that impinge upon the target unit area within unit time is given as

$$I = nv. \tag{15.3}$$

This quantity is called the intensity of the beam.

The **rate of neutron reactions** R, defined as the number of neutron interactions per unit time and volume, can be now obtained by combining Eq. (15.1) with Eq. (15.3),

$$R = \frac{\Upsilon}{Z} = \sigma N n v, \tag{15.4}$$

where Z is the thickness of the target. The product

$$\Sigma = \sigma N \tag{15.5}$$

is called the **macroscopic cross section** and represents the effective area for a given reaction and neutron energy of all nuclei per unit volume of target.

In laboratory conditions, when the microscopic cross section is to be determined, all neutrons should have the same speed. However, in a nuclear reactor the speeds and directions of neutrons crossing a certain point in space can vary. Even though the directions of neutrons do not influence the reaction rate, the effect of neutron speed can be significant. Suppose $n(v)$ is the density of neutrons of speed v per unit speed interval, then $n(v)dv$ is the number of neutrons having speed in the interval from v to $v + dv$. Thus, the total reaction rate for neutrons with speeds varying from v_{min} to v_{max} is found as,

$$R_{tot} = \int_{v_{min}}^{v_{max}} \Sigma(v) n(v) v dv. \tag{15.6}$$

In reactor calculations it is more convenient to express the neutron speed v in terms of its kinetic energy $E = m_n v^2 / 2$, where $m_n \approx 1.675 \times 10^{-27}$ kg is the rest mass of the neutron. When released in fissions, neutrons usually have high kinetic energies, in the range of few million electron volts. These neutrons are called fast neutrons, since they move with high speeds. After several scattering collisions with various nuclei in the reactor, neutrons lose their kinetic energy until they are in a mechanical equilibrium at the medium temperature. Such neutrons are called the thermal neutrons, and at ordinary temperatures, they have average energy of about 0.035 eV. To facilitate computations, it is a common practice to use energy limits of zero and infinity. In addition, the product $n(v)vdv$ can be identified as the total distance travelled in unit time by all neutrons present in unit volume and having speed in the range from v to $v + dv$. Thus the total path travelled by neutrons of all energies per unit time and volume, referred to as the **total neutron flux** or just *neutron flux*, is obtained as,

$$\phi = \int_0^\infty n(E) v(E) dE. \tag{15.7}$$

The total reaction rate in a polyenergetic neutron system may be written in terms of the neutron flux and the average macroscopic cross section $\overline{\Sigma}$ as,

$$R = \overline{\Sigma}\phi, \tag{15.8}$$

where,

$$\overline{\Sigma} = \frac{\int_0^\infty \Sigma(E)n(E)v(E)\mathrm{d}E}{\int_0^\infty n(E)v(E)\mathrm{d}E}. \tag{15.9}$$

The total neutron flux given by Eq. (15.7) may, in general, be a function of the spatial location and the time $\phi(x,y,z,t) = \phi(\mathbf{r},t)$, where (x,y,z) are neutron Cartesian coordinates, $\mathbf{r}$ is the neutron location vector, and t is the time. The spatial variability of the neutron flux may primarily be caused by a spatially non-uniform distribution of the neutron density, but also by non-uniform distribution of the neutron energy. Both these quantities may also vary with time. Similarly, the macroscopic cross section may vary in space due to non-uniform distribution the atomic mass density N. Thus, the reaction rate is,

$$R(\mathbf{r},t) = \overline{\Sigma(\mathbf{r})}\phi(\mathbf{r},t). \tag{15.10}$$

The results obtained for the reaction rates are general and apply to a broad range of reaction types occurring in nuclear reactors. In particular, if $\overline{\Sigma}$ represents the macroscopic cross section for the fission reaction, the reaction rate will determine the time and space distribution of the reactor power. The major reactions occurring in nuclear reactors are discussed below.

Elastic Scattering

During an elastic scattering, usually denoted as (n,n), a neutron and nucleus collide with no change in the nucleus structure. For example, the elastic scattering of neutrons on carbon isotope $^{12}_6C$ (a graphite form of carbon can exist in a graphite-moderated nuclear reactor) is,

$$^1_0n + ^{12}_6C \rightarrow ^1_0n + ^{12}_6C. \tag{15.11}$$

Due to the collision, the nucleus recoils from its original location and the energy of the scattered neutron is reduced. Considering the momentum and energy conservation for the neutron-nucleus system, the energy of the neutron after collision E' depends on the neutron energy before the collision E, the atomic mass number of the nucleus A, and the angle θ at which the neutron is scattered as,

$$E' = \frac{E}{(A+1)^2}\left[\cos\theta + \sqrt{A^2 - \sin^2\theta}\right]^2. \tag{15.12}$$

When neutron is directly scattered backwards ($\theta = \pi$), it suffers the largest possible energy loss and,

$$(E')_{min} = \alpha E, \tag{15.13}$$

where α is called the **collision parameter** defined as,

$$\alpha = \left(\frac{A-1}{A+1}\right)^2. \tag{15.14}$$

When estimating the average number of collisions that are required for a neutron to be moderated from an initial energy E_0 to some final energy E_N, where N is the total number of collisions, it is useful to introduce the average logarithmic energy loss during one collision defined as

$$\xi \equiv \overline{\ln\left(\frac{E}{E'}\right)}. \tag{15.15}$$

The **average logarithmic energy loss** from E_0 to E_N can be represented as follows

$$\overline{\ln\left(\frac{E_0}{E_N}\right)} = \overline{\ln\left(\frac{E_0}{E_1}\right)} + \overline{\ln\left(\frac{E_1}{E_2}\right)} + ... + \overline{\ln\left(\frac{E_{N-1}}{E_N}\right)} = N\xi, \tag{15.16}$$

where we used the fact that the average of a sum is equal to a sum of the averages. Thus, we get the following expression for the total number of collisions that are required to moderate neutron from E_0 to E_N

$$N = \frac{\overline{\ln\left(\dfrac{E_0}{E_N}\right)}}{\xi}. \tag{15.17}$$

The **average logarithmic energy loss during one collision** can be found as

$$\xi \equiv \overline{\ln\left(\frac{E}{E'}\right)} = 1 - \frac{(A-1)^2}{2A}\ln\left(\frac{A+1}{A-1}\right), \tag{15.18}$$

where A is the mass number of the target nucleus. It can be seen that the parameter ξ decreases with the mass number. This indicates that the number of collisions that are needed to slow down a fast neutron, as given by Eq. (15.17), increases with increasing nuclide mass.

Example 15.1:

Find the number of collisions that are needed on average to moderate a fast neutron with energy 2 MeV to the thermal neutron energy of 0.035 eV in deuterium and graphite moderators.

Solution

Taking $A = 2$ for deuterium in Eq. (15.18) we find the parameter $\xi = 1 - (2-1)^2/(2 \cdot 2)\ln((2+1)/(2-1)) \cong 0.725$. Using Eq. (15.17) we find the required average number of collisions as $N = \ln(2 \times 10^6/0.035)/0.725 \cong 24.6$. For graphite we have $A = 12$, which gives $\xi \cong 0.158$ and $N \cong 113.2$. Thus the required number of collisions to moderate a fast neutron with energy of 2 MeV to the thermal neutron energy of 0.035 eV for deuterium and graphite moderators is 25 and 113, respectively.

The **average cosine of the scattering angle** $\overline{\cos\theta} \equiv \overline{\mu}_0$ describes the preferred direction of neutron after collision and is often used in the analysis of neutron slowing down. It can be shown that this value is given as,

$$\overline{\mu}_0 = \frac{\int_0^{4\pi} \cos\theta\, d\Omega}{\int_0^{4\pi} d\Omega} = \frac{2\pi \int_0^{\pi} \cos\theta \sin\theta_c d\theta_c}{4\pi} = \frac{2}{3A}, \tag{15.19}$$

Where θ and θ_c are the scattering angles in the laboratory and the center-of-mass system[1], respectively.

The microscopic cross section for elastic scattering σ_s is a function of the energy of the incident neutron, and three distinct regions can be identified. For low-energy region the potential scattering takes place and the cross section is given by

$$\sigma_s = 4\pi R^2, \tag{15.20}$$

where R is the nuclear radius that can be estimated from Eq. (1.10). In the intermediate range of energies there is a region of multiple resonances, where the microscopic cross section has high values within narrow energy intervals. At still higher energies the resonances are so close to each other that they can no longer be resolved and in this region σ_s is a smooth and slowly varying function of energy. The elastic scattering cross section for ^{12}C is shown in Fig. 15.1.

---

[1] A reference coordinate system which moves with the velocity of the center of mass of the neutron and nucleus, so that the total momentum of the nucleus-neutron system is zero. The elastic scattering event is usually isotropic when analyzed in the center-of-mass system.

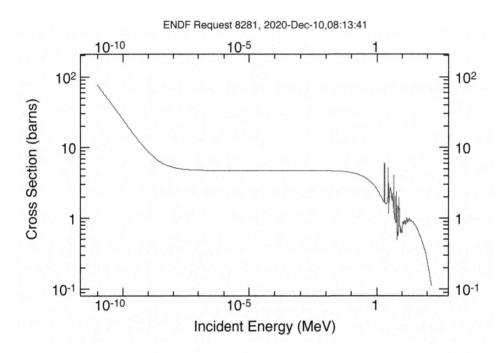

Figure 15.1 Elastic scattering cross section of ^{12}C (From `https://www-nds.iaea.org`).

Inelastic Scattering

Inelastic scattering differs from elastic scattering in that the target nucleus is left in an excited state. This process occurs only when the neutron has sufficient energy to place the target nucleus in the first excited state. This means that the microscopic cross section for inelastic scattering σ_{in} is zero up to some threshold energy. This threshold energy decreases with increasing mass number. In particular, it is about 4.8 MeV for ^{12}C, while it drops to only 44 keV for ^{238}U. Heavy nuclei, such as for uranium isotopes, have many excited states well below 1 MeV; thus, inelastic scattering can contribute to the initial slowing down of neutrons in a reactor. Since the nucleus is left in an excited state, the energy of the emitted neutron can be significantly less than the energy of the incident neutron. If the compound nucleus decays by emission of two or more neutrons, the reactions are referred to as (n,2n), (n,3n), etc. Their corresponding microscopic cross sections are denoted $\sigma_{n,2n}$, $\sigma_{n,3n}$, and so on.

Neutron Radiative Capture

In the first stage of many reactions, the neutron combines with the target nucleus to form an excited compound nucleus. The term **neutron radiative capture**, often simply referred to as the radiative capture, is usually restricted to those cases where the excited compound nucleus decays by the emission of gamma rays. For example, taking $^{238}_{92}$U as the target nucleus, radiative capture leads to the formation of $^{239}_{94}$Pu as follows,

$$^{1}_{0}n + {}^{238}_{92}U \longrightarrow ({}^{239}_{92}U)^{*} \xrightarrow{\beta^{-}} {}^{239}_{93}Np \xrightarrow{\beta^{-}} {}^{239}_{94}Pu, \qquad (15.21)$$

which indicates that the process of formation of ^{239}Pu takes two beta decays: first into neptunium-239 and then into plutonium-239.

The **radiative capture cross section** σ_γ can be divided into three neutron energy E regions: the low-energy region, in which $\sigma_\gamma \sim 1/\sqrt{E}$, the intermediate-energy region with multiple isolated resonances at the same energies as the resonances in σ_s, and the high-energy region, where σ_γ drops smoothly to very small values. The radiative capture cross section of $^{238}_{92}$U is shown in Fig. 15.2.

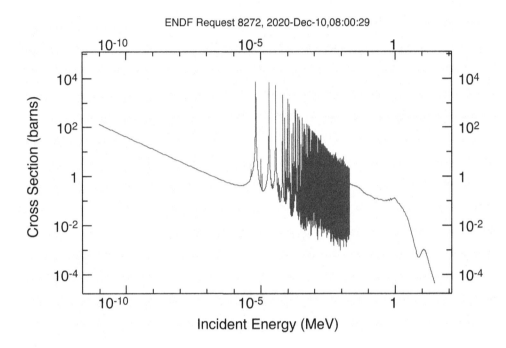

Figure 15.2 Radiative capture cross section of $^{238}_{92}$U (From `https://www-nds.iaea.org`).

Nuclear Fission

We introduced the basic characteristics of the *nuclear fission* reaction in §1.3.4, where we discussed the energy of nuclear reactions, and in §6.3, where we discussed fission products, fission neutrons, and the energy released in fission reactions.

The fission reaction is only possible for fissile and fissionable nuclides. A nuclear fission reaction for $^{235}_{92}$U is as follows

$$^{1}_{0}n + ^{235}_{92}U \rightarrow (^{236}_{92}U)^* \rightarrow \text{fission reaction products}, \tag{15.22}$$

which indicates that the fission process has two stages: first a creation of a compound nucleus $(^{236}_{92}U)^*$ and next a fission of the compound nucleus into two or more fragments. For example, a typical fission reaction is

$$^{1}_{0}n + ^{235}_{92}U \rightarrow (^{236}_{92}U)^* \rightarrow ^{140}_{54}Xe + ^{94}_{38}Sr + 2^{1}_{0}n + 200 \text{ MeV}. \tag{15.23}$$

Fission products are unstable, since they have too high number of neutrons. Less than 1% of the fission products decay by the delayed emission of neutrons. Most of decays are through beta emission, accompanied by one or more gamma photons. Usually more than one decay is required to attain stability. For the fission fragments in Eq. 15.23 we have

$$^{140}_{54}\text{Xe} \xrightarrow{\beta} {}^{140}_{55}\text{Cs} \xrightarrow{\beta} {}^{140}_{56}\text{Ba} \xrightarrow{\beta} {}^{140}_{57}\text{La} \xrightarrow{\beta} {}^{140}_{58}\text{Ce}. \tag{15.24}$$

and

$$^{94}_{38}\text{Sr} \xrightarrow{\beta} {}^{94}_{39}\text{Y} \xrightarrow{\beta} {}^{94}_{40}\text{Zr}. \tag{15.25}$$

Equation 15.23 shows only one example of the more than 40 different fragment pairs that result from fission, which in general can have atomic mass numbers between 72 and 160. Nearly all of the fission products fall into two broad groups: the light group with mass numbers between 80 and 110, and the heavy group with numbers between 125 and 155.

The microscopic cross sections of nuclei for neutron-induced fission σ_f resemble radiative capture cross in their dependence on the energy of the incident neutron. The microscopic fission cross section for $^{235}_{92}\text{U}$ is shown in Fig. 15.3. Nuclei that lead to fission following the absorption of low-energy neutrons are called **fissile**. For certain nuclei, however, fission is possible when a high-energy neutron is absorbed. Such nuclei are said to be **fissionable**. The principle **fissile isotopes** are ^{233}U, ^{235}U, and ^{239}Pu, whereas ^{238}U is an example of the **fissionable isotope**.

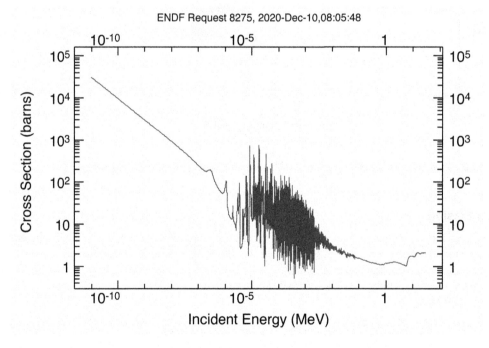

Figure 15.3 Fission cross section of $^{235}_{92}\text{U}$ (From https://www-nds.iaea.org).

The probability that a neutron that is captured in a fissile nuclide causes a fission is just $\sigma_f/(\sigma_f + \sigma_\gamma) = 1/(1 + \sigma_\gamma/\sigma_f) = 1/(1 + \alpha)$, where the **capture-to-fission ratio** α is defined as,

$$\alpha \equiv \frac{\sigma_\gamma}{\sigma_f}. \tag{15.26}$$

For reactor calculations it is convenient to define the parameter η, frequently referred to as the **reproduction factor**, which is equal to the number of fast neutrons released in fission per thermal neutron absorbed by a fissile nucleus. It can be obtained as a product of the average number of neutrons liberated in fission v and the probability that a captured neutron causes a fission,

$$\eta = \frac{v\sigma_f}{\sigma_f + \sigma_\gamma} = \frac{v}{1+\alpha}. \tag{15.27}$$

The reproduction factor plays an important role in fuel conversion and breeding. It is evident that η must be greater than unity to have in chain reaction excess neutrons for transmutation of fertile isotopes into fissile isotopes. This very important aspect of nuclear power sustainability potential is further discuss in §15.4.1.

15.1.2 NEUTRON FLUX

The knowledge of neutron distribution in the reactor core is very important for prediction of its over-all features like power level and power distribution. In general, exact description of positions and velocities of all neutrons is neither possible nor necessary. Even though the general form of equations are known (integro-differential Boltzmann equations), their simplified form known as diffusion theory approximation is employed. Further, it can be assumed that all neutrons have the same speed which leads to the 1-group diffusion theory approximation. This kind of approximation leads to a quite crude model, nevertheless due to its simplicity it can be used for first-step analyses of nuclear reactors. As a most straightforward continuation, one can consider multi-group diffusion theory from which two-group approximation (where only fast and slow or thermal neutrons are considered) gained a considerable popularity and is often employed for practical analysis of nuclear reactors.

Neutron Balance Equation

The most general method that can be used to describe the spatial and temporal distribution of neutrons is based on the neutron transport equation, also called the Boltzmann equation. The neutron transport theory is beyond the scope of the present book. A concise introduction to this topic can be found in [82]. In this section we employ a one-speed diffusion approximation, which represents the simplest and most widely used mathematical description of the neutron distribution in nuclear reactors.

The net rate of change of the neutron density n can be obtained from the neutron conservation principle, in which the neutron sources and losses are considered. Assuming that neutrons follow Fick's law of diffusion, the one-speed neutron balance equation can be written as,

$$\frac{\partial n(\mathbf{r},t)}{\partial t} = D\nabla^2 \phi(\mathbf{r},t) - \Sigma_a \phi(\mathbf{r},t) + S(\mathbf{r},t), \tag{15.28}$$

where $\phi(\mathbf{r},t)$ is the neutron flux at location $\mathbf{r}$ and time t. The first term on the right-hand side of the equation represents escape (or leakage) of neutrons from a volume element. The term describes the neutron diffusion process with D as the diffusion coefficient. The second term represents the neutron loss due to absorption, where Σ_a is the macroscopic absorption cross section for the one-speed neutrons in the system. The term $S(\mathbf{r},t)$ represents neutron sources end is equal to the rate of neutron production per unit volume and unit time. The neutron source term depends on circumstances, but in nuclear reactors it results from nuclear fissions.

The diffusion equation can be derived from the Boltzmann equation by using a number of simplifying approximations. In particular it can be shown that the diffusion coefficient D is given as,

$$D = \frac{1}{3(\Sigma_t - \Sigma_s \overline{\mu}_0)} \equiv \frac{1}{3\Sigma_{tr}}, \tag{15.29}$$

where $\Sigma_t = \Sigma_a + \Sigma_s$ is the total macroscopic cross section, Σ_s is the macroscopic scattering cross section, and $\overline{\mu}_0$ is the *average cosine of the scattering angle* for collisions in the laboratory system. The reciprocal of the total macroscopic cross section has an important physical interpretation and represents the average total distance λ a neutron will travel before undergoing any collision. Thus,

$$\lambda = \frac{1}{\Sigma_t}, \tag{15.30}$$

where λ is referred to as the **mean free path**. The term $\Sigma_{tr} \equiv \Sigma_t - \Sigma_s\overline{\mu}_0)$ is called the **transport cross section**, and its reciprocal is the **transport mean free path**, λ_{tr},

$$\lambda_{tr} = \frac{1}{\Sigma_{tr}}. \tag{15.31}$$

The differential balance equation Eq. (15.28) will provide a unique solution only when proper initial and boundary conditions are specified. These conditions should be derived from the physical nature of the problem. One of these conditions is concerned with the boundary between the diffusion medium (such as the nuclear reactor core) and the surrounding medium, such as air, which from the neutron transport point of view can be treated as a vacuum.

A widely used vacuum boundary condition, derived from neutron transport theory, is as follows,

$$\phi(r,t)\,|_{b+d} = 0, \tag{15.32}$$

where b represents the physical boundary of the diffusion medium and d is the **extrapolation length** given as,

$$d = 0.7104\lambda_{tr}. \tag{15.33}$$

Since d is usually very small, of the order of few centimeters, it can be neglected in many typical applications with the diffusing medium dimensions of few meters.

Theory of Homogeneous Critical Reactor

The reactor core has to contain a fissile material in which neutron absorptions can lead to fissions and the production of more neutrons. Since fission reactions lead to release of nuclear fission energy, a coolant is needed to remove the heat from the core. The thermal reactor has to contain a moderator material to moderate neutrons from high energies to low energies in order to increase the probability of fissions. Since the reactor core has to be kept in steady-state condition during normal operation, a neutron absorbing material is needed to control the speed of the chain reaction. All these reactor materials must be separated and supported by an internal core structure made of high quality construction materials.

Depending on the size of fuel elements containing the fissile material and the length of the neutron mean free path, neutrons can participate in a varying number of collisions in each of the reactor materials. If the neutron passes through several materials between two consecutive collisions, the reactor is said to be **homogeneous**. If, instead, neutrons undergo several consecutive collisions in each material, the reactor is said to be **heterogeneous**.

In the homogeneous reactor, the medium properties can be calculated based on volume fractions and atomic densities of all component materials. Assuming the one-speed diffusion approximation for a finite homogeneous multiplying medium, the neutron balance equation can be written as,

$$\frac{1}{v}\frac{\partial\phi(\mathbf{r},t)}{\partial t} = D\nabla^2\phi(\mathbf{r},t) - \Sigma_a\phi(\mathbf{r},t) + v\Sigma_f\phi(\mathbf{r},t). \tag{15.34}$$

Assuming a bare reactor in a vacuum, the boundary condition for the neutron flux at the extrapolated boundary $\tilde{\mathbf{b}} = \mathbf{b} + \mathbf{d}$ is as follows,

$$\phi(\tilde{\mathbf{b}},t) = 0, \tag{15.35}$$

were **b** is the physical boundary and **d** is the extrapolation length vector normal to the boundary surface, pointing outward from the medium, and having the magnitude given by Eq. (15.33). The initial condition is given as,

$$\phi(\mathbf{r},0) = \phi_0(\mathbf{r}), \tag{15.36}$$

where $\phi_0(\mathbf{r})$ is the known neutron flux distribution at time $t = 0$.

We seek a solution of the following form,

$$\phi(\mathbf{r},t) = R(\mathbf{r})T(t). \tag{15.37}$$

Substituting Eq. (15.37) into Eq. (15.34) and dividing by DRT yields,

$$\frac{1}{\text{v}D}\frac{1}{T}\frac{dT}{dt} + \frac{\Sigma_a - \text{v}\Sigma_f}{D} = \frac{1}{R}\nabla^2 R. \tag{15.38}$$

Since the terms on the left-hand side is time dependent only, it can be equal to the space-dependent term on the right-hand side of the equation only when both sides of the equation are equal to a constant. Denoting this constant $-B^2$, we obtain the following set of differential equations,

$$\frac{1}{\text{v}D}\frac{1}{T}\frac{dT}{dt} + \frac{\Sigma_a - \text{v}\Sigma_f}{D} = -B^2, \tag{15.39}$$

and

$$\frac{1}{R}\nabla^2 R = -B^2. \tag{15.40}$$

Since the time-dependent equation is a first-order ordinary differential equation, its solution can be easily found as,

$$T(t) = Ce^{-\text{v}\left(\Sigma_a - \text{v}\Sigma_f + DB^2\right)t}, \tag{15.41}$$

where C is the integration constant.

Solution of Eq. (15.40) depends on the shape of the multiplying medium under consideration. Assuming a spherical shape in which the solution depends on the radial coordinate r only and applying substitution $R(r) = U(r)/r$, Eq. (15.40) becomes,

$$\frac{d^2U}{dr^2} + B^2U = 0. \tag{15.42}$$

The general solution of this equation is,

$$U(r) = C_1 \sin Br + C_2 \cos Br, \tag{15.43}$$

and thus the general solution of Eq. (15.40) has the following form,

$$R(r) = C_1 \frac{\sin Br}{r} + C_2 \frac{\cos Br}{r}, \tag{15.44}$$

where C_1 and C_2 are proper integration constants. Since for $r = 0$ the solution should be limited, we require that $C_2 = 0$, and the solution becomes,

$$R(r) = C_1 \frac{\sin Br}{r}. \tag{15.45}$$

Assuming that the extrapolated radius of the sphere is $\tilde{r}_0 = r_0 + d$, where r_0 is the physical radius of the sphere and d is the extrapolation length, the boundary condition that Eq. (15.45) should satisfy is,

$$\sin B\tilde{r}_0 = 0. \tag{15.46}$$

This equation is satisfied by an infinite number of B-values,

$$B_n = \frac{n\pi}{\tilde{r}_0}, \quad n = 1, 3, ..., \tag{15.47}$$

which provides an infinite number of solutions for Eq. (15.45),

$$R_n(r) = C_n^* \frac{\sin B_n r}{r}. \tag{15.48}$$

Values given by Eq. (15.47) and functions given by Eq. (15.48) are called **eigenvalues** and **eigenfuctions** of Eq. (15.40), respectively. Thus, the solution of Eq. (15.34) together with boundary condition given by Eq. (15.35) formulated for a spherical reactor with extrapolated radius $\tilde{r}_0$ is as follows,

$$\phi(r,t) = \sum_{n=1,3,...} C_n e^{-v(\Sigma_a - v\Sigma_f + DB_n^2)t} \frac{\sin B_n r}{r}, \tag{15.49}$$

where constants $C_n = C \cdot C_n^*$ can be found from the given distribution of neutron flux at time $t = 0$. It can be seen that the neutron flux will not grow in time when for each n the following is valid,

$$\Sigma_a - v\Sigma_f + DB_n^2 \geqslant 0, \tag{15.50}$$

or,

$$B_n^2 \geqslant \frac{v\Sigma_f - \Sigma_a}{D}. \tag{15.51}$$

However, since the eigenvalues are ordered as $B_1^2 = (\pi/\tilde{r}_0)^2 < B_2^2 < B_3^2 < ...$, the asymptotic solution of Eq. (15.49) will not grow in time when the following is valid,

$$B_1^2 = \left(\frac{\pi}{\tilde{r}_0}\right)^2 \geqslant \frac{v\Sigma_f - \Sigma_a}{D}. \tag{15.52}$$

In particular, the asymptotic steady-state solution is obtained when $B_1^2 = (\pi/\tilde{r}_0)^2 = (v\Sigma_f - \Sigma_a)/D$ and it is equal to $\phi(r,t) = C_1(\sin \pi r/\tilde{r}_0)/r$. The reactor in which the steady-state prevails is said to be **critical** and equation,

$$B_1^2 = \frac{v\Sigma_f - \Sigma_a}{D}, \tag{15.53}$$

is referred to as the **criticality condition**.

The first eigenvalue B_1^2 is known as **geometric buckling**, because, according to Eq. (15.40) it is equal to $-\nabla^2 R/R$, which is the measure of bending or curvature of the spatial distribution of the neutron flux. The geometry buckling, denoted B_g^2, depends only on the reactor geometry. Similarly, the right-hand side of Eq. (15.53) is known as **material buckling** and it is denoted B_m^2. Thus, the criticality condition can be simply written as $B_g^2 = B_m^2$. When the neutron flux is increasing in time, the reactor is said to be **supercritical**, and the corresponding condition is $B_g^2 < B_m^2$. For $B_g^2 > B_m^2$ the neutron flux decreases with time and the reactor is said to be **subcritical**. Geometric bucklings and the corresponding neutron flux profiles for selected core geometries are given in Fig. 15.4.

The criticality condition given by Eq. (15.53) requires that for a given shape and size of the reactor core the material buckling is precisely determined and set equal to the geometry buckling. Any deviation from this equality will cause the reactor power to increase or decrease in time. The rate of this reactor power change is determined by the first **time eigenvalue** λ_1 given by,

$$\lambda_1 = v\Sigma_a - vv\Sigma_f + vDB_g^2 = v\Sigma_a \left(1 + L^2 B_g^2\right) \left(1 - \frac{v\Sigma_f/\Sigma_a}{1 + L^2 B_g^2}\right), \tag{15.54}$$

Geometry		Geometry buckling B_g^2	Flux profile
Slab		$\left(\frac{\pi}{\tilde{a}}\right)^2$	$\cos\left(\frac{\pi x}{\tilde{a}}\right)$
Infinite cylinder		$\left(\frac{v_1}{\tilde{r}_0}\right)^2$	$J_0\left(\frac{v_1 r}{\tilde{r}_0}\right)$
Sphere		$\left(\frac{\pi}{\tilde{r}_0}\right)^2$	$\frac{1}{r}\sin\left(\frac{\pi r}{\tilde{r}_0}\right)$
Parallelepiped		$\left(\frac{\pi}{\tilde{a}}\right)^2 + \left(\frac{\pi}{\tilde{b}}\right)^2 + \left(\frac{\pi}{\tilde{c}}\right)^2$	$\cos\left(\frac{\pi x}{\tilde{a}}\right)\cos\left(\frac{\pi y}{\tilde{b}}\right)\cos\left(\frac{\pi z}{\tilde{c}}\right)$
Finite cylinder		$\left(\frac{v_1}{\tilde{r}_0}\right)^2 + \left(\frac{\pi}{\tilde{H}}\right)^2$	$J_0\left(\frac{v_1 r}{\tilde{r}_0}\right)\cos\left(\frac{\pi z}{\tilde{H}}\right)$

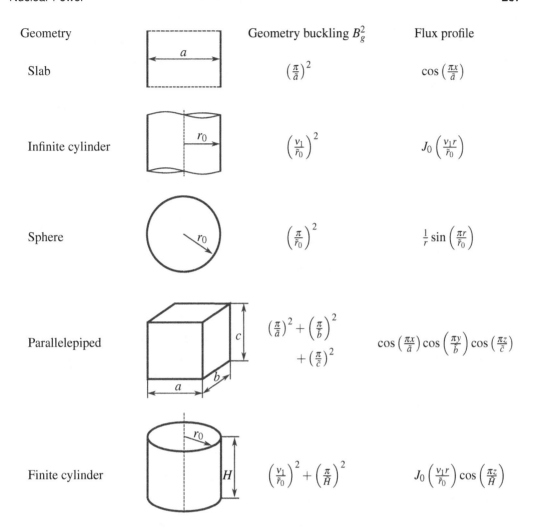

Figure 15.4 Common core shapes with their geometric bucklings and critical flux profiles: $\tilde{a} = a + d$, $\tilde{b} = b + d$, $\tilde{c} = c + d$, $\tilde{r}_0 = r_0 + d$, $\tilde{H} = H + 2d$, $v_1 \approx 2.4048$, and d is the extrapolation length.

where we have defined the **neutron diffusion length** L

$$L \equiv \sqrt{D/\Sigma_a}, \tag{15.55}$$

which is a useful quantity in reactor physics. It can be shown that L is essentially a measure of how far the neutrons will diffuse from a source before they are absorbed. Now the criticality condition can be expressed as,

$$\lambda_1 = 0 \quad \Leftrightarrow \quad k \equiv \frac{v\Sigma_f/\Sigma_a}{1 + L^2 B_g^2} = 1, \tag{15.56}$$

where k is the **multiplication factor** characterizing the chain reaction. It can be interpreted as the ratio of the rate of neutron production to the rate of neutron loss (absorption and leakage) in reactor. The relative deviation of k from one is known as the **reactivity** ρ,

$$\rho = \frac{k-1}{k}. \tag{15.57}$$

During reactor normal operation the multiplication factor is very close to one and the reactivity can be approximated as $\rho \approx k - 1$.

15.1.3 THE NEUTRON CYCLE IN THERMAL REACTOR

To analyze the physical processes that determine the neutron multiplication factor, we consider a homogeneous, uranium-fuelled, thermal-neutron reactor. Usually the fuel consists of uranium and plutonium with a wide variety of isotopic compositions. The natural uranium contains ^{238}U (99.274%), ^{235}U (0.720%), and ^{234}U (0.0.006%). The fractional abundance of a specific isotope of an element is expressed in atom percentage (at%), whereas man-made fraction of fissile ^{235}U which is higher than in naturally occurring uranium is usually expressed as weight fraction (wt%), and is referred to as the **uranium enrichment**.

The ^{235}U atom fraction for uranium is defined as follows,

$$e_a(\text{at}\%) = \frac{\text{No. of atoms } ^{235}U}{\text{No. of atoms U}} \times 100. \tag{15.58}$$

The uranium enrichment as a weight fraction is given as,

$$e_w(\text{wt}\%) = \frac{\text{No. of grams } ^{235}U}{\text{No. of grams U}} \times 100. \tag{15.59}$$

Example 15.2:

Find atomic number densities of ^{235}U and ^{238}U in the UO_2 fuel, knowing that the uranium enrichment is e_w(wt%) and the fuel mass density is ρ.

Solution

The weight fraction of ^{235}U in UO_2 fuel can be found as,

$$w_5 = \frac{N_5 A_5}{N_5 A_5 + N_8 A_8 + 2(N_5 + N_8)A_{Ox}}, \tag{15.60}$$

where A_5, A_8 and A_{Ox} are atomic weights of ^{235}U, ^{238}U, and oxygen, respectively, and N_A is Avogadro's constant. The number density of the ^{235}U can be written as,

$$N_5 = \frac{w_5 \rho N_A}{A_5} = \frac{\rho N_A}{A_5 + 2A_{Ox} + (A_8 + 2A_{Ox})(N_8/N_5)}. \tag{15.61}$$

From the definition of the uranium enrichment we have,

$$\frac{e_w}{100} = \frac{N_5 A_5}{N_5 A_5 + N_8 A_8} = \frac{1}{1 + \frac{A_8}{A_5}\frac{N_8}{N_5}} \tag{15.62}$$

. Thus, the ratio N_8/N_5 can be found as,

$$\frac{N_8}{N_5} = \left(\frac{100 - e_w}{e_w}\right)\left(\frac{A_5}{A_8}\right). \tag{15.63}$$

Finally, we obtain the number densities of the ^{235}U and ^{238}U as follows,

$$N_5 = \frac{\rho N_A}{A_5} \frac{1}{1 + 2\frac{A_{Ox}}{A_5} + \left(1 + 2\frac{A_{Ox}}{A_8}\right)\left(\frac{100 - e_w}{e_w}\right)}, \tag{15.64}$$

$$N_8 = \frac{\rho N_A}{A_8} \frac{1}{\left(1 + 2\frac{A_{Ox}}{A_5}\right)\frac{e_w}{100 - e_w} + 1 + 2\frac{A_{Ox}}{A_8}}, \tag{15.65}$$

and the number density of the oxygen is found as,

$$N_{Ox} = 2(N_5 + N_8). \tag{15.66}$$

Since the number of fissions in a critical reactor has to be constant, not all neutrons produced in fissions lead to additional fissions. Some of the neutrons leak out of the reactor and others are absorbed without fission. Thus it is instructive to analyze the fission chain reaction from the point of view of neutron. We can assume that a given neutron is "born" in a fission event and is then scattering about the reactor until it meets its "death" in either an absorption reaction or by leaking out of the reactor. All these processes, schematically illustrated in Fig. 15.5, are frequently referred to as the "neutron life cycle".

Let us assume that initially there are N_0 fast neutrons in a thermal reactor. These fast neutrons will induce fissions in both fissile and fissionable material. The net increase of fast neutrons due to fissions by fast neutrons is expressed in terms of the **fast fission factor** in such way that after fast fissions, the total number of fast neutrons increases to $N_0\varepsilon$, where ε is the fast fission factor.

In addition to causing fissions, fast neutrons leak out from the reactor. This process is expressed in terms of the **nonleakage probability for fast neutrons** P_{FNL}. Thus, the number of neutrons that start slowing down is $N_0\varepsilon P_{FNL}$.

While slowing down, the neutron energies can coincide with resonances in the capture cross sections of the reactor materials. The probability that a neutron is not captured during the slowing-down process is referred to as the **resonance escape probability** and denoted p. The number of neutrons that reach thermal energies is then given as $N_0\varepsilon P_{FNL}p$. In general, p-factor is not easy to calculate. An approximate, empirical correlation for a uranium-fuelled homogeneous system with $e_a \ll 1$ is [8]

$$p = \exp\left[-\frac{2.73}{\xi}\left(\frac{N_8}{\Sigma_s}\right)^{0.514}\right],$$
(15.67)

where ξ is the average logarithmic energy loss for the mixture of moderator and fuel, N_8 is the atomic number density of $^{238}\mathrm{U}$, and Σ_s is the macroscopic cross section for the moderator-fuel mixture in barns per unit volume.

Thermal neutrons can still participate in non-productive processes that do not lead to fissions. One of such processes is, similarly as for the fast neutrons, the neutron leakage out from the reactor. Defining the **nonleakage probability for thermal neutrons** P_{TNL} in a similar way as for the fast neutrons, the total number of thermal neutrons that remain in the reactor is $N_0\varepsilon P_{FNL}pP_{TNL}$. These neutrons will have to be absorbed in the reactor materials, including fuel, moderator, construction materials, and fission products.

The conditional probability for absorption in fuel is referred to as the **thermal utilization factor** and denoted f. For a system consisting of a homogeneous mixture of fuel and moderator only, the thermal utilization factor can be found as,

$$f = \frac{\Sigma_{a,F}}{\Sigma_{a,F} + \Sigma_{a,M}},$$
(15.68)

where $\Sigma_{a,F}$ is the macroscopic absorption cross section for the fuel and $\Sigma_{a,M}$ is the macroscopic absorption cross section for the moderator.

Thus, the number of thermal neutrons absorbed in fuel is $N_0\varepsilon P_{FNL}pP_{TNL}f$. Finally, recalling the definition of the reproduction factor η, the number of fast neutrons in next generation after thermal fission will be $N_0\varepsilon P_{FNL}pP_{TNL}f\eta$. For a homogeneous uranium-fuelled reactor, the reproduction factor is given approximately by,

$$\eta = 2.42\left[\frac{e_a\sigma_{f,5}}{(100-e_a)\sigma_{a,8}+e_a\sigma_{a,5}}\right],$$
(15.69)

where $\sigma_{f,5}$ is the microscopic fission cross section for $^{235}\mathrm{U}$, $\sigma_{a,5}$ is the microscopic absorption cross section for $^{235}\mathrm{U}$, $\sigma_{a,8}$ is the microscopic absorption cross section for $^{238}\mathrm{U}$, and e_a is uranium enrichment in at%.

The number of neutrons will stay constant in a reactor only when the following condition is satisfied,

$$k_{\text{eff}} \equiv \eta \varepsilon p f P_{FNL} P_{TNL} = 1,$$ (15.70)

where k_{eff} is referred to as the **effective multiplication factor**.

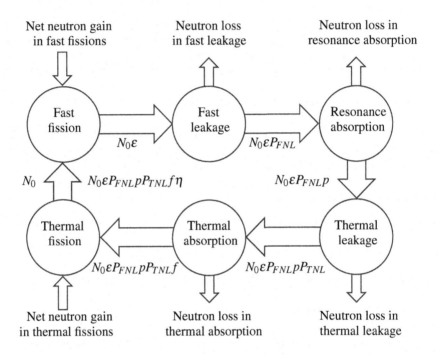

Figure 15.5 Neutron balance and chain fission reactions in a thermal reactor; N_0—initial number of neutrons, ε—fast fission factor, P_{FNL}—nonleakage probability for fast neutrons, p—resonance escape probability, P_{TNL}—nonleakage probability for thermal neutrons, f—thermal utilization factor, η—reproduction factor.

Based on Eq. (15.70), known as the **six-factor formula**, the criticality condition can be expressed in terms of a product of six factors that determine k_{eff}. Two of these factors, namely P_{FNL} and P_{TNL} account for the possibility that neutrons can leak out from a reactor of finite size. For an infinite medium, the nonleakage probabilities are equal to unity, and the corresponding multiplication factor is referred to as the **infinite medium multiplication factor**. It is found from the following **four-factor formula**,

$$k_\infty \equiv \eta \varepsilon p f.$$ (15.71)

In general, calculation of factors in Eq. (15.70) is quite difficult. In practice a more complex iterative approach is followed to determine the criticality condition. Nevertheless the four-factor and six-factor formulas are very useful to help understanding the various mechanisms involved in nuclear reactor operation. They can also support estimates that need to be made in the preliminary reactor design. For example, it can be deduced that the choice of fuel determines the values of η and ε, whereas f and p can be varied considerably by changing the relative location and mass of fuel and moderator. In particular, they can be both increased by a heterogeneous design of the reactor in which fuel elements are surrounded by moderator rather than design containing a homogeneous mixture of fuel and moderator.

15.2   REACTOR ANALYSIS AND DESIGN

The task of reactor analysis and design is to determine set of system parameters which will ensure safe, reliable, and economical reactor operation. Typically the process is initiated with preliminary studies to identify the range over which system parameters can be varied while still preserving the required safety margins. The final best choice of system parameters is obtained in many steps, in which increasingly detailed studies are performed. Multiple libraries with material data and specialized computer code packages are used to support the design process.

The design activities include the nuclear core analysis in order to determine the criticality conditions, the power distribution, the reactivity control, and the fuel depletion. Both steady-state and transient behaviors of the core are taken into account. Other areas of reactor core analyses deal with the thermal and mechanical core analyses, including degradation of materials in the reactor environment.

All reactor designs are subject to thorough studies to ensure that they are satisfying the safety standards imposed by safety authorities. Typically a safety analysis report is required for every nuclear power plant to be licensed for operation.

15.2.1   STEADY-STATE REACTOR PHYSICS

The purpose of steady-state reactor physics is to determine the power distribution throughout a reactor core during normal operation. The most typical calculations start from engineering specifications and a nuclear data library, and employ various numerical approaches to solve the neutron transport equation.

The neutron balance equation in its full form is known as the Boltzmann transport equation. In general this equation is multidimensional, since it considers the distribution of neutrons in space, energy, and time. Thus deterministic methods to solve the Boltzmann equation throughout the reactor core are computationally very expensive and impractical. Instead simplified versions of the Boltzmann transport equation, such as the neutron diffusion equation, are used. Nowadays reactor core analysis and design can be performed using nodal two-group diffusion methods [82].

Raw cross-section data sets provided by measurements need to be significantly processed to make them useful for nuclear engineering applications. In particular, before the data sets can be used with required degree of confidence, they must be evaluated. In addition, to facilitate their usage, the data sets are stored in standard formats. One of the best known standard nuclear data libraries is called the Evaluated Nuclear Data File (ENDF/B) library, created in 1966 at Brookhaven National Laboratory [42]. The 8th major release of the library, referred to as ENDF/B-VIII.0, was published in 2018 [12]. Such reliable and useful data sets are obtained in several organizations around the world, including JEFF in Europe [69], JENDL in Japan [76], CENDL in China [32], and BROND in Russia [10].

A standard experimental nuclear reaction database, known as EXFOR, is used worldwide for exchange of information, including nuclear reaction data, bibliography, sources of data, and history of the data sets. EXFOR format is compatible with the internal storage and retrieval system of five major nuclear data centers: NDS (IAEA)[2], NEADB (OECD)[3], NNDC (USA)[4], JCPRG (Japan)[5] and CDFE (Russia)[6].

---

[2]http://www-nds.iaea.org/exfor/

[3]http://www.oecd-nea.org/janisweb/search/exfor/

[4]http://www.nndc.bnl.gov/exfor/

[5]http://www.jcprg.org/exfor/

[6]http://www.cdfe.sinp.msu.ru/exfor/

15.2.2 THERMAL-HYDRAULIC DESIGN

Determination of core temperature distributions and heat transfer rates plays an important role in reactor design. Most of the power reactors have as a goal to produce steam with high enough parameters to be used in an associated steam thermal cycle. High thermal efficiency of the thermal cycle requires steam at high temperature. However, the maximum achievable temperature in a reactor core is strictly limited by the fuel failure and the release of radioactive material into the coolant. One of the main goals of the thermal-hydraulic design is to assure that the reactor efficiency and safety constraints are met. The design calculations are performed with dedicated computer codes, which solve heat transfer equations using power distributions obtained from the nuclear core analysis.

15.3 REACTOR KINETICS AND DYNAMICS

The emission of delayed neutrons during a nuclear fission is strongly affecting the transient behavior of a nuclear reactor. Delayed neutrons are emitted in a chain of beta-decays of certain fission products, referred to as the **precursors of delayed neutrons**. There are about 40 different precursors of delayed neutrons that are produced in fissions of ^{235}U. Usually all these precursors are lumped into six groups characterized by yield fractions and decay constants.

Reactor kinetics is the study of time-dependent behavior of a nuclear reactor due to changes in the reactivity. In the **point kinetics model** it is assumed that the transient behavior of the reactor can be described by a single parameter, such as the total number of neutrons in the core or the total power of the core. Using parameters of the critical reactor as a reference, the normalized point kinetic equations can be written as follows,

$$\frac{dx}{dt} = -\frac{(\beta-\rho)}{\Lambda}x + \frac{1}{\Lambda}\sum_{i=1}^{6}\beta_i y_i + \frac{\rho}{\Lambda} + \frac{S}{n_e}, \qquad (15.72)$$

$$\frac{dy_i}{dt} = \lambda_i(x - y_i), \quad i = 1,...,6, \qquad (15.73)$$

where $x = (n - n_e)/n_e$ is a normalized deviation of the number of neutrons in the reactor core from the equilibrium condition for the critical reactor, n—actual number of neutrons, n_e—equilibrium number of neutron in the critical reactor, $y_i = (C_i - C_{ie})/C_{ie}$—normalized deviation of precursor-i concentration from the equilibrium, C_i—actual concentration of precursor i, C_{ie}—concentration of precursor i at equilibrium, S—external neutron sources, $\beta = \sum_{i=1}^{6}\beta_i$—total yield of delayed neutrons, β_i—yield of delayed neutrons of i-th group, λ_i—decay constant of i-th group of precursors, Λ—average neutron generation time[7].

The **six-group point kinetics model** given by Eq. (15.72) and Eq. (15.73) consists of seven linear ordinary differential equations with variable coefficients. If we further assume that the coefficients in the equations are constant, the set of equations can be solved analytically. In particular, a simple **one-group point kinetics model** with constant coefficients can be formulated as follows,

$$\frac{dx}{dt} = -\frac{(\beta-\rho)}{\Lambda}x + \frac{\beta y + \rho}{\Lambda} + \frac{S}{n_e}, \qquad (15.74)$$

$$\frac{dy}{dt} = \lambda(x - y). \qquad (15.75)$$

Here we assumed the averaged decay constant λ as follows,

$$\lambda = \beta\left(\sum_{i=1}^{6}\frac{\beta_i}{\lambda_i}\right)^{-1}. \qquad (15.76)$$

---

[7]The name results from the fact that Λ represents the average time between two birth events in successive neutron generations.

Equations (15.74) and (15.75) can be solved analytically for various changes of the reactivity. In particular, assuming a step change of reactivity as,

$$\rho = \begin{cases} 0 & t < 0 \\ \rho_0 & t \geq 0 \end{cases},$$ (15.77)

and the following initial conditions,

$$x(0) = 0, \quad y(0) = 0,$$ (15.78)

the solution for the relative power change is as follows,

$$x(t) = \left(\frac{\rho_0}{\Lambda} + \frac{S_0}{n_e}\right)\left[\frac{\lambda}{s_1 s_2} + \frac{s_1 + \lambda}{s_1(s_1 - s_2)}e^{s_1 t} + \frac{s_2 + \lambda}{s_2(s_2 - s_1)}e^{s_2 t}\right],$$ (15.79)

where,

$$s_{1,2} = \frac{-\left(\frac{\beta}{\Lambda} - \frac{\rho_0}{\Lambda} + \lambda\right) \pm \sqrt{\left(\frac{\beta}{\Lambda} - \frac{\rho_0}{\Lambda} + \lambda\right)^2 + 4\frac{\lambda \rho_0}{\Lambda}}}{2}.$$ (15.80)

The solution given by Eq. (15.79) indicates that the reactor power will increase indefinitely if any of the roots s_1 and s_2 is greater than zero. Inspecting Eq. (15.80) reveals that this will always be the case when $\rho_0 > 0$. The role of the delayed neutrons (represented by β in Eq. (15.80)) is to reduce the rate at which the power increases.

In reality, the reactivity is a function of reactor power. When the reactor power increases, the reactivity changes due to the temperature changes of the fuel and the moderator. The **temperature coefficient of reactivity** is defined as,

$$\alpha_T \equiv \frac{\partial \rho}{\partial T} = \frac{\partial}{\partial T}\left(\frac{k-1}{k}\right) = \frac{1}{k^2}\frac{\partial k}{\partial T} \approx \frac{1}{k}\frac{\partial k}{\partial T}.$$ (15.81)

One of the fundamental properties of nuclear fuel is that when its temperature increases, the resonance escape probability, given by Eq. (15.67) for the uranium-fuelled homogeneous reactor, decreases. This is caused by the so called **Doppler effect**, which in thermal reactors is due primarily to epithermal capture resonances in the nonfissionable fuel isotopes (^{232}Th, ^{238}U, ^{240}Pu). Thus, a more realistic expression for the reactivity change is as follows,

$$\delta \rho = \delta \rho_C(t) + \alpha_T \delta T + \cdots,$$ (15.82)

where $\delta \rho_C(t)$ is the reactivity change introduced by the control system, δT is the temperature change in the reactor core and α_T is the corresponding temperature reactivity coefficient. Typical values of the temperature coefficient of reactivity for fuel (also called the **Doppler coefficient**) in light water reactors are from -1 pcm/K to -4 pcm/K.[8]

The combined response of a reactor power, fuel and coolant temperatures, and system pressure including the effect of reactivity feedbacks is the subject of **reactor dynamics**. In addition to the Doppler coefficient, the effects of other reactivity coefficients for coolant temperature and coolant void, and due to thermal expansion of the core, are investigated.

---

[8]1 pcm/K corresponds to reactivity change of 1 pcm (per cent mille = 10^{-5}) per temperature change of one kelvin.

15.4   FUEL COMPOSITION CHANGES

During reactor operation the composition of the fuel changes in both space and time. These changes can be traced by fuel burnup calculations which provide a picture of the core composition, thermal power distribution, effective multiplication factor, and other reactor parameters as a function of time. The purpose of burnup calculations is to determine the best arrangement of fuel in the core from the point of view of heat removal and the fuel economy.

Two aspects of fuel burnup are described in more detail in this section. The first one is concerned with fuel conversion and breeding that are closely related to the sustainability potential of the nuclear power. The second aspect is related to the fuel poisoning that is affecting the reactivity in nuclear reactor.

15.4.1   FUEL CONVERSION AND BREEDING

During reactor operation, the composition of the reactor core slowly changes with time. The changes are slow enough to keep the reactor in critical state by the control system adjustments. This means that the composition changes of the core can be studied in time intervals during which a constant neutron flux can be assumed.

It is extremely important to be able to accurately model the nuclide composition in the core during reactor operation. This composition is strongly affecting the core multiplication as well as flux and power distributions. For sustainability reasons, it is important to follow the process of conversion and breeding of new fuel.

The differential equations governing the fuel composition changes are called the Bateman equations. These equations can be formulated for all nuclides to describe the time change of their atomic densities during the depletion time step. Assuming properly averaged cross sections and fluxes, the Bateman equations are as follows,

$$\frac{dN_k}{dt} = \phi \sum_{i=s}^{m} N_i \sigma_{f,i} y_{i \to k} + \phi \sum_{z=r}^{q} N_z \sigma_{c,z} \gamma_{z \to k} + \sum_{j=n}^{p} N_j \lambda_j \alpha_{j \to k} - \lambda_k N_k - \phi N_k \sigma_{a,k}, \qquad (15.83)$$

where: N_k—time dependent atomic density of nuclide k during the depletion time step, ϕ—neutron flux during the depletion time step, $\sigma_{f,i}$—microscopic fission cross section of nuclide i,—$\sigma_{c,z}$—microscopic capture cross section of nuclide z including (n, xn) reactions,—$\sigma_{a,k}$—microscopic absorption cross section of nuclide k, λ_j, λ_k—radioactive decay constants of nuclides j and k, $y_{i \to k}$—yield of nuclide k due to fission of nuclide i, $y_{z \to k}$—probability that a neutron capture in nuclide z produces nuclide k, $\alpha_{j \to k}$— probability that the radioactive decay of nuclide j produces nuclide k, k—nuclide index for which the equation is written, i—summation index over all precursor nuclides m yielding nuclide k through neutron-induced fission events, $i = s, ..., m$, z—summation index over all precursor nuclides q yielding nuclide k through neutron capture events, $z = r, ..., q$, j—summation index over all precursor nuclides p yielding nuclide k through radioactive events, $j = n, ..., p$, s, r, n—specific first nuclides undergoing fission, capture, and radioactive decay events, respectively.

The main reason for development of fast reactors has been to achieve fuel breeding which could provide clean energy resource for our planet for several millennia. Fuel breeding is achieved by conversion of a fertile isotope into a fissile isotope by neutron capture (n, γ). The two naturally occurring fertile isotopes are $^{238}_{92}U$ and $^{232}_{90}Th$, which can be converted into fissile isotopes $^{239}_{94}Pu$ and $^{233}_{92}U$, respectively.

A conversion chain of $^{238}_{92}U$ into $^{239}_{94}Pu$, together with generation of americium and curium, is shown in Fig. 15.6. The changes of concentration of the most important nuclides in the chain can be described with the Bateman equations. The mechanism of generation of ^{239}Pu (nuclide with index $k = 2$ in the figure) involves several stages. By capturing a neutron, ^{238}U ($k = 1$) becomes ^{239}U that rapidly beta-decays ($T_{1/2} = 23.5$ min) into ^{239}Np. This nuclide is then undergoing a consecutive beta decay ($T_{1/2} = 2.35$ d) and transforms into ^{239}Pu ($k = 2$). Similarly as ^{235}U, ^{239}Pu is a fissile nuclide. Due to additional neutron captures, nuclides ^{240}Pu ($k = 3$), ^{241}Pu ($k = 4$), and ^{242}Pu ($k = 5$) are produced. Finally, ^{241}Am ($k = 7$) and ^{243}Am ($k = 6$) are resulting from beta-decays of ^{241}Pu ($T_{1/2} = 14.3$ y) and ^{243}Pu ($T_{1/2} = 4.98$ h), respectively.

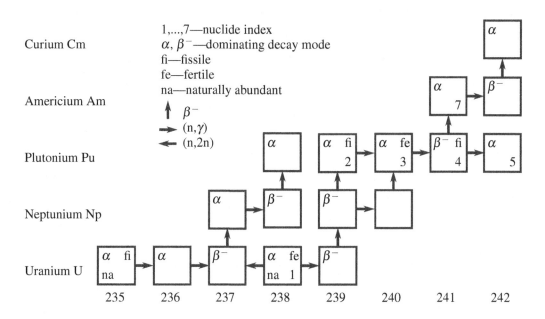

Figure 15.6 Uranium-plutonium conversion chain.

A conversion chain of ^{232}Th into $^{233}_{92}$U, together with generation of neptunium, is shown in Fig. 15.7. Initially, due to the neutron capture, ^{232}Th ($k = 1$) transforms into ^{233}Th, which beta-decays ($T_{1/2} = 22.2$ min) into ^{233}Pa. Next, ^{233}U ($k = 2$) is produced by the consecutive beta-decay of ^{233}Pa ($T_{1/2} = 27.0$ d). Further neutron captures lead to the following nuclides: ^{234}U ($k = 3$), ^{235}U ($k = 4$), ^{236}U ($k = 5$), and ^{237}U. Finally, ^{237}Np ($k = 6$) is produced by beta-decay of ^{237}U ($T_{1/2} = 8.75$ d).

15.4.2 FISSION PRODUCT POISONING

Fission products are of concern for reactor operation since they contain materials which are parasitic absorbers of neutrons. In particular, two isotopes have significant microscopic cross section for absorption: ^{135}Xe and ^{149}Sm. Both these isotopes behave differently, since xenon is unstable and with time is removed from fission products due to decay, whereas samarium is stable and accumulates in the fission products. Due to high microscopic cross section for absorption of these isotopes, their presence in the reactor core has a strong impact on the thermal utilization factor, and thus on the effective multiplication factor and the reactivity.

Xenon Poisoning

The neutron absorption cross section of ^{135}Xe is about 2.6×10^6 barns. It is produced directly by fission with insignificant yield, but it is a product of tellurium ^{135}Te decay chain,

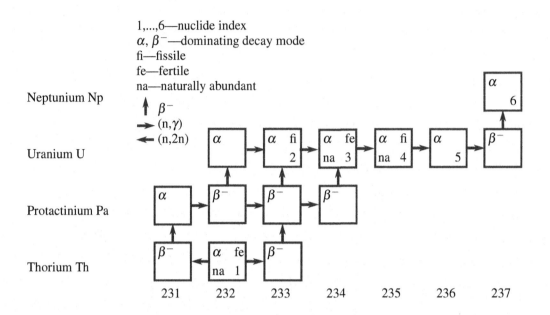

Figure 15.7 Thorium-uranium conversion chain.

$$\,^{135}_{52}\text{Te} \xrightarrow[19\,\text{s}]{\beta^-} \,^{135}_{53}\text{I} \xrightarrow[6.57\,\text{h}]{\beta^-} \,^{135}_{54}\text{Xe} \xrightarrow[9.10\,\text{h}]{\beta^-} \,^{135}_{55}\text{Cs} \xrightarrow[2.3\times10^6\text{y}]{\beta^-} \,^{135}_{56}\text{Ba}. \tag{15.84}$$

The half-life of ^{135}Te is so short that it can be assumed that ^{135}I is produced directly from fission. ^{135}I is not a strong neutron absorber, but it decays to form ^{135}Xe. Thus, the half-life of ^{135}I plays an important role in the nuclear reactor poisoning by ^{135}Xe.

The rate of change of ^{135}I concentration depends on its production due to fission and its removal due to decay and neutron absorption,

$$\frac{\mathrm{d}I}{\mathrm{d}t} = \gamma_I \Sigma_f \phi - \lambda_I I - \sigma_{a,I} I \phi, \tag{15.85}$$

where I is the atomic number density of ^{135}I, γ_I - yield of ^{135}I from fission, λ_I—decay constant of ^{135}I, $\sigma_{a,I}$—microscopic absorption cross section for ^{135}I, Σ_f—macroscopic fission cross section, and ϕ—neutron flux density. Similarly, the rate of change of concentration of ^{135}Xe is given as,

$$\frac{\mathrm{d}X}{\mathrm{d}t} = \gamma_X \Sigma_f \phi + \lambda_I I - \lambda_X X - \sigma_{a,X} \phi X, \tag{15.86}$$

where X—atomic number density of ^{135}Xe, γ_X—effective yield of ^{135}Xe from fission, λ_X—decay constant of ^{135}Xe, and $\sigma_{a,X}$—microscopic absorption cross section for ^{135}Xe. For a critical reactor operating at a constant power, the equilibrium concentrations of ^{135}I (I_0) and ^{135}Xe (X_0) are found as,

$$\frac{\mathrm{d}I}{\mathrm{d}t} = 0 = \gamma_I \Sigma_f \phi - \lambda_I I_0 \Rightarrow I_0 = \frac{\gamma_I \Sigma_f \phi}{\lambda_I}, \tag{15.87}$$

and

$$\frac{\mathrm{d}X}{\mathrm{d}t} = 0 = \gamma_X \Sigma_f \varphi + \lambda_I I_0 - \lambda_X X_0 - \sigma_{a,X} \varphi X_0 \Rightarrow X_0 = \frac{\gamma_X \Sigma_f \varphi + \lambda_I I_0}{\lambda_X + \sigma_{a,X} \varphi}, \tag{15.88}$$

where the neutron absorption by ^{135}I is neglected as small in comparison with other terms.

After shutdown of a nuclear reactor that was operating for a long time at steady-state, the concentrations of ^{135}I and ^{135}Xe are given by the following set of differential equations and initial conditions,

$$\frac{dI}{dt} = -\lambda_I I, \quad I(0) = I_0, \tag{15.89}$$

$$\frac{dX}{dt} = \lambda_I I - \lambda_X X, \quad X(0) = X_0. \tag{15.90}$$

The time change of ^{135}Xe concentration is found as,

$$X(t) = \frac{\phi \Sigma_f (\gamma_X + \gamma_I)}{\lambda_X + \phi \sigma_{a,X}} e^{-\lambda_X t} + \frac{\phi \Sigma_f \gamma_I}{\lambda_I - \lambda_X} \left(e^{-\lambda_X t} - e^{-\lambda_I t} \right) \tag{15.91}$$

with maximum value at,

$$t_{max} = \frac{1}{\lambda_X - \lambda_I} \ln \left[\frac{\lambda_X}{\lambda_I} \left(1 + \frac{\lambda_I - \lambda_X}{\lambda_I} \frac{X_0}{I_0} \right) \right], \tag{15.92}$$

where t_{max} is time in seconds after the reactor shutdown. For typical thermal power reactors, the peak xenon time is approximately equal to 11.6 h.

Samarium Poisoning

Samarium-149 is the second most important fission product poison because of its high thermal neutron absorption cross section of 4.1×10^4 barns. This isotope appears in the decay chain of ^{149}Nd, which is a fission product with a significant yield. The decay chain that leads to creation of ^{149}Sm is as follow,

$$^{149}_{60}\text{Nd} \xrightarrow[1.73\text{ h}]{\beta^-} {}^{149}_{61}\text{Pm} \xrightarrow[53.1\text{ h}]{\beta^-} {}^{149}_{62}\text{Sm}, \tag{15.93}$$

where ^{149}Sm is stable.

The behavior of ^{149}Sm can be analyzed from a simplified model, in which the promethium is assumed to be created directly from fissions, since the ^{149}Nd half-life time (1.73 h) is sufficiently shorter than the ^{149}Pm half-life time (53.1 h). Thus the rate of change of the ^{149}Pm is as follows,

$$\frac{dP}{dt} = \gamma_P \Sigma_f \phi - \lambda_P P, \tag{15.94}$$

where γ_P is the effective yield of ^{149}Pm, P—atomic number density of ^{149}Pm, Σ_f—macroscopic cross section for fission, ϕ—the neutron flux, and λ_P—the decay constant of ^{149}Pm.

The rate of formation of the ^{149}Sm is as follows,

$$\frac{dS}{dt} = \gamma_S \Sigma_f \phi + \lambda_P P - \sigma_{a,S} \phi S. \tag{15.95}$$

Here S is the atomic number density of ^{149}Sm, γ_S—the effective yield of ^{149}Sm, and $\sigma_{a,S}$—the microscopic cross section for absorption of thermal neutrons by ^{149}Sm. Since the yield of ^{149}Sm is nearly zero, the equation simplifies to,

$$\frac{dS}{dt} = \lambda_P P - \sigma_{a,S} \phi S. \tag{15.96}$$

In all reactors except those operating at very low flux, the concentrations of ^{149}Pm and ^{149}Sm come into equilibrium in a few days' time. These equilibrium concentrations are found from Eqs. (15.94) and (15.95) by placing the time derivatives equal to zero, and are found to be

$$P_0 = \frac{\gamma_P \Sigma_f \phi}{\lambda_P},$$ (15.97)

and

$$S_0 = \frac{\gamma_P \Sigma_f}{\sigma_{a,S}}.$$ (15.98)

After reactor shutdown, the concentrations are described with the following equations,

$$\frac{\mathrm{d}P}{\mathrm{d}t} = -\lambda_P P,$$ (15.99)

$$\frac{\mathrm{d}S}{\mathrm{d}t} = \lambda_P P,$$ (15.100)

and the initial conditions are as follows,

$$P(0) = P_0, \quad S(0) = S_0.$$ (15.101)

Here P_0 and S_0 are concentrations of promethium and samarium at equilibrium, respectively. The post-shutdown concentration of samarium is found as

$$S(t) = S_0 + P_0 \left(1 - e^{-\lambda_P t}\right).$$ (15.102)

At $t \to \infty$, the samarium concentration approaches $S_0 + P_0$ as the promethium decays into samarium with time constant approximately equal to 78 h.

15.5   REACTOR TYPES

There are five main nuclear reactor types that are currently operable and additional six types which are under development, as shown in Fig. 15.8 and Fig. 15.9, respectively. Light water reactors and other thermal spectrum reactors are most commonly used for power generation and for naval propulsion. Fast spectrum reactors are under development.

15.5.1   CURRENTLY OPERABLE REACTORS

The most common type of reactor for power generation is the **pressurized water reactor** (PWR). PWRs use ordinary water as both coolant and moderator. The primary cooling circuit, which flows through the core and the steam generator, contains water at very high pressure to avoid boiling in the core. The secondary circuit is used to generate steam in the steam generator. The steam is used to drive the turbine in the same way as in the fossil fired power plants.

Only a single external circuit is used in the **boiling water reactor** (BWR). The pressure in the circuit is about 7–7.5 MPa, allowing water to boil in the core. The reactor is designed to operate with 12–15% steam fraction by mass in the top part of the core. After passing through steam separators and dryers, the steam is dry enough to drive the turbine. Since the turbine is a part of the reactor circuit, it can be contaminated with short-lived radioactivity that is always present in the primary circuit. Thus the turbine must be shielded and radiological protection must be provided during maintenance.

The **pressurized heavy water reactor** (PHWR) has been developed since the 1950s in Canada as the CANDU, and from 1980s also by India. PHWRs are using heavy water as moderator and natural (or slightly enriched) uranium as fuel. Because of its satisfactory slowing-down power and small capture cross section for neutrons, heavy water is an excellent moderator and reflector. Thanks to this, a natural-uranium reactor is smaller and requires considerably less fuel if heavy water is the moderator rather than graphite.

The **gas cooled reactor** (GCR) has been developed in the UK, first as the *Magnox* reactor and next as the **advanced gas cooled reactor** (AGR). AGRs use carbon dioxide as the coolant, graphite as the moderator, and enriched UO_2 as the fuel. The carbon dioxide circulates through the core, reaching 650 °C, and through steam generators placed inside the concrete and steel pressure vessel. The steam high temperature yields a high thermal efficiency of about 41%.

The light water graphite-moderated reactor (LWGR) has been developed in Soviet Union as the RBMK. It employs 7 meter long vertical pressure tubes running through graphite moderator. Similarly as in BWRs, water flowing through the core is allowed to boil at 290 °C and about 6.9 MPa. Low enriched UO_2 is used as the fuel. Since moderation is provided by graphite, the boiling of coolant reduces its density without inhibiting the fission reaction, which can lead to positive reactivity feedback.

15.5.2 ADVANCED REACTORS

The six types of advanced reactors which are under development, frequently referred to as the **generation IV reactors**, are shown in Fig. 15.9. The development of generation IV reactors was initiated by US Department of Energy in the year 2000, when the Generation IV International Forum (GIF) was launched. The main goals for the generation IV reactors include reduction of radiative wastes as well as improvement of safety, economy and proliferation resistance.

The **sodium cooled fast reactor** (SFR) is a fast-neutron spectrum system designed for management of high-level wastes and, in particular, management of plutonium and other actinides. The primary coolant system can either be arranged in a pool layout, or in compact loop layout. Sodium core outlet temperatures are typically 530–550 °C. A large margin to coolant boiling is achieved by design, which is an important safety feature of SFRs. To improve safety, a secondary sodium system acts as a buffer between the reactor core sodium and the working fluid (water/steam or nitrogen) in the balance-of-plant system.

The **molten-salt reactor** (MSR) uses the molten-salt fuel mixture to produce the fission power. MSRs are fueled with uranium or plutonium fluorides dissolved in a mixture of molten fluorides, with Na and Zr fluorides as the primary option. MSRs have unique inherent safety features thanks to fail-safe drainage, passive cooling, and a low inventory of volatile fission products in the fuel. The reactor can use ^{238}U or ^{232}Th, as a fertile material dissolved as fluorides in the molten salt. Due to the thermal or epithermal spectrum of the fluoride MSR, ^{232}Th achieves the highest conversion fractions.

The **lead-cooled fast reactor** (LFR) is a fast-neutron spectrum system with closed fuel cycle, and liquid Pb or Pb-Bi alloy as a coolant. Various options are considered, including a long refuelling interval battery ranging from 50–150 MWe, a modular system from 300–400 MWe, and a large monolithic plant at 1200 MWe. This type of reactor has a potential to operate with natural circulation and at higher temperatures than liquid sodium allows. Energy conversion systems can employ supercritical Brayton or Rankine cycles, or process heat applications such as hydrogen production and desalination.

The **gas-cooled fast reactor** (GFR) is a fast-spectrum, helium-cooled and closed fuel cycle system. Versatile applications, such as electricity production, hydrogen generation, or process heat are feasible due to the high outlet temperature of the helium coolant.

15.6 NUCLEAR FUEL CYCLE

The nuclear fuel cycle starts with the mining of uranium and ends with the disposal of nuclear waste. There are many techniques to mine uranium ore, with underground mining being the most common due to the depths of most deposits. In general, surface mining is used where deposits are closer to the surface than 120 m. Increasingly popular is in situ leach mining, where oxygenated groundwater is circulated through a very porous orebody to dissolve the uranium oxide and bring it

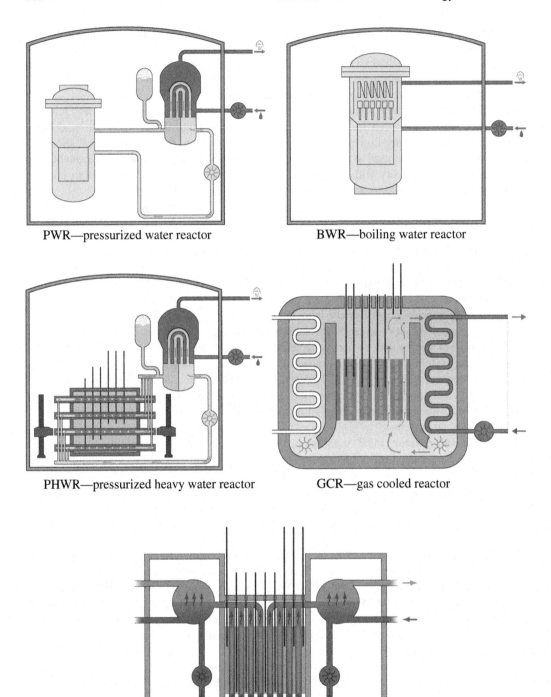

PWR—pressurized water reactor   BWR—boiling water reactor

PHWR—pressurized heavy water reactor   GCR—gas cooled reactor

LWGR—light water graphite-moderated reactor

Figure 15.8 Main reactor types currently in commercial operation. Source: Wikimedia Commons.

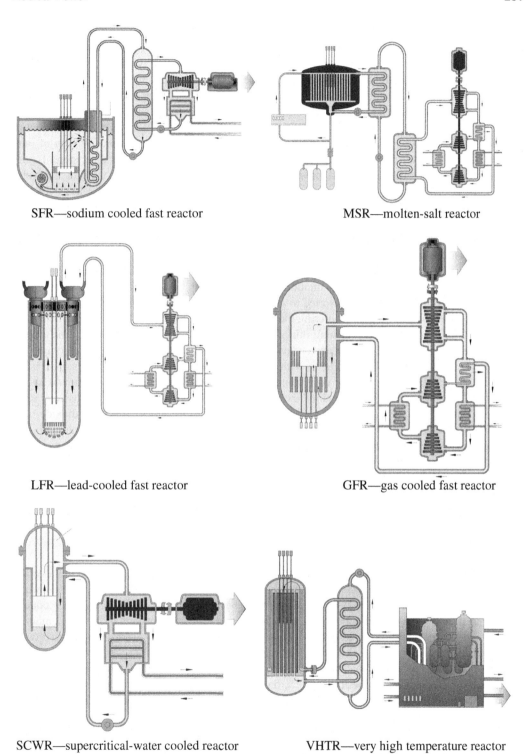

SFR—sodium cooled fast reactor   MSR—molten-salt reactor

LFR—lead-cooled fast reactor   GFR—gas cooled fast reactor

SCWR—supercritical-water cooled reactor   VHTR—very high temperature reactor

Figure 15.9 Generation IV reactors. Source: Wikimedia Commons.

to the surface. The ore is brought to a mill where it is crushed to a fine powder and further processed with concentrated acid, alkaline, or peroxide solutions to convert into U_3O_8 (triuranium octoxide), a form of uranium referred to as **yellowcake**. Actually modern yellowcake typically contains 70% to 90% of U_3O_8 with some additions of uranium dioxide (UO_2) and uranium trioxide (UO_3). During this initial process, the concentration of uranium increases from as little as 0.1% in the original ore to more than 80% in the yellowcake.

The concentration of the fissile $^{235}_{92}U$ needs to be increased from 0.72%, typical for the natural uranium, to between 3.5% and 5%. The process of isotope separation and concentration of one isotope relative to others is called enrichment. The uranium enrichment requires three steps. In the first step uranium dioxide is converted into uranium hexafluoride (UF_6), which is a gas at relatively low temperature. In the next step, the isotope separation takes place. Since fluorine has only one isotope, UF_6 molecules differ in mass about 1% only due to different atomic masses of $^{235}_{92}U$ and $^{238}_{92}U$. Using various methods, such as based on centrifuges or using the laser enrichment, the gaseous uranium hexafluoride is separated into two streams: one being enriched to the required level, and the other stream depleted in $^{235}_{92}U$. In the final third step, the enriched uranium hexafluoride is de-converted into uranium dioxide.

A fuel manufacturing plant receives uranium either as uranium hexafluoride or uranium trioxide. These uranium forms need to be converted to uranium dioxide prior to pellet fabrication. Ceramic pellets are formed by pressing and sintering the uranium dioxide at a high temperature. The cylindrical pellets are then encased into long thin zircaloy tubes which serve as cladding. The empty space in the tube is filled with helium and an end cap is sealed on. After inspection of the completed fuel rods, they are assembled into bundles, with appropriate spacers according to the reactor type.

On average, a 1000 MWe reactor core contains 75 tonnes of low enriched uranium and about 27 tonnes of fresh enriched fuel is required each year. During fission process of $^{235}_{92}U$, fission products are produced and accumulated in fuel rods. Due to irradiation, sintered uranium dioxide fuel pellets may undergo marked structural changes, primarily as a result of the high internal temperature and the large temperature gradients. Thermal stresses lead to radial cracks and changes in the grain structure. In addition, other forms of fuel degradation take place such as swelling and densification, effectively limiting the level of fuel burn-up to about 40 to 55 GWd/t, depending on the fuel enrichment. Burn-up in GWd/t (gigawatt-days thermal per tonne of heavy metal) is the conventional measure for oxide fuels, and 60 GWd/t is equivalent to about 6.5% of the original uranium atoms to fission directly or indirectly via transformation to fissile plutonium[9]. For metallic fuels, the atomic percent metrics is used, and a new light water reactor fuel is targeting 21 atomic percent burn-up [62].

After 18–36 months of operation, the used fuel is removed from the reactor. The used fuel usually have about 3% of fission products and minor actinides, 1% of $^{235}_{92}U$, 0.6% of fissile plutonium (almost 1% of plutonium in total) and 95% of $^{238}_{92}U$. Several hundred different fission products are produced during the reactor operation. Due to heat generation and emission of radiation, mainly from fission products and minor actinides, the used fuel is unloaded into a storage pool filled with water immediately adjacent to the reactor. Used fuel is held in such pools for several months or even years until it becomes less radioactive and can be moved to a storage that is cooled by naturally-circulating air. Depending on the local policy, the used fuel may be transferred to central storage facilities, before it is either reprocessed in order to recycle most of it, or prepared for permanent disposal.

The purpose of reprocessing of used fuel is to remove fission products and to recover uranium and plutonium. In the first stage, fuel assemblies are chopped into sections from which the spent material is leached with hot nitric acid. Next, separation and purification of materials is performed. Various

---

[9]Plutonium fission provides about one-third of the energy output in light water reactors and more than half in CANDU reactors. Even though in CANDU reactors using natural uranium burn-up is only 7.5 GWd/t, due to plutonium fission, this is equivalent to almost 50 GWd/t for enriched fuel.

complex solvent-extraction separation processes can be used for separation of compounds or metal complexes. The purex process, using n-tributyl phosphate as the extractant, is typical of solvent extraction procedures employed in the treatment of used fuel. Reprocessing enables recycling of the uranium and plutonium into fresh fuel, and produces a significantly reduced amount of waste. Typically the amount of waste from a 1000 MWe reactor can be as low as 750 kg per year.

15.7 NUCLEAR POWER SAFETY

Adequate safety level of nuclear power plants can be achieved through a combination of proper design, manufacture, construction, and operation of their systems and components. The most important general safety consideration in design is to protect individuals, society, and environment from any harm due to radiological hazards. The radiation protection has as a goal to ensure that any radiation exposure is kept below prescribed limits and as low as reasonably achievable. In case of any accidents, means should be envisioned to mitigate their radiological consequences.

Typical safety design of a nuclear power plant is based on deterministic analyses complemented by probabilistic assessments. The deterministic approach follows the safety concept of **defense in depth** which usually comprises four levels. The first level is based on conservative design, quality assurance, and surveillance activities to prevent departures from normal operation. The second level of protection is provided to ensure the integrity of the fuel cladding and the reactor coolant pressure boundary. To this end detection of deviations from normal operation is used. For example, the reactor core should be designed in such way that fuel pellets do not release an excessive amount of radioactive fission products, and they do not challenge the integrity of the fuel cladding. Any failure of fuel cladding can be detected due to increase of the radioactivity level of coolant, and further propagation of the failure can be stopped. The third level of defense is provided by engineered safety features and protective systems that ensure mitigation of accidents and consequently prevent their evolution into severe accidents. Finally, the fourth level of protection is the containment together with systems preserving its integrity in case of a severe accident.

In addition to the confinement of radioactive material, two other safety functions are of fundamental importance and should be considered in the design of the reactor core. Firstly, a reliable and robust system for control of reactivity should be designed. Such system enables the power level and the power distribution to be maintained within safe operating limits. Secondly, continuous and reliable removal of heat from the core should be ensured so that the peak fuel temperature and the cladding temperature have sufficient margins to their limit safety values.

A probabilistic safety approach is used to analyze low probability events with multiple failures of varying levels of severity, including the total loss of safety-grade systems. The procedures are based on so called fault trees and event trees, more extensively discussed in §17.1.2. A system **fault tree** is a logic diagram which represents the component failure modes that, when combined, lead to the system failure. Correspondingly, system **event tree** is a graphical means of identifying the various possible consequences of the initiating event, showing the evolution of a series of events in time. A successful construction of an event tree provides a qualitative analysis of what happens after an initiating event. Various methods can be used when a quantitative analysis is desired. One of them is to assign numerical values from fault tree analysis to different state in nuclear systems.

A comprehensive reactor safety study based on the use of fault trees and event trees was published in 1975 and is commonly known as WASH-1400 or the Rasmussen Report [70]. The study includes analysis of several thousand potential accident scenarios. Many of these scenarios could be identified as of no serious concern due to either very low probability or insignificant consequences. The study did, however, identify a small number (about 20) of cases with the major sources of risk. Another important aspect that the study indicated was the need for an adequate data base that would allow for reduction of uncertainties.

The accidents in the Three Mile Island in 1979, Chernobyl in 1986, and Fukushima in 2011 contributed to the current emphasis on the use of probabilistic techniques for the analysis of nuclear systems, even though reactor design characteristics and scenarios in these three accidents were entirely different [58].

15.8 FUSION REACTORS AND OTHER TECHNOLOGIES

At the present time, commercial-scale electricity production from nuclear energy is possible only in fission reactors. However, since the late 1940s, there have been sustained efforts at trying to harness the fusion energy as a source of useful power. This is mainly because the potential energy yield by a deuterium-based fusion is enormous. Deuterium has an abundance of 0.015% of all hydrogen isotopes on Earth. Thus, in the 1.386×10^9 km^3 of Earth's water, there are about 1.4×10^{43} deuterium atoms. If all these deuterium atoms could be fused in D–D reactions, about 1.8×10^{31} J of energy could be obtained (see Problem 1.10 for the Q value of such reaction). At the world's energy consumption rate in 2017 (about 4×10^{20} J), this amount of energy source would provide our energy needs for about 45 billion years.

15.8.1 POTENTIAL FUSION REACTIONS

Although many fusion reactions are possible (as shown in §1.3.4), only one **D-T fusion reaction** has high practical potential:

$$^2_1\text{H} + ^3_1\text{H} \rightarrow ^4_2\text{He} \ (3.5 \text{ MeV}) + ^1_0\text{n} \ (14.1 \text{ MeV}), \quad Q = 17.6 \text{ MeV}. \tag{15.103}$$

In addition, two other **D-D fusion reactions** are of interest:

$$^2_1\text{H} + ^2_1\text{H} \rightarrow ^3_2\text{He} \ (0.82 \text{ MeV}) + ^1_0\text{n} \ (2.45 \text{ MeV}), \quad Q = 3.27 \text{ MeV}, \tag{15.104}$$

$$^2_1\text{H} + ^2_1\text{H} \rightarrow ^3_1\text{H} \ (1.01 \text{ MeV}) + ^1_1\text{H} \ (3.02 \text{ MeV}), \quad Q = 4.03 \text{ MeV}. \tag{15.105}$$

The D–T reaction requires tritium which does not occur naturally on Earth except in very low quantities formed in cosmic ray collisions. However, the tritium can be manufactured by reacting neutrons with lithium isotopes ^6Li and ^7Li. Reaction D–T has the disadvantage that most of the energy is carried off by the neutron, which makes the energy harder to capture. The reaction is considered as the most promising one since it has a low Coulomb barrier, thus requires the lowest temperature, the lowest density, and the shortest confinement time to proceed. At the same time, out of all possible fusion reactions, the D–T reaction has the highest energy release.

15.8.2 FUSION POWER DENSITY

As discussed earlier the fusion reactions require plasma with ionized nuclei and electrons heated to high temperatures to overcome the repulsive Coulomb forces between reactants. The fusion power density generated by plasma can be found as

$$P_{fus} = R_{fus}Q_{fus}, \tag{15.106}$$

where R_{fus} is the reaction rate density expressed in units reactions·m^{-3}·s^{-1}, Q_{fus} is the energy release per fusion reaction, expressed in MeV per reaction, and P_{fus} is the fusion power density expressed in MeV·m^{-3}·s^{-1}. This can be converted to a more common unit of power density W·m^{-3}·s^{-1} using relationship 1 MeV·s$^{-1} \approx 1.6 \times 10^{-13}$ W.

For two sets of particles moving with relative speed v_r and number densities n_1 and n_2, the fusion reaction rate between the particles is given as,

$$R_{fus} = \sigma_{12}(v_r)n_1n_2v_r. \tag{15.107}$$

Here $\sigma_{12}(v_r)$ is a proportionality factor called cross section, with unit of an area. Commonly used unit, similarly to fission reactions, is barn (b), where $1\text{ b} = 10^{-24}\text{ cm}^2 = 10^{-28}\text{ m}^2$.

In a general case reacting particles have distribution of energies well-described by the Maxwellian distribution and are moving in various directions. Averaging the reaction rate formula given by Eq. (15.107) over all velocities and directions, the following **reaction rate parameter** is defined,

$$\langle \sigma v \rangle_{12} \equiv \int_{\mathbf{v}_1} \int_{\mathbf{v}_2} \sigma(v_r) v_r F_1(\mathbf{v}_1) F_2(\mathbf{v}_2) d\mathbf{v}_1 d\mathbf{v}_2, \tag{15.108}$$

where $v_r = |\mathbf{v}_1 - \mathbf{v}_2|$ and $F_1(\mathbf{v}_1)$, $F_2(\mathbf{v}_2)$ are normalized velocity distribution functions for the two sets of particles. Assuming Maxwellian energy distributions of two sets of particles with masses m_1 and m_2, the reaction rate parameter is obtained as [2],

$$\langle \sigma v \rangle_{12} = \left[\frac{8}{\pi \mu k_B^3 T^3} \right]^{3/2} \int_0^{\infty} \sigma(E) E \exp(-E/k_B T) dE. \tag{15.109}$$

Here $\mu = m_1 m_2/(m_1 + m_2)$ is the reduced mass of the system, k_B is the Boltzmann constant, and T is the plasma temperature in K.

The reaction rate parameter $\langle \sigma v \rangle_{12}$ is expressed in units $\text{m}^3 \cdot \text{s}^{-1}$ and is typically presented for various reaction types as plots or tables in terms of the **kinetic temperature** $k_B T$.

15.8.3 PLASMA CONFINEMENT METHODS

One fundamental question in fusion science is to determine conditions required for a net energy output from fusion. Energy is needed to heat the plasma up to the temperature required for fusion and to compensate for energy losses in the hot plasma. For D-T fusion, ion energies should be higher than 370 keV to overcome the Coulomb barrier. This seems to be not much compared to the energy that is obtained from the fusion reaction. For example, using a medium energy accelerator for injection of deuterons of 500 keV into tritiated target would lead to energy multiplication $E_{out}/E_{in} = 17.6/0.5 \approx 35$. However, this high energy multiplication ratio is practically unachievable. In reality deuterons entering the target lose energy through the processes of ionization and heating the target. Experiments and theory show that they are far more likely to scatter with additional energy loss by bremsstrahlung radiation. A more promising approach is to heat up a mixture of deuterium and tritium to high-enough temperature to cause ionization and high speeds of fuel ions. Because the activation of fusion reactions occurs here due to random thermal motions of the reacting nuclei, this process is called **thermonuclear fusion**.

The critical technical requirement for the thermonuclear fusion is the sustainment of a sufficiently stable high temperature plasma for sufficiently long period of time. The plasma is said to be at the **critical ignition temperature** when the power generated is equal to the plasma power loss. It can be shown that for the D-T reaction the critical ignition temperature is about 3×10^7 K, whereas for the D-D reaction it is as high as 6×10^8 K. Confinement of fuel ions is thus needed to maintain these conditions within the required reaction volume.

Material confinement fusion is impossible, because all known materials melt below 5000 K, and fusion plasma requires a temperature of 10^8 K. **Inertial confinement fusion** involves compression of a small fusion fuel pellet about 1 to 3 mm in diameter to high density and temperature by external laser or ion beams. The beam induces an inward directed momentum of the outer layers of the pellet and sends a shock wave into the center. This shock wave compresses the fuel to high density and heats it to temperature sufficiently high that fusion reactions occur.

One of the most effective means of plasma confinement involves the use of magnetic fields. Solenoidal fields belong to oldest and most widely used **magnetic confinement** devices in plasma physics research. However, these devices suffer from severe ion leakages through the ends and thus provide little prospect for use as fusion reactors. The ion leakages can be reduced by squeezing the magnetic field lines at the ends. Such squeezed field configuration is referred to as a **magnetic mirror** since it is able to reflect charged particles. However, to eliminate end leakage entirely, the obvious solution is to connect the ends by turning a solenoidal field into a toroidal field. The most widely pursued device using the toroidal magnetic field topology is **tokamak**[10], introduced by Sakharov and Tamm in 1968.

15.8.4 FUSION PERFORMANCE CRITERIA

Performance of fusion reactors can be described by different parameters. In general, plasma density and temperature must be high enough to produce nuclear fusion. For D-T fusion reaction, the following **Lawson criterion** is formulated to describe the conditions that are required for ignition:

$$n\tau_E \geq L \equiv \frac{12k_BT}{\langle \sigma v \rangle E_\alpha}, \tag{15.110}$$

where $E_\alpha \cong 3.5$ MeV is α particle energy after the fusion reaction and n is the number density of protons in plasma. The **energy confinement time** τ_E is defined as

$$\tau_E = \frac{\text{energy content of the plasma}}{P_{loss}} = \frac{1}{P_{loss}}(n+n_e)\frac{3}{2}k_BT = \frac{3nk_BT}{P_{loss}}, \tag{15.111}$$

where we have used the fact that the average kinetic energy of the electrons and ions with a Maxwellian distribution is $3k_BT/2$ and the number density of protons n is equal to the number density of electrons n_e. The above expression was proposed by Lawson who argued that, in a plasma that has ignited, the fusion power density that goes into heating the plasma (P_{heat}) must exceed the power density lost to the environment (P_{loss}).

As fusion research matured, a more useful figure-of-merit was found to be the following **triple product** criterion

$$n\tau_E T \geq \frac{12k_BT^2}{\langle \sigma v \rangle E_\alpha}. \tag{15.112}$$

Both the Lawson and the triple product criteria estimate the condition required for ignition, when the plasma can be maintained in steady state or its temperature is increasing without external heating.

In steady-state fusion reaction, the **fusion gain factor** Q is defined as the ratio of the total power density produced by fusion P_{fus} to the heating that must be externally supplied P_{heat}:

$$Q \equiv \frac{P_{fus}}{P_{heat}}. \tag{15.113}$$

$Q = 1$ is termed **breakeven** and $Q \to \infty$ when **ignition** is attained.

15.8.5 ITER

In the early 1990s, an international effort was undertaken to construct a new generation of tokamak reactors that were intended to produce more fusion energy than the energy needed to create the plasma. Construction of the first such reactor called ITER[11] began in 2010 at Cadarache in the south of France. The ITER project aims to initiate D-T fusion reactions before 2030. The reactor is designed to produce 500 MW of fusion power for extended period of time of roughly 8 minutes. The produced fusion power will be 10 times more than is needed to maintain the plasma ($Q = 10$).

---

[10]The name tokamak is an acronym of a Russian phrase that translates to *toroidal chamber in magnetic coils.*

[11]The name associated with various acronyms such as "International Thermonuclear Energy Reactor" or simply Latin for "the way."

15.8.6   OTHER TECHNOLOGIES

Nuclear fission and fusion reactors are using thermodynamic cycle to produce electricity. Technologies are also under development for directly converting nuclear radiation or the thermal energy produced by the radiation into electrical energy. Over the past several decades, a few such converters have been developed.

Radioisotope thermoelectric generators (RTGs) provide electrical power using decay heat from radionuclides. They use thermoelectric cells to convert thermal energy into electrical energy. Many such devices of varying designs and power capacities have been made and tested for a variety of terrestrial and space applications. They are fuelled mainly by ^{210}Po (with design life of 90 days), ^{238}Pu (with design life of 5 years), ^{90}Sr (with design life of more than 2 years), and ^{242}Cm (with design life of 90 days). The electric power of RTGs is varying from 0.001 to 500 W. Tests are carried out to use small nuclear reactors to heat a liquid coolant that heats the hot junction of the thermoelectric cells.

Radioisotope thermionic generators use a thermionic electrical generator and the decay heat of radioisotopes to generate electricity. **Thermionic electrical generators** use two closely spaced metal plates of different temperatures to directly convert thermal energy into electricity. The hotter plate (called emitter or cathode) is kept at high-enough temperature to boil electrons from its surface into the gap between the plates. The low-temperature plate (called collector or anode) collects the electrons producing current through an external electrical load. The efficiency of this converter is limited by the ideal Carnot efficiency and high emitter temperatures are required, typically in excess of 1400 K, to achieve high conversion efficiency. Typical conversion efficiencies are between 1 and 10% and the potential differences are from 0.3 to 1.2 V per cell. Higher voltage can be obtained by connecting several cells in series.

PROBLEMS

PROBLEM 15.1

Derive Eq. (15.12), assuming that the ratio of the nucleus mass to the neutron mass is approximately equal to the atomic mass number of the nucleus A.

PROBLEM 15.2

Derive an approximate relationship between the uranium enrichment expressed in at% and the enrichment expressed in wt%.

PROBLEM 15.3

Find the reproduction factor of uranium fuel with enrichment 5wt%.

PROBLEM 15.4

Find the logarithmic energy decrement ξ for water.

PROBLEM 15.5

Calculate the average logarithmic energy decrement ξ for a homogeneous water-moderated and uranium-fuelled reactor with moderator-to-fuel ratio $N_M/N_F = 3$, where N_M and N_F are atomic densities of moderator and fuel, respectively.

PROBLEM 15.6

Calculate the thermal utilization factor f for a homogeneous water-moderated and uranium-fuelled reactor with uranium enrichment $e_a = 3.5\text{at}\%$ and moderator-to-fuel ratio $N_M/N_F = 3$, where N_M and N_F are atomic densities of moderator and fuel, respectively.

PROBLEM 15.7

Consider an infinite, homogeneous reactor fuelled with natural uranium and moderated by heavy water. What is the optimal ratio of moderator to fuel, to obtain the highest infinite multiplication factor?

PROBLEM 15.8

Calculate the reactivity ρ for an infinite homogeneous water-moderated and uranium-fuelled reactor with uranium enrichment $e_w = 3.5\text{wt}\%$ and moderator-to-fuel ratio $N_M/N_F = 3$, where N_M and N_F are atomic densities of moderator and fuel, respectively. Assume $\varepsilon = 1$.

Part III

External Effects

16 Energy and Environment

Human energy transformation is strongly interconnected with the natural environment. On the one hand, all energy technologies are using natural resources, either directly as fuel, or indirectly as raw materials that are needed for manufacturing of machines and infrastructure. On the other hand, energy transformation facilities are always involving use of land, water, and air, which can lead to pollution or other negative impacts on the natural environment.

In this chapter we discuss some of the most important effects of human energy transformation on the natural environment. Recent anthropogenic emissions of CO_2 have strong influence on the concentration of this gas in the Earth's atmosphere. Since CO_2 is an important atmospheric greenhouse gas, its concentration plays a significant role in Earth's climate changes.

While the basic science described in this chapter is well-established, many aspects of Earth's climate are still beyond the reach of current science, as described in §3.1. Nevertheless, an intensive development of new models that can be used for estimation of the evolution of Earth's climate is ongoing. This topic is covered by many good references, dealing with basic aspects of atmospheric physics [40], [5], and the climate change [83].

16.1 CLIMATE

Climate is a long-term averaged weather in a given area. The classical period used for describing a climate is 30 years, as defined by the World Meteorological Organization (WMO)[1]. The climate description contains information on such parameters as the average temperature in different seasons or over the whole year, rainfall, and sunshine. Global climate indicators are published by WMO on a yearly basis since 1994. The "State of the Global Climate" reports provide most recent trends of such indicators as, between others, the global mean temperature, the atmospheric concentrations of greenhouse gases, the ocean heat content, the sea level, the marine heatwaves, the ocean acidification, and the status of the cryosphere [89].

Understanding the mechanisms that are governing climate changes is vital for further development of human civilization. Such recent phenomena as increasing CO_2 concentration, resulting in increased greenhouse effect, and ocean acidification, cause several undesired high impact events. Such high impact events as heat waves, cold waves, fires, draughts, and flooding constitute key risks to achieve several of the sustainable development goals. In particular, they can lead to rising poverty, food insecurity, health issues, water scarcity, damaged infrastructure, rising inequalities, displacement, ecosystem collapse, biodiversity loss, and conflicts.

The anomaly of Earth's surface temperature during 1850–2019 is shown in Fig. 16.1, where, as the reference, the average temperature during 1961–1990 is used. Over that period of time, many different methods have been used for measuring temperature, which could lead to different systematic errors and uncertainties. Nevertheless, with taking into account all major sources of uncertainties and systematic errors, it is possible to obtain a reasonably accurate global average temperature history since 1850. From Fig. 16.1 it is evident that since that year until 2019, the global mean temperature has increased by slightly more than 1 K. Most of this increase, about 0.8 K, has occurred since 1980s. It can be observed that the temperature development during recent 40 years significantly departs from the previous, more moderate temperature changes. In next sections we discuss possible coupling mechanisms between the global average temperature and the global average atmospheric concentration of carbon dioxide and other greenhouse gases.

---

[1]https://public.wmo.int

DOI: 10.1201/9781003036982-16

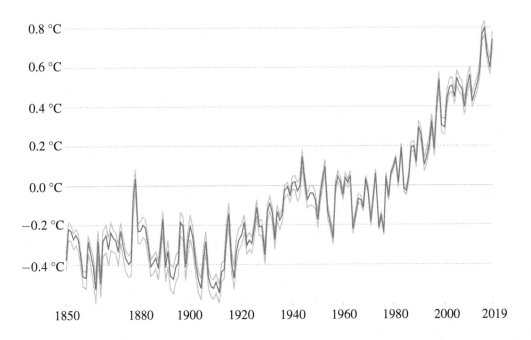

Figure 16.1 Global average land-sea temperature anomaly during 1850–2019 relative to the 1961–1990 average temperature, showing median average temperature change as well as upper and lower 95% confidence intervals. Source: Hadley Centre (HadCRUT4). Reproduced after OurWorldInData.org.

16.2 GREENHOUSE EFFECT

To understand the greenhouse effect, we first assume that Earth can be treated as a black body, which perfectly absorbs incoming solar radiation at $I_\odot = 1366$ W/m^2 and radiates it back into space at a temperature $T_\oplus$. This temperature can be obtained from the energy balance equation for Earth. Since the only way of energy transfer between Earth and the space is radiation, the radiative equilibrium requires that,

$$\pi R_\oplus^2 I_\odot = 4\pi \sigma R_\oplus^2 T_\oplus^4, \tag{16.1}$$

where σ is the Stefan-Boltzmann constant and $R_\oplus$ is Earth's radius. Thus, the equilibrium surface temperature is obtained as,

$$T_\oplus = \left(\frac{I_\odot}{4\sigma} \right)^{1/4} \cong 278.6 \text{ K} \cong 5.5 \text{ }°C. \tag{16.2}$$

This simple model provides a reasonable estimate of the equilibrium temperature of Earth, but we need a more accurate estimate of the present temperature to predict the future trends.

The two most important effects that have been neglected in the simple model are the surface reflectivity, or albedo, and the greenhouse effect. The **albedo** of an object is defined as the fraction of incident electromagnetic energy that is reflected by the object. If Earth was completely covered in ice, its albedo would be about 0.84, but it would drop to 0.14 if Earth was covered by a dark green forest canopy. Earth's albedo has been measured using satellites since the late 1970s, and its global average value is

$$a_\oplus \cong 0.3. \tag{16.3}$$

This value is significantly higher than the albedo of an open ocean, which is 0.06. This difference primarily exists because of the contribution of clouds.

If the effect of albedo is taken into account, the net energy gained by Earth by radiation is $\pi R_{\oplus}^2 I_{\odot} - a_{\oplus} \pi R_{\oplus}^2 I_{\odot} = (1 - a_{\oplus}) \pi R_{\oplus}^2 I_{\odot}$. Thus the surface temperature of Earth, assuming a realistic value of the albedo, but still neglecting the greenhouse effect, would be,

$$T_s = \left[\frac{(1 - a_{\oplus}) I_{\odot}}{4\sigma} \right]^{1/4} \cong 254.8 \text{ K} \cong -18.3 \text{ °C}. \tag{16.4}$$

Both this estimate and the one found earlier in Eq. (16.2), which ignored albedo, yield much colder Earth's surface temperature than the currently measured temperature. The underestimation of the surface temperature is due to the neglect of the greenhouse effect.

The **greenhouse effect** (also known as the *heat trapping phenomenon*) occurs when a material that absorbs incoming solar radiation is covered by a layer of another material that is transparent to the incoming radiation but that absorbs the emitted infrared radiation. At equilibrium the absorbing layer reradiate the same amount of energy in all directions, including the absorbing material. In the context of the global Earth's system, the absorbing material is the surface of Earth and the covering layer is the atmosphere. The majority of the gases present in the atmosphere are transparent to outgoing infrared radiation. However, such gases as carbon dioxide (CO_2), methane (CH_4), water vapor (H_2O), nitrous oxide (N_2O), ozone (O_3), chlorofluorocarbons (CFCs) and hydrofluorocarbons (HCFCs and HFCs), referred to as the **greenhouse gases**, absorb certain frequencies of infrared radiation.

We can extend our simple model to take into account the greenhouse effect by adding an additional layer that represents the atmosphere, as shown in Fig. 16.2. We can write two energy

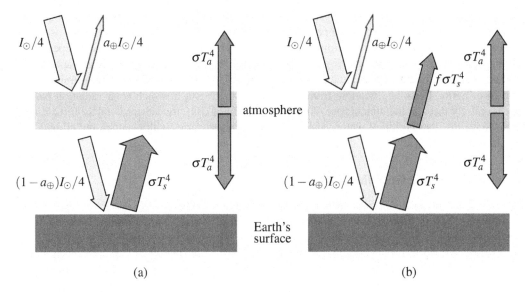

Figure 16.2 Energy flow in a simplified model of Earth with albedo $a_{\oplus}$ and a single layer atmosphere that transmits all solar radiation and: (a) absorbs all infrared radiation; (b) transmits fraction f of the infrared radiation.

conservation equations: one for the radiative equilibrium at Earth's surface and one for the radiative equilibrium of Earth's atmosphere. Assuming that temperatures of the surface and the atmosphere are T_s and T_a, respectively, the equations are as follows. According to Fig. 16.2(b), the radiative equilibrium for the entire Earth's system requires

$$\sigma T_a^4 + f \sigma T_s^4 = \frac{(1 - a_{\oplus}) I_{\odot}}{4}, \tag{16.5}$$

where parameter f represents the fraction of infrared radiation that is transmitted through the atmosphere without absorption. For $f = 0$, as shown in Fig. 16.2(a), all infrared radiation is absorbed, whereas for $f = 1$ no radiation is absorbed, which corresponds to the previously considered case with no greenhouse effect. The radiative equilibrium at Earth's surface requires

$$\sigma T_s^4 = \frac{(1 - a_\oplus)I_\odot}{4} + \sigma T_a^4. \tag{16.6}$$

The solution of these two equation is

$$T_s = \left[\frac{(1 - a_\oplus)I_\odot}{2(1 + f)\sigma}\right]^{1/4}. \tag{16.7}$$

If we assume that no infrared radiation is transmitted through the atmosphere ($f = 0$), the surface temperature is obtained as $T_s = (0.7I_\odot/2\sigma)^{1/4} \cong 303.1$ K $\cong 29.9$ °C. If, on the contrary, we assume that all infrared radiation is transmitted through the atmosphere ($f = 1$), the surface temperature is $T_s = (0.7I_\odot/4\sigma)^{1/4} \cong 254.8$ K $\cong$ -18.3 °C. As we can see, this result is identical with previously obtained surface temperature when neglecting the greenhouse effect. It can be shown that $f = 0.226$ (that is when 22.6% of infrared radiation is transmitted out through the atmosphere without absorption) yields $T_s \cong 288$ K $\cong 15$ °C.

In reality, the fraction of infrared radiation that is transmitted through the atmosphere without absorption depends on the absorption properties of many components present in the atmosphere. The diatomic molecules N_2 and O_2 cannot easily absorb photons and are transparent to most incoming solar radiation and to outgoing infrared radiation. In the troposphere, below roughly 15 km altitude, water vapor, CO_2, and other greenhouse gases absorb both incoming solar radiation and outgoing infrared radiation. Water vapor, which stays within the bottom 3 km of the atmosphere, is the strongest absorber of the radiation. Absorption responsible for greenhouse effect depends on not only temperature and concentration of emission gases but also wavelength. Water vapor has a number of absorption bands that capture the incoming solar radiation and outgoing infrared radiation from Earth. There is a window from about 8 μm to 12 μm in which solar radiation passes almost entirely through the atmosphere. Absorption bands of CO_2 are centered at 15, 4.3, 2.7, and 2 μm. Since the distribution of water vapor in the atmosphere varies with weather and climate, its modeling is complicated.

16.3 EARTH ENERGY IMBALANCE

Human-induced emissions of greenhouse gases contribute to composition changes of the Earth's atmosphere. This, in turn, causes a radiative imbalance at the top of the atmosphere which is driving global warming. Understanding the heat gain of the Earth is fundamental to assess how this affects rising surface temperature and warming oceans, atmosphere, and land. A recent comprehensive study attempted to address this challenging issue and estimated heat gain in the atmosphere, cryosphere and land over the period 1960–2018 [75]. The study obtains a total heat gain of 358±37 ZJ over the period 1971–2018, which is equivalent to a global heating rate of 0.47±0.1 W/m^2. This effect is accelerating in recent years and the global heating rate of 0.87±0.12 W/m^2 is reported during 2010–2018. Stabilization of climate requires that the Earth energy imbalance reduces to approximately zero. This would be possible if the amount of CO_2 in the atmosphere was reduced from 410 to 353 ppm.

The majority (89%) of the heat gained by the Earth system during 1971–2018 has been absorbed by oceans. Land has absorbed 6% and the remaining heat is distributed between the cryosphere (4%) and the atmosphere (1%). Whereas the heat absorption in recent years (2010–2018) in oceans and atmosphere has increased by 1% each, an opposite trend is noted in cryosphere and land, where the heat absorption has decreased by 1% each.

16.4 CO_2 CONCENTRATION

After water vapor, carbon dioxide is the next most important greenhouse gas. Water evaporates from the oceans and land, and precipitates from the atmosphere back to the surface. This process is quite rapid and typically the average water vapor content of the atmosphere comes into equilibrium with the climate within about two weeks. For other greenhouse gases, however, the time scale over which the atmospheric concentration reaches equilibrium with the climate is much longer. In particular CO_2 is likely to be the primary driver of climate change over the time scale of the next one or two centuries.

The concentration of carbon dioxide has varied significantly over the Earth's 4.54 billion year history. Most probably it was present in the first atmosphere, shortly after Earth's creation, but its concentration increased mainly by outgassing from volcanism and other processes. About 500 million years ago, during the Cambrian period, the concentration was as high as 4 000 ppm. It dropped to about 180 ppm during an alternating series of glacial and interglacial periods that began 2.58 million years ago, and is still ongoing.

Measurements of atmospheric CO_2 levels taken during the past half century indicate a steady increase in concentration of this gas in the atmosphere. Global average atmospheric concentration of CO_2 measured at high resolution using preserved air samples from ice cores is shown in Fig. 16.3. During the past 800,000 years the concentration of CO_2 oscillated between 170 and 300 ppm. In modern time the level of 300 ppm was exceeded in the beginning of 1900s and the concentration is constantly increasing since then.

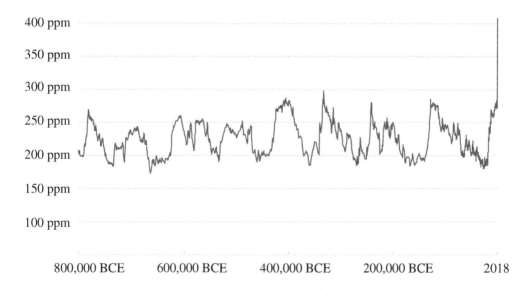

Figure 16.3 Global average long-term atmospheric concentration of carbon dioxide measured at high resolution using preserved air samples from ice cores. Source: EPICA Dome C CO_2 (2015) and NOAA (2018). Reproduced after OurWorldInData.org.

There are many natural sources of atmospheric CO_2, including volcanic outgassing, the combustion of organic matter, wildfires, and the respiratory processes of living aerobic organisms. Anthropogenic sources of CO_2 include the burning of fossil fuel for heating, power generation, and transport.

Example 16.1:

Calculate the total mass of carbon dioxide in the atmosphere assuming its current concentration of 415.24 ppm[2] and knowing that the total mean atmospheric mass is 5.1480×10^{18} kg.

Solution

Note that CO_2 concentration in the atmosphere is quoted on a per volume basis, in parts per million by volume (ppmv $\equiv$ ppm). We need to find the concentration in parts per million by mass (ppmm). Since the molecular mass of CO_2 is 44.01 amu and the average atmospheric molecular mass is 28.9647 amu, we find

$$1 \text{ ppm} \cong \frac{44.01}{28.9647} \text{ ppmm} \cong 1.519 \text{ ppmm}. \tag{16.8}$$

Thus, the mass of CO_2 in the atmosphere is

$$5.1480 \times 10^{18} \times 415.24 \times 1.519 \times 10^{-6} \text{ kg } CO_2 \cong 3\ 247 \text{ Gt } CO_2. \tag{16.9}$$

16.5 GREENHOUSE GAS EMISSIONS

Providing good estimates of greenhouse gas emissions from energy sector requires reliable and extensive coverage on domestic and traded energy. The understanding of emissions in the late 20th and 21st centuries has significantly improved as compared to previous long-term reconstructions. The Intergovernmental Panel on Climate Change (IPCC) provides clear guidelines on methodologies and best practice for measuring and monitoring greenhouse gas estimates at the national level[3].

Greenhouse gas emissions are converted to **carbon dioxide equivalents** (CO_2eq) by multiplying each gas by its 100-year global warming potential value: the amount of warming one tonne of the gas would create relative to one tonne of CO_2 over a 100-year time-scale. Greenhouse gases include carbon dioxide (CO_2), methane (CH_4), nitrous oxide (N_2O), and smaller traces of so-called F-gases, such as hydrofluorocarbons (HFCs), chlorofluorocarbons (CFCs), and sulfur hexafluoride (SF_6). In 2016, CO_2 was the largest contributor to greenhouse gases accounting for 74.4% of total emissions, followed by CH_4 (17.3%), N_2O (6.2%) and F-gases (2.1%)[4]. The world annual greenhouse gas emissions during years 1990–2016 are shown in Fig. 16.4.

The world CO_2 emissions are resulting mainly from burning coal (39.4%), oil (33.9%) and gas (21.0%), followed by cement production (4.2%), gas flaring[5] (1.3%), and from other industries (0.2%). The world averaged emissions per capita reached 4.72 tonnes in 2019. Emissions per capita

---

[2]www.co2.earth, accessed 2021-02-13.

[3]*IPCC Good Practice Guidance and Uncertainty Management in National Greenhouse Gas Inventories*, available online at https://www.ipcc-nggip.iges.or.jp/public/gp/english/6_Uncertainty.pdf, accessed 2021-02-23.

[4]https://ourworldindata.org/greenhouse-gas-emissions, accessed 2021-02-24.

[5]Intentional burning of natural gas, often on oil or gas extraction sites.

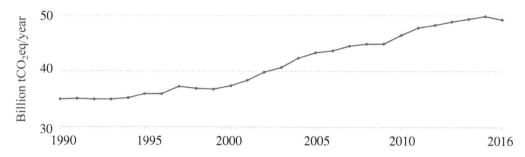

Figure 16.4 World annual greenhouse gas emissions. Source: CAIT Climate Data Explorer via Our World in Data.

significantly vary in different regions and countries, as shown in Table 16.1. The distribution across different fuel sources is very dependent on energy production and mix in a given country. In China and Poland the coal is dominant, whereas oil is prevailing in US and Norway, and gas in Russia.

Table 16.1
CO_2 Emissions per Fuel and Capita in Selected Countries in 2019 (tCO_2eq)

Country	Coal	Oil	Gas	Cement	Flaring	Other	Total
Brasil	0.31	1.48	0.31	0.09	0.01	0	2.2
China	5.05	1.06	0.41	0.58	0	0	7.1
Denmark	0.64	3.56	1.04	0.2	0.04	0.07	5.55
EU-27	1.66	2.85	1.74	0.17	0.05	0.08	6.55
France	0.42	3.01	1.33	0.11	0.05	0.06	4.98
Germany	2.82	3.23	2.07	0.16	0.02	0.1	8.4
India	1.22	0.49	0.09	0.11	0.01	0	1.92
Italy	0.52	2.41	2.41	0.13	0.04	0.05	5.56
Japan	3.43	3.3	1.72	0.2	0.01	0.07	8.73
Netherlands	1.94	3.07	4.34	0.01	0.06	0.08	9.5
Norway	0.79	3.99	2.5	0.14	0.41	0.07	7.9
Poland	4.83	2.31	0.93	0.2	0.13	0.12	8.52
Russia	2.71	2.72	5.55	0.15	0.26	0.12	11.51
Spain	0.47	3.07	1.52	0.21	0.08	0.06	5.41
Sweden	0.82	2.95	0.19	0.16	0.08	0.05	4.25
United Kingdom	0.4	2.51	2.39	0.06	0.07	0.04	5.47
United States	3.33	7.12	5.19	0.13	0.22	0.08	16.07
World	1.86	1.6	0.99	0.2	0.06	0.01	4.72

Source: https://ourworldindata.org/emissions-by-fuel, retrieved 2021-02-24.

The energy sector is by far the biggest contributor to global greenhouse gas emissions, which reached 49.4 billion tonnes CO_2 eq. in 2016. This sector was responsible for 73.2% of the emissions, where 24.2% resulted from the energy (electricity and heat) use in industry, 17.5% from the energy use in commercial and residential buildings, and 16.2% in transport. The remaining 15.3% was due to unallocated fuel combustion, fugitive emissions[6] from energy transformations, and energy use in the agriculture and fishing. The two other sectors contribution with significant emissions are

---

[6]Fugitive emissions are the often accidental leakages of methane to the atmosphere during oil and gas extraction and transportation, from damaged or poorly maintained pipes. This also includes flaring.

agriculture, forestry and land use (18.4%), cement and chemical process industry (5.2%), and waste (3.2%).

While the greenhouse gas emissions should be reduced in all sectors whenever possible, it is clear that the most significant reductions can be achieved in the electricity and heat production sector, followed by the transport sector. A complete de-carbonization of the electricity and heat sector would result in reduction of emissions by as much as 40%. A complete electrification of the transport sector would lead to emission reduction by about 16% at the most, providing that the electricity used in transport is completely emission-free. With the current contribution of fossil fuels in the electricity production, the reduction of emissions in the transport sector would be much less efficient.

Emission intensities of selected electricity supply technologies are shown in Table 16.2. Electricity generated in power plants with pulverized coal as fuel causes the highest emissions among all currently commercially available technologies presented in the table, with life-cycle median emissions of 820 gram of CO_2 equivalent per kWh (gCO$_2$eq/kWh). Hydropower, nuclear, solar (CSP), and wind technologies cause the lowest emissions of 24, 12, 27, and 11 gCO$_2$eq/kWh, respectively. Pre-commercial technologies based on the carbon dioxide capture and storage (CCS) are expected to have emissions in a range of 160–220 gCO$_2$eq/kWh.

16.6 AIR POLLUTION

Air pollution is one of the world's largest health and environmental problems. The Institute for Health Metrics and Evaluation (IHME) in its Global Burden of Disease (GBD) study provides estimates of the number of deaths attributed to the range of risk factors for disease. The study shows that the fourth largest risk factor for attributable death in 2019 was air pollution, accounting for 6.67 million deaths (11.3% of all female and 12.2% of all male deaths) [31].

Air pollution results from the release into the atmosphere of various gases and particles at rates that exceeds the natural capacity of the environment to dissipate and dilute or absorb them. **Outdoor air pollution** originates from natural and anthropogenic sources. The natural sources contribute substantially to local air pollution in arid regions more prone to forest fires and dust storms, however, the contribution from human activities far exceeds natural sources, and include:

- Fuel combustion from motor vehicles.
- Heat and power generation in oil and coal power plants and boilers.
- Industrial facilities such as manufacturing factories, mines, and oil refineries.
- Municipal and agricultural waste sites and waste incineration or burning.
- Residential cooking, heating, and lighting with polluting fuels.

The pollutants with the strongest evidence of health effects are particulate matter (PM), ozone (O_3), nitrogen dioxide (NO_2) and sulphur dioxide (SO_2).

Sulphur dioxide (SO_2) and nitrogen oxide (NO_x) emissions and effects

The largest sources of SO_2 and NO_x in the atmosphere are fuel combustion processes in the power generation sector, in industry, and in the commercial and residential sectors. In addition, significant NO_x emissions are caused by the transport sector. Nitric oxide (NO) accounts for the majority of NO_x emissions. Nitrogen dioxide (NO_2) is to some extend emitted directly as well, but it mainly results from subsequent oxidation of NO.

Exposure to high concentrations of SO_2 and NO_x can harm human respiratory system and make breathing difficult. SO_2 can harm trees and plants by damaging foliage and decreasing growth. After oxidation to form H_2SO_4, it is responsible for acid rains, which harm sensitive ecosystems such as lakes and forests.

Table 16.2
Emissions of Selected Electricity Supply Technologies (gCO$_2$eq/kWh)

Options	Direct Emissions Min/Med/Max	Infrastructure Emissions[a] Typ. Value	Biogenic Emissions[b] Typ. Value	Methane Emissions Typ. Value	Lifecycle Emissions[b] Min/Med/Max
Currently Commercially Available Technologies					
Coal–PC (Pulverized Coal)	670/760/870	9.6	0	47	740/820/910
Gas–Combined Cycle	350/370/490	1.6	0	91	410/490/650
Biomass–cofiring	n.a.[c]	-	-	-	620/740/890[d]
Biomass–dedicated	n.a.[c]	210	27	0	130/230/420[e]
Geothermal[f]	0	45	0	0	6/38/79
Hydropower[g]	0	19	0	88	1/24/2200
Nuclear	0	18	0	0	3.7/12/110
Concentrated Solar Power[h]	0	29	0	0	8.8/27/63
Solar PV–rooftop	0	42	0	0	26/41/60
Solar PV–utility	0	66	0	0	18/48/180
Wind onshore	0	15	0	0	7/11/56
Wind offshore	0	17	0	0	8/12/35
Pre-Commercial Technologies					
CCS–Coal–Oxyfuel	14/76/110	17	0	67	100/160/200
CCS–Coal–PC	95/120/140	28	0	68	190/220/250
CCS–Coal–IGCC	100/120/150	9.9	0	62	170/200/230
CCS–Gas–Combined Cycle	30/57/98	8.9	0	110	94/170/340
Ocean[i]	0	17	0	0	5.6/17/28

Source: Schlömer S. et al., 2014: Annex III, Technology-specific Cost and Performance Parameters. In: *Climate Change 2014: Mitigation of Climate Change. Contribution of Working Group III to the Fifth Assessment Report of the Intergovernmental Panel on Climate Change*, Cambridge University Press, Cambridge, United Kingdom and New York, NY, USA.

[a] Including supply chain.

[b] Including albedo effect.

[c] Direct emissions from biomass combustion at the power plant are positive and significant, but are reduced due to the CO$_2$ absorbed by growing plants. The net emissions depend on the chemical carbon content of biomass and the power plant efficiency.

[d] Indirect emissions for co-firing are based on relative fuel shares of biomass from dedicated energy crops and residues (5–20%) and coal (80–95%).

[e] Lifecycle emissions from biomass are for dedicated energy crops and crop residues.

[f] Includes both flash steam and binary cycle power plants.

[g] Includes both run-of-the-river and reservoir hydropower.

[h] Includes both CSP with storage as well as CSP without storage.

[i] Includes both tidal and wave energy conversion technologies.

Household air pollution is one of the leading causes of disease and premature death in the developing world. It results from cooking fires and burning fuels such as dung, wood and coal in inefficient stoves or open hearths. It includes particulate matter, methane, carbon monoxide, polyaromatic hydrocarbons and volatile organic compounds. Burning kerosene in simple wick lamps also produces significant emissions of fine particles and other pollutants.

Particulate matter (PM$_{2.5}$, PM$_{10}$) and volatile organic compound (VOC) emissions and effects

Particulate matter (PM) is the term for a mixture of solid and liquid particles suspended in air, considered as an atmospheric pollutant. It is mainly composed of ions, reactive gases, organic compounds, metals, and particle carbon core. Some particles are emitted from human activities such as fuel combustion for power generation, industry, transport, waste incineration and agriculture, and domestic heating. Others are emitted from natural sources such as sea salt, natural dust and pollen.

PM$_{10}$ and PM$_{2.5}$ are particles with equivalent diameters below 10 and 2.5 micrometers, respectively. All such particles pose great health hazard, because they can get deep into the lungs and even into the bloodstream.

Volatile organic compounds (VOCs) are defined as any organic chemical compounds that under normal conditions are gaseous or can vaporize and enter the atmosphere. They include both human-made and naturally occurring chemical compounds, and are often divided into two categories: methane and non-methane VOCs (NMVOCs). The largest sources for VOCs are fuel combustion for transport (both directly as gasoline or indirectly as automobile exhaust gas), power generation, industry, and commercial and residential sector.

Exposure to high concentrations of VOCs has several negative health and environmental impacts such as production of ground level ozone (smog), global warming (methane), and fauna and flora degradation.

The life cycle emissions of NO$_x$ and SO$_2$ for electricity generation from selected technologies are shown in Table 16.3.

Table 16.3

Emissions of NO$_x$ and SO$_2$ for Electricity Generation from Selected Electricity Supply Technologies

Option	Electricity Output		Fuel Input	
	NO$_x$ (kg/MWh$_{out}$)	SO$_2$ (kg/GJ$_{in}$)	NO$_x$ (kg/MWh$_{out}$)	SO$_2$ (kg/GJ$_{in}$)
Hard coal	0.3–3.9	0.03–6.7	0.028–0.352	0.003–0.596
Lignite	0.2–1.7	0.6–7.0	0.025–0.161	0.047–0.753
Natural gas	0.2–3.8	0.01–0.32	0.037–0.277	0.0002–0.044
Oil	0.5–1.5	0.85–8.0	0.081–0.398	0.112–0.698
Biomass	0.08–1.7	0.03–0.94	0.007–0.128	0.004–0.094
Nuclear	0.01–0.04	0.03–0.038	-	-
Hydro	0.004–0.06	0.001–0.03	-	-
Solar	0.15–0.4	0.12–0.29	-	-
Wind	0.02–0.11	0.02–0.09	-	-

Source: [85]

16.7 WATER USE AND CONTAMINATION

Water is essential for cooling thermal power plants (coal, natural gas, nuclear, solar thermal, and biomass). It is also fundamental for hydro, wave, tidal, and marine current power. Almost all energy transformation systems require water at some point during their life cycles. Three types of water impacts can be distinguished: consumption, withdrawal, and contamination. **Water consumption** refers to its evaporation, transpiration, or incorporation during the system operation. This type of water use is typical for thermal power plants using cooling towers. In addition, hydropower can result in significant water consumption through evaporation from reservoirs. Although the consumed water remains in the global water cycle, its amount available for other purposes is reduced. **Water withdrawal** refers to the amount of water that is temporarily diverted from its natural source. Withdrawal for cooling can affect aquatic ecosystems when effluent water is returned at a higher temperature than that of its source. Withdrawal for hydropower can affect aquatic ecosystem by increasing dissolved oxygen content and by changing water flow pattern. Water can be **contaminated** due to many processes, such as fossil and nuclear fuel mining and processing. For some renewable technologies, such as solar PV cells, upstream mining and manufacturing operations can lead to the release of hazardous water pollutants from industrial effluents.

Energy usage causes diverse water pollution problems, such as ocean and freshwater acidification, eutrophication, and oil spills. **Acidification** is a chemical change of the environment in which inflow of hydrogen ions H^+ (protons) is greater than their outflow or neutralization. More hydrogen ions lower the pH (pH = -log$[H^+]$, where $[H^+]$ is the hydrogen ion concentration expressed in mol/liter) and increase acidity. It is now commonly accepted that the ongoing **ocean acidification**, that is the observed decrease in the pH of the Earth's oceans, is caused by the uptake of CO_2 from the atmosphere [13]. When water (H_2O) and CO_2 mix, they combine to form carbonic acid (H_2CO_3), which subsequently releases hydrogen ions (H^+) that bond with other molecules. The main cause of ocean acidification is the burning of fossil fuels. The ocean absorbs about 30% of the CO_2 that is released in the atmosphere, and as levels of atmospheric CO_2 increase, so do the levels in the ocean. It is estimated that during the past 200 years, ocean water has become 30% more acidic, and its pH value decreased from 8.25 to 8.14. As the most affected surface layers gradually mix into deep water, the entire ocean is affected. These relative quick change of ocean chemistry is adversely affecting marine life, which cannot adopt fast enough.

Like oceans, freshwater is also acidifying. The main reason for freshwater acidification is atmospheric depositions and soil leaching of SO_x and NO_x. An additional contribution is due to acid rains, formed when SO_x and NO_x react with water, oxygen, and oxidants within the clouds [51]. Even though it is difficult to quantify the effects of CO_2 concentration levels in fresh water, there has been a clear increase of this concentration in the last century due to anthropogenic influences [88].

In all petroleum-handling processes there is a certain probability of an oil spill, which is the release of a liquid petroleum hydrocarbon into the environment. Oil spills at sea are generally much more damaging then those on land, since they can spread over large areas in a thin oil slick which can cover beaches with a thin coating of oil. Crude oil and refined fuel spills from tanker ship accidents have damaged vulnerable ecosystems in, between others, Alaska[7], the Gulf of Mexico[8], and the Galapagos Islands. Spilled oil can also contaminate drinking water supplies.

16.8 LAND USE

Land use assessments of electric power technologies are increasingly important. However, understanding the spatial impacts of these technologies still presents methodological challenges. On the one hand, the current electric power system is heterogeneous and includes a range of fuel and power technology mixes. The main difficulty is to develop a balanced assessment method that takes into account the differences between various technologies in the mix. On the other hand, because energy

---

[7]Exxon Valdez oil spill occurred on March 24, 1989 near Tatitlek, Alaska, where 37 000 tons of crude oil was spilled.

[8]About 780 000 m^3 of crude oil was discharged during the Deepwater Horizon oil spill that occurred on April 20, 2010.

mix, conversion efficiency, local natural environment, and social average productivity vary by country, land use studies of the same electric power technology may yield significantly different results. As an example, estimates of the primary energy consumption associated with the manufacture of an E-40 wind turbine[9] in Germany and Brazil were 0.43 and 0.11 MJ/kWh, respectively, [59]. This discrepancy is directly influencing the land use assessment by wind power in the two countries.

Land use alone fails to completely reflect the real land impacts of different energy transformation technologies. Some technology options rarely change the land's geographical properties and pose fewer threats such as soil nutrient loss, functional degradation, and irreversible land transformation. Proper impact interpretation and comparison require more information, including data on the properties and conditions of the land required, the duration and reversibility of land transformation, and the quality of its use.

The land use for fossil-fired plants is dominated by upstream impacts of mining or extraction and transportation of fuel. Furthermore, the direct land use of surface coal mining is significantly higher than that of underground coal mining. The life cycle land use for coal (averaged for hard coal and lignite) and natural gas (averaged for combined cycle and steam turbine), excluding carbon capture and storage, is 180–5 700 and 340 m^2/GWh, respectively.

For renewable technologies, with the exception for bioenergy, the largest use of land occurs during the operation stage. Rooftop-mounted photovoltaic panels have almost negligible additional land use. Photovoltaic power stations and wind turbine farms can be deployed on pasturelands, brownfields, or even deserts. Their land use is relatively static and there is no further increase of land use after they are built. The land use for hydropower is highly site-specific and depends partially on the amount of land flooded in reservoir creation. Run-of-river hydropower requires less land use than does reservoir hydropower. Bioenergy production from dedicated feedstock has the highest land requirements of 12 600 m^2/GWh. It is followed by wind (1 000–2 000 m^2/GWh), hydropower (151–4 100 m^2/GWh), concentrated solar thermal power (250–900 m^2/GWh), geothermal (160–900 m^2/GWh), and solar photovoltaic (50–750 m^2/GWh).

For nuclear power, fuel mining and power plant site use account for two-thirds of life-cycle land use, and the land coverage of a nuclear power plant is usually larger than that of an equivalent coal-fired power plant, partially owing to safety considerations in the design of their layouts. The life cycle land use for nuclear power is in a range from 130 to 1 200 m^2/GWh [65].

16.9 MINERAL USE

Energy transformation technologies rely on a variety of minerals. Clean energy technologies generally require more minerals than fossil fuel-based counterparts. An electric car uses five times as much minerals as a conventional car and an onshore wind farm requires eight times as much minerals as a gas-fired plant of the same capacity. Lithium, cobalt and nickel are needed to manufacture batteries with greater charging performance and higher energy density. Copper is essential for increasing use of electricity from remotely located wind farms and solar panels. Some rare earth elements such as neodymium are needed for powerful magnets that are vital for wind turbines and electric vehicles. Even in fossil fuel-based technologies more extensive use of minerals is needed to achieve higher efficiency. For example, the modern most efficient coal-fired power plants require more nickel than the least efficient ones in order to manufacture steel that allow for higher combustion temperatures.

Table 16.4 contains estimations of amounts of minerals used in selected power generation technologies. The key minerals used for solar PV, wind power, natural gas, coal, and battery storage are presented. The major minerals required for the manufacture and operation of wind turbines

---

[9]Horizontal-axis wind turbine with rated power of 500 kW

are copper, zinc, nickel, and molybdenum. In addition such bulk commodities like iron ore, aluminium, limestone, and carbon are used. Wind turbines use steel for the towers, nacelle structural components, and the drivetrain, accounting for about 80% of the total weight. Some turbine generator designs use direct-drive magnets, which contain the rare earth metals such as neodymium and dysprosium. Such direct-drive permanent magnet generators simplify the design by eliminating the gearbox, and this design solution is attractive for offshore applications because it reduces maintenance.

A typical crystalline silicon PV panel, which is currently the dominant technology, contains about 76% glass in the panel surface, 10% polymer in the encapsulant and back-sheet foil, 8% aluminium in the frame, 5% silicon, 1% copper in the interconnectors, and less than 0.1% silver in the contact lines.

Lithium ion batteries are used in almost all electric vehicles and are also important for stationary energy-storage applications. Such batteries are made of two electrodes (anode and cathode), current collectors, a separator, electrolyte, a container, and sealing parts. The anode is usually made of graphite, with a copper foil current collector. The cathode is typically a layered transition metal oxide, with an aluminium foil current collector. A porous separator and electrolyte are placed in between the electrodes, and all components are housed in an aluminum container. The most common lithium ion batteries for electric vehicle applications are nickel-manganese-cobalt, lithium-iron phosphate, nickel-cobalt-aluminium, and lithium-manganese oxide [33].

Table 16.4
Minerals Used in Selected Power Generation Technologies

Mineral	Offshore Wind (kg/MW)	Onshore Wind (kg/MW)	Solar PV (kg/MW)	Natural Gas (kg/MW)	Coal (kg/MW)	Battery Storage[a] (kg/MWh)
Copper	11498.6	1730.5	2809.8	1085.3	783.9	-
Nickel	663.7	556.6	1.3	-	655.5	284.4
Manganese	56.4	56.4	-	-	6.3	284.4
Cobalt	-	-	-	1.8	1.8	284.4
Silicon	-	-	6620.3	-	-	-
Chromium	293.6	788.5	-	2.0	281.0	-
Molybdenum	127.8	127.8	-	-	57.0	-
Zinc	5454.0	5451.3	25.1	-	-	-
Rare earth	105.9	69.5	-	-	-	-
Graphite	-	-	-	-	-	829.4
Lithium	-	-	-	-	-	142.2
Other	97.3	91.2	31.3	15.6	215.3	-

Source: IEA, Minerals used in selected power generation technologies, IEA, Paris; retrieved on 2021-03-09 from https://www.iea.org/data-and-statistics/charts/minerals-used-in-selected-power-generation-technologies.
[a] Calculated from mineral use for a medium-size lithium-ion battery pack with capacity 42.2 kWh, using 12 kg of nickel, manganese and cobalt, 35 kg of graphite, and 6 kg of lithium, as estimated by German Raw Material Agency (DERA) in 2018.

PROBLEMS

PROBLEM 16.1

Derive a more realistic model of Earth's atmosphere than the one shown in Fig. 16.2(a) by assuming that the atmosphere contains two layers with different temperatures.

PROBLEM 16.2

Develop further the model from Problem 16.1 and show that an n-layer atmosphere gives a surface temperature $T_s = (n+1)^{1/4} \times 255 \text{ K}$.

PROBLEM 16.3

Calculate the total mass of CO_2 that needs to be removed from the atmosphere to reduce the CO_2 concentration from 417 to 250 ppm.

PROBLEM 16.4

During 2018 the global emission of CO_2 into the atmosphere was 31.7×10^9 tonnes. What would be the corresponding increase of the CO_2 concentration during that year if this emission was the only source of CO_2 and all emitted CO_2 stayed in the atmosphere?

17 Risks, Safety, and Cost Analysis

Energy transformation has both positive and negative impacts on societies. On the one hand, access to abundant, affordable, secure, safe, and clean energy is beneficial for humans. On the other hand, extraction, transportation, and use can have negative consequences to the health, environment, and economics of societies. In the previous chapter we discussed the environmental impacts of energy systems. This chapter deals with such aspects as risk, safety, and costs associated with energy systems.

Any complex technology, and most of modern energy systems belong to this category, is prone to events that can lead to certain undesired consequences. Risk analysis techniques are needed to assess both the frequency and the consequences of such events, while safety analysis techniques are used for preventing the occurrence of such events.

Any energy system, if it is to be viable for an extended period of time, must be able to sell a product (electricity, heat, or another energy form) for more than its cost. Since the existence of the system depends on this principle, an economic assessment of any project is an essential part of the decision to pursue it. Risk and safety analyses are discussed in sections §17.1 and §17.2, respectively, whereas section §17.3 deals with the cost analysis, and in particular with calculation of the levelized cost of energy.

17.1 RISK ANALYSIS

Many different definitions of **risk** have been proposed. In simple terms risk is the possibility of loosing something that humans value, such as health, well-being, property, or the environment. In general, the descriptions and definitions of risk are different for different areas of application, such as business, economics, safety, and security. The International Organization for Standardization (ISO) Guide 77[1] provides basic vocabulary to develop common understanding on risk management concepts and terms across different applications.

Similarly to risk definitions, there are many different metrics that are used for its quantitative description. A metric frequently used for complex engineering systems describes risk $\mathscr{R}$ as a set of triplets

$$\mathscr{R} = \langle s_i, p_i, c_i \rangle \quad \text{for } i = 1, 2, ..., N, \tag{17.1}$$

where s_i is a sequence of events called a **scenario**, p_i is the probability of the scenario, c_i is the consequence of the scenario, and N is the number of scenarios chosen to describe the risk.

Risk is the combination of the predicted frequency of an undesired initiating event and the predicted damage such an event might cause if the ensuing follow-up events were to occur. In general, the damage from an accident sequence i can be analyzed with a continuum of outcomes between x and $x + dx$ such that $\int_X^\infty \mathscr{D}_i(x)dx$ represents the total damage exceeding the magnitude X. Thus for an initial event i that causes a variety of predicted consequences of type j, the risk is given as,

$$\mathscr{R}_i(\geq X) = \mathscr{F}_i \sum_j \int_X^\infty \mathscr{D}_{ij}(x)dx, \tag{17.2}$$

where $\mathscr{F}_i$ is the frequency of an event sequence i and $\mathscr{D}_{ij}(x)$ is the damage due to the initial event i and consequence type j. As can be seen $\mathscr{R}_i(\geq X)$ is the complementary cumulative distribution

---

[1]ISO Guide 73:2009, Risk management—Vocabulary, available at www.iso.org.

DOI: 10.1201/9781003036982-17

function for accident sequence i. The practical approach to evaluate risk of energy systems is presented in the following sections.

17.1.1 RISK OF ENERGY SYSTEMS

All energy systems entail risks to the environment and to the health and welfare of people. However, our information about them is subject to great uncertainty, and their comparison is difficult. For example, for centralized energy production and distribution systems, risks and benefits are not shared equally, since the person who receives the benefit generally does not suffer the risk.

A comparison of energy-related risks performed in [86] used the following three bases:

1. Energy-related risks of a given kind were compared with risks arising from background effects of the same kind; for example, the risks of cancer from the emissions of nuclear power plants can be compared to the average risk of cancer in the general population or the hypothetically estimated cancer risk associated with exposure to natural background radiation.
2. Cross comparisons were made among alternative energy technologies, systems, or strategies with respect to similar kinds of risks; for example, comparison of the relative risks to ecosystems from coal combustion and hydropower.
3. Energy-related risks were compared to more familiar risks; for example, fatalities from nuclear reactor accidents were compared to fatalities from commercial airline accidents.

17.1.2 PROBABILISTIC RISK ASSESSMENT

Probabilistic risk assessment (PRA) is a systematic technique for investigating the transformation of an undesired *initiating event* into a set of possible outcomes and their consequences. This analysis is often based on the *event tree* analysis. A *fault tree* is used to represent the combination of component failures that could result in the system failure, treated as the *top event* of the tree.

Fault Tree Analysis

A system **fault tree** is a logic diagram that connects component failure modes that combine to produce a failed state of the system called the top event. A fault tree construction starts with a precise definition of the **top event**. All subordinate fault trees are connected by logical connective functions, called gates, such as AND and OR. Thus the final fault tree consists of successive layers of events, connected by gates, that are possible contributors to the top failure event. A simplified fault tree diagram for an example electrical system is shown in Fig. 17.1. The electrical system consists of the offsite AC power and a diesel generator that would be used if the offsite power failed. Both these components are connected in parallel to a 4 kV AC bus which through a switcher room provides emergency electric power. Such systems are present in nuclear power plants and their availability is vital during emergency situations such as the loss-of-coolant accident.

From the fault tree it is seen that the top event T, which represent a failure of AC, results from several fault events such as: loss of offsite power (event X_3), loss of diesel generator (event X_4), failure of 4-kV AC bus (event X_2), and failure of AC due to malfunction in the switcher room (event X_1). The probability of loss of electric power is then

$$P_1 = P(T) = P(X_1 \cup X_2 \cup X_3 \cap X_4) \equiv P(X_1 + X_2 + X_3 X_4). \tag{17.3}$$

Here for brevity of equations we write $X_3 X_4 \equiv X_3 \cap X_4$ to represent the intersection of two events X_3 and X_4. Similarly $X_1 + X_2 \equiv X_1 \cup X_2$ is the union of events X_1 and X_2.

Using the product axiom for probabilities, the probability $P(X_3 X_4)$ can be found as,

$$P(X_3 X_4) = P(X_3 \mid X_4) P(X_4) = P(X_4 \mid X_3) P(X_3). \tag{17.4}$$

Basic events:

X_1 = failure of AC due to malfunction in switcher room

X_2 = failure of 4 kV AC bus

X_3 = loss of offsite power

X_4 = loss of diesel generator

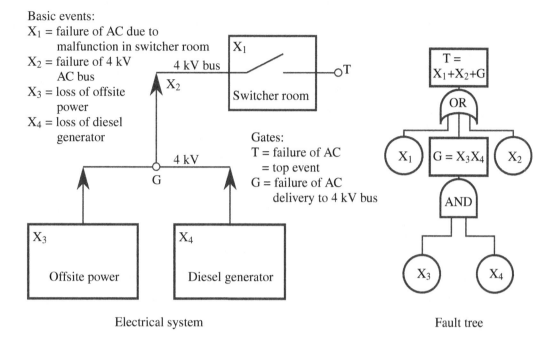

Gates:

T = failure of AC = top event

G = failure of AC delivery to 4 kV bus

Electrical system   Fault tree

Figure 17.1 Simplified electric system and its fault tree. Source [58]

Here the conditional probability $P(X_3 \mid X_4)$ is defined as the probability of event X_3 (loss of offsite power) given event X_4 (loss of diesel generator) has occurred. In the special case when the two events are independent (that is the loss of offsite power and the loss of generator do not depend of each other, and this is the case for properly designed systems) we have $P(X_3 \mid X_4) = P(X_3)$ and $P(X_3 X_4) = P(X_3)P(X_4)$.

For probability of the union of events, the following is valid,

$$P(X_1 + X_2) = P(X_1) + P(X_2) - P(X_1 X_2). \tag{17.5}$$

Assuming that events X_1 (failure of AC due to malfunction in switcher room) and X_2 (failure of 4 kV AC bus) are independent, we can write,

$$P(X_1 + X_2) = P(X_1) + P(X_2) - P(X_1)P(X_2). \tag{17.6}$$

Here the term $P(X_1 X_2) = P(X_1)P(X_2)$ is subtracted from the sum of the two events considered independently, $P(X_1) + P(X_2)$, to exclude the possible double counting resulting from the intersection of the two events. Using Eq. (17.4) and Eq. (17.6) in Eq. (17.3), we arrive at the following expression for the probability of loss of electric power,

$$P_1 = P(X_1) + P(X_2) - P(X_1)P(X_2) + P(X_3)P(X_4) - P(X_1)P(X_3)P(X_4)$$
$$- P(X_2)P(X_3)P(X_4) + P(X_1)P(X_2)P(X_3)P(X_4). \tag{17.7}$$

A fault tree analysis, as shown above, can be complicated by common cause failures. We can consider an improvement of the electric system shown in Fig. 17.1 by introducing additional three diesel generators. The probability term $P(X_4)$ could be then replaced by $[P(X_4)]^4$ in Eq. (17.7), and since this probability is a small number, the resulting probability of the loss of electricity power P_1 would be significantly reduced. However, this result wouldn't be valid if all four diesel generators can fail due to a common cause, such as flooding by a tsunami wave. If this possibility is overlooked, the probability of the top event would be significantly underestimated.

Event Tree Analysis

In complex energy systems, event trees provide step-by-step risk analysis technique. The **event tree** contains a number of paths representing sequences of events that follow an undesired **initiating event**, and lead to a predicted undesired consequence.

An example of a simplified event tree for a loss-of-coolant accident (LOCA) in a light water reactor is shown in Fig. 17.2. For a given initiating event (for LOCA it is a pipe break) which can occur with a predicted frequency λ, the availability or failure of the hierarchy of relevant safety systems (electric power, emergency core cooling, fission product removal, containment integrity) are considered sequentially.

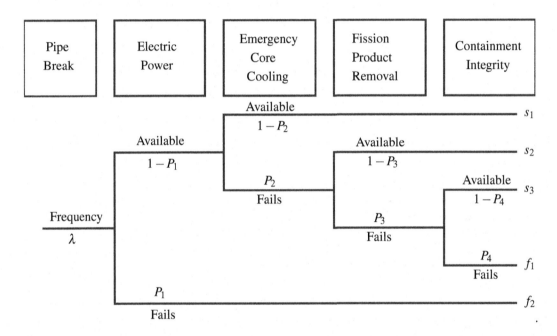

Figure 17.2   Simplified event tree for a loss-of-coolant accident. Source [58].

The conditional probabilities of each of the system failure events are obtained by fault tree analysis. For example, the failure probability of the electric system was derived in the previous section and is given by Eq. (17.7). Similar derivations are used to determine conditional probabilities P_2 through P_4. Multiplication of the conditional probabilities for the branches involved in a sequence then gives the probability of the sequence, providing that the failure events are independent. For example, the undesired outcome of loss of containment integrity occurs by the system failure sequence given as $f_1 = \lambda(1 - P_1)P_2P_3P_4$. This corresponds to an available electric power followed by sequential failures of emergency core cooling, fission product removal and containment integrity. In most cases, however, failure events are dependent. For example, electric system failure implies also failure of emergency cooling system, if the system is based on pumps driven by electric motors. As a consequence, the fission product removal and containment integrity will fail during the progression of the accident. This sequence reduced to a single line of the event tree shown in the lower portion of Fig. 17.2, and contribute with the failure frequency equal to $f_2 = \lambda P_1$. The total frequency of system failures resulting in unacceptable release of fission products is then $f = f_1 + f_2$.

17.2 HAZARDS IN ENERGY SYSTEMS

A common dictionary definition of hazard is: "a risk or source of danger". As a result, it is common to use all these terms synonymously. However, for the purpose of safety risk management, hazard, risk, and danger mean distinctly different things. Hazard is a condition with the potential of causing an undesired consequence, and danger is exposure to a hazard.

Practically all forms of energy are hazardous to humans. For example, electricity can harm people by electric shock or by exposure to an electrical arc. Hydraulic or pneumatic potential energy, when released in an uncontrolled manner, may crash or stuck individuals by moving machinery, equipment or other items. Thermal energy of objects with high or low temperatures or radiation may cause such common injuries as burn, scales, dehydration, and frostbite. Identifying hazards in energy systems is the first important step in the process of safety risk management. Once the hazards are identified, the associated risk can be assessed.

Risk and safety analysis helps to predict what might happen when an undesired initiating event occurs and safety systems fail to perform their intended function. To start with, the fault tree and the event tree analysis is applied as a useful risk analysis technique. In the next step safety analysis is performed with an objective to design the components of a system under consideration so that undesired initiating events do not occur or, if they do, that dedicated systems are activated to prevent or mitigate any undesired consequences. Types of initiating events and the practical application of the safety analysis depends on the specific energy system.

The overall risk of any energy system may be divided into two categories: occupational and public risk. Occupational risk is incurred by those involved in the process of producing and operating an energy system, whereas public risk is incurred by everyone else.

Comparison of the relative risk of energy systems is not straightforward and requires a substantial research effort. To make the comparison fair, we have to find the total risk to human health divided by the net energy production. In addition, we must consider all life-cycle risk components. In particular, we need to include all the processes required for components of a system in order to determine the overall risk. For example, risks associated with mining the coal, sand, copper, iron, uranium, and other raw materials that are required, as well as the risk due to fabricating them into glass, copper tubing, fuel rods, and other necessary components would be included. To this would be added the risk associated with transporting material, manufacturing components, and the more obvious risk of construction and operating the coal-fired plant, solar panel, or nuclear power plant [49].

Many studies have been performed to evaluate the health burdens of different forms of commercial power generation. The health effects of energy systems can most easily be assessed by a bottom-up approach, in which emissions and risks from each stage of the power generation cycle are measured and tracked to the endpoints at which they cause harm to individuals. The effects are calculated for a given power station using specified fuel sources. This approach is characterized by the so-called impact pathway, in which emissions from a source are traced through as they disperse into the environment, after which the effect of the dispersed pollutants are estimated [63].

A major set of studies called the ExternE program provided data on health effects of electricity generation in Europe using lignite, coal, oil, gas, biomass, and nuclear. The study showed that air pollution-related effects expressed in terms of fatalities per TWh of generated electricity are dominating and several orders of magnitude greater than effects due to accidents. The health effect of lignite is the most severe one with 32.6 fatalities per TWh, followed by coal (24.5), oil (18.4), biomass (4.63), gas (2.8), and nuclear (0.052) [63].

An assessment of energy accident risks across biofuels, biomass, geothermal, hydropower, hydrogen, nuclear power, solar energy and wind energy over the period 1950–2014 shows 686 accidents resulting in 182,794 human fatalities and \$265.1 billions in property damage [81]. Eight of the top ten accidents with the greatest fatalities were all related to hydropower, the other two being Chernobyl (second in the top ten) and Fukushima (seventh). Hydropower alone is responsible for 177,665 fatalities, followed by nuclear with 4803 fatalities and wind with 126 fatalities.

A normalized risk expressed per TWh of energy produced was estimated for a smaller sub-sample of accidents over the period 1990–2013. The estimation shows that wind energy had the highest accident frequency per TWh (0.0917), followed by solar energy (0.0136), biomass (0.0099), biofuels (0.0072), nuclear (0.0009), and hydro (0.0002). Wind power had also the highest risk in terms of fatalities per TWh (0.035), followed by hydro (0.0235), solar (0.0190), biomass (0.0164), nuclear (0.0097), and biofuels (0.0048). The three technologies with the greatest property damage per TWh were nuclear (greater than $3 million), wind ($235,000), and geothermal and biomass ($150,000) [81].

17.2.1 SOLAR POWER

Solar energy is one of the safest energy systems with 0.0190 fatalities per TWh of produced electricity as reported over the period 1990–2013. Its seven accidents constituted only about 1 percent of all accidents by frequency and only two accidents killed more than one person: an explosion at a polysilicon manufacturing plant in Japan in 2014, and when a solar thermal water heater fell on a crowd of people in South Korea in 2013 when it was being installed. Solar energy had no accidents with damage greater than $10 million. For solar photovoltaics, all accidents involved installers [81].

The solar photovoltaic industry has grown rapidly in recent years and requires more and more employees to install the photovoltaic panels. According to a report released by the International Renewable Energy Agency, in 2018 almost 11 million people globally worked in the renewable energy industry. Around 3.6 million of those employees worked for companies engaged in photovoltaics and the conversion of solar energy to electricity [50].

For solar energy there are four main hazards that have potential of causing undesired consequencies:

•   Lifting solar panels, leading to such undesired consequences as herniated discs, rotator cuff tears, and hip and low-back strains.
•   Trips and falls during construction, leading to such undesired consequences as broken, fractured, or shattered bones, severe back, neck, and head trauma, internal injuries, puncture injuries, and deaths.
•   Electrical shocks, leading to such undesired consequences as thermal burns, muscle, nerve, and tissue damages, falls from a surprise shock, and death.
•   Fall from a ladder, leading to such undesired consequences as fractures or sprains, puncture injuries, back, neck, and head trauma, and deaths.

17.2.2 WIND POWER

Among all low carbon technologies, wind power causes the most accidents per TWh (0.092) as well as the most deaths per TWh (0.035). Most accidents impacted either maintenance personnel or people living close to or onsite at the wind farm. The most serious accident in terms of deaths occurred in 2012 when a bus collided with a truck transporting a wind turbine tower in Sao Paulo State, Brazil, killing 17 people [81].

The wind energy sector is still relatively new, with wind turbine technology constantly progressing in tower design and component technology. The wind turbine manufactures tend not to publicize failure data. It could be argued that the hazards found within a wind farm are not too different from those that exist in other industries today. However, considering the sometimes unique and extreme conditions in which these hazards are found (e.g. isolated, remote and difficult-to-reach areas and extreme weather conditions), the new combination of these hazards and the inexperience of some of the workers in this sector, it is possible that these hazards may not be controlled or managed appropriately [29].

Accident data in the wind energy sector are hard to find and usually the information available is not very comprehensive. However, it is clear that with an increasing number of wind turbines built, more accidents is occurring. The Caithness Windfarm Information Forum[2] gathers worldwide information on wind turbine related accidents, which is gathered through press reports or official information releases. Until 31 March 2021, they reported 2840 accidents, leading to 126 occupational fatalities, and 93 public fatalities. By far the biggest number of found incidents was due to blade failure, which can arise from a number of possible sources, and results in either whole blades or pieces being thrown from the turbine. Pieces of a blade are documented as travelling up to 1.5 km. The second most common accident is the wind turbine fire. In stormy and dry weather this creates a wide-area fire risk due to scattering of the burning debris. Additional accident causes include structural failures (damage to turbines and tower collapse), ice throw, and transport [21].

Main hazards for those working with wind turbines include: falling, in particular when the turbine is being installed; working in confined spaces such as tower, nacelle, hub, and blades; electrocution; fires; and moving parts. The last two hazards are also affecting the community.

17.2.3 HYDROPOWER

Hydropower, after wind power, is the second riskiest low-carbon energy technology with 0.0235 fatalities per TWh. However, the statistics is largely influenced by a single disaster that occurred on August 8, 1975, in the Shimantan hydroelectric facility on the Ru River, in the Henan Province of China. Due to the dam failure, 1670 million tons of water was released in just 5 hours, creating a massive tidal wave that killed 26,000 people immediately. Another 145,000 people died of fatal injuries and subsequent epidemics and famine [80].

During commissioning of hydroelectric facility, the potential community hazards are those common to large industrial and infrastructure projects. Additional community hazards that may be associated with hydropower projects are as follows:

- Dam failure, including the potential failure of coffer dams during construction, that can lead to extensive downstream flooding with potentially catastrophic consequences, including loss of life and destruction of property.
- Draining around or downstream of the project.
- Destabilizing slopes surrounding the reservoir due to changes in loads occurring during reservoir filling and during variation in reservoir levels.
- General health issues in large reservoirs with stagnant water, promoting water-borne diseases.

17.2.4 COMBUSTION-BASED THERMAL POWER

Combustion-based thermal power heavily rely on fuel that has to be mined or harvested. Mining of fossil fuels is hazardous and has negative environmental effects. Surface mines affect negatively the landscape. Underground mines generally affect the landscape less, but the ground above the mine tunnels can collapse, and acidic water can drain from abandoned underground mines. In addition, methane gas that occurs in coal deposits can explode if it concentrates underground.

Coal and lignite mining has negative occupational health effects partly due to work accidents and partly due to fatal diseases that miners develop, such as pneumoconiosis, progressive massive fibrosis, emphysema, chronic bronchitis, and accelerated loss of lung function [63].

At the power generation stage the main effects arise from the emissions of primary small particles ($PM_{2.5}$) and the creation of secondary small particles (PM_{10}). In general, the following emissions result from coal burning:

- Sulfur dioxide (SO_2), which contributes to acid rains and respiratory illnesses.

---

[2]www.caithnesswindfarms.co.uk.

- Nitrogen oxides (NOx), which contribute to smog and respiratory illnesses.
- Particulates, which contribute to smog, haze, and respiratory illnesses and lung disease.
- Carbon dioxide (CO_2), which is the primary greenhouse gas produced from burning coal.
- Mercury and other heavy metals, which have been linked to both neurological and developmental damage in humans and animals.
- Fly ash and bottom ash, which are residues created when power plants burn coal.

The health effects from oil and gas are lower than those from coal, mainly because the effects from primary and secondary particles are much smaller.

Biomass can be transformed into useful energy by combustion or by thermochemical or biochemical conversion to liquid (ethanol, methanol) or gaseous fuels (methane, hydrogen). However, its usefulness as a major energy resource is limited by the inherent inefficiency of photosynthesis, which captures no more than a small percentage of solar energy reaching the Earth's surface. The power yield of even the most productive cultivated crops is therefore little higher than 1 W m^{-2}. This is an order of magnitude lower than direct solar capture through photovoltaic or thermal systems, and up to four orders of magnitude lower than fossil-fuel combustion [63].

17.2.5 GEOTHERMAL POWER

Geothermal is an inherently low-risk system from an accident standpoint. Over the period 1900-2013 it presented no reported fatalities, no accidents totalling more than $1 million, and only $800,000 in total damages across four accidents. The largest of these, with $300,000 in damages, related mostly to medical expenses from when an employee at geothermal facility in Canada fell into a pool of heated water and was badly burned [81].

There are, however, several hazards specific to the geothermal power. The most important hazards are those associated with drilling, sudden well discharge of CO_2 or H_2S with suffocating effect, catastrophic failure from the use of binary cycle plants, and the hydrothermal eruption, when superheated water trapped below the surface of the earth rapidly converts from liquid to steam, violently disrupting the confining rock.

17.2.6 NUCLEAR POWER

Nuclear energy accidents are relatively infrequent compared to other low carbon technologies, but they cause the highest property damage per TWh. According to [81], 172 accidents with 4803 fatalities and 240,854.3 million 2013US$ in damage occurred within the nuclear sector over the period 1950–2014. By far the highest damage of $162,650.7 million was caused by the Fukushima Daiichi nuclear power plant accident in 2011 in Japan. The second most costly damage, estimated to $15,500 million, occurred in 1995 in Japan when the fast breeder reactor Monju suffered fire caused by leaking sodium.

Nuclear is among the safest electricity-generating technologies in terms of fatalities per TWh. In a study from 2007, which includes accidents in Three Mile Island reactor in Pennsylvania, United States, and in Chernobyl reactor, Ukraine, and takes into account occupational deaths from mining, transport and other routes of exposure, the over-all fatalities rate is 0.052 per TWh [63]. A more recent study from 2016 spans the period from 1990 to 2013 (thus it only includes Fukushima accident) provides a slightly lower rate of 0.0097 fatalities per TWh [81].

The majority of conventional hazards present on a nuclear power plant are common with other forms of industry. These include, e.g., fire, dropped loads, electric shock, or vehicle movements. The only unique element to a nuclear power plant is the nuclear hazard associated with the use of nuclear fuel.

Nuclear hazards are part of radiological hazards and are mainly due to the potential for an uncontrolled nuclear fission reaction resulting from the use of uranium fuel. The hazards that could lead to a radiological release are each stringently regulated in accordance with national and international legislation. These regulations enforce the use of a wide range of measures to ensure that a nuclear power plant operation is safe. In general, the types and amount of protection required are chosen according to the hazard type and its potential impact. For example, if a hazard can affect the health of the public, a wide variety of protection is specified to ensure this will not happen. Less protection will be adopted if the hazard can only have a financial impact.

To reduce risks, the nuclear industry adopts some key methodologies such as elimination of hazards by design and providing backup for important equipment. For the acceptable risk level, the principle *as low as reasonably achievable* (ALARA) is followed. The combined methodology is known as the principle of **defense in depth** (DID) and typically consists of the following steps:

- Design nuclear power plant in such way that identified hazards cannot lead to undesired consequences.
- Operate nuclear power plant within safe operating parameters.
- Fit primary equipment with redundancy.
- Design diverse and segregated backup equipment.
- Design engineered safety features and on site actions to mitigate event consequences.
- Provide emergency procedures to reduce risk to people and environment.

In general, regulating nuclear power safety is a national responsibility. However, it is obvious that international standards are needed to enhance safety globally. Such international safety standards are established by the International Atomic Energy Agency (IAEA). The IAEA safety standards reflect an international consensus on what constitutes a high level of safety for protecting people and the environment from harmful effects of ionizing radiation. They fall into three main categories: safety fundamentals, safety requirements, and safety guides.

Safety guides provide recommendations and guidance on how to comply with the safety requirements. They provide international good practices to help users to achieve high levels of safety. For example, a specific safety guide is concerned with deterministic safety analysis for nuclear power plants [46]. This particular guide focuses on neutronic, thermalhydraulic, fuel, and radiological analysis. It covers aspects of the analysis of releases of radioactive material up to and including the determination of the source term for releases to the environment for anticipated operational occurrences and accident conditions.

17.3 COST ANALYSIS

The purpose of cost analysis is to determine if any investment is sound, and in particular, to verify whether its benefits outweigh the costs. This is particularly true for a project such as power plant which entails a large investment. Cost analysis provides also a basis for comparing projects, which involves comparing the total expected cost of each option against its total expected benefits.

An economic assessment of any project is an essential part of the decision to pursuit it. It should always accompany an engineering analysis to determine the design and efficiency of the plant, and the time required for its construction. Economic and engineering analyses should be employed in a balanced fashion in order to ensure the success of the project. Thus, a comprehensive understanding of both disciplines is desirable.

17.3.1 CALCULATION METHODS

Traditional cost-benefit analysis, as applied to energy decisions, does not usually include all important effects. This is particularly true for effects concerned with sociological, political, and even environmental risks. For example, a comparison of loss of natural beauty with the advantage of local business development is not included because there is no consensus about the models that should be applied. Still a political judgement is applied when comparing costs and benefits of particular groups or geographical regions. For example, some regions might bear more than their share of the adverse consequences of rapidly developed mining, but the economic benefits would not return to the affected communities in the same proportion.

In spite of the above-mentioned difficulties and limitations, there is ongoing effort to provide a transparent and intuitive metric of energy cost that can be used in policy making, modeling and public discussions. This metric is called in general the levelized cost of energy, and can be regarded as the minimum constant price at which energy must be sold in order to break even over the lifetime of the energy-transformation project. If the energy is sold as electricity, the concept of the levelized cost of electricity (LCOE) is used. The LCOE allows comparison of different methods of electricity generation on a consistent basis. Basic concepts used in derivation of the LCOE such as the time value of money, capital costs, and fuel cycle expenses are discussed below.

Time Value of Money

The time value of money is a basic financial principle describing how money in the present is worth more than an equal amount in the future. The relationship between the present time value PV and the future value FV is as follows,

$$PV = \frac{FV}{(1+i)^n},$$
(17.8)

where i is the interest rate per period and n is the number of periods.

The two most important cases include borrowing money and investing money. Let us first consider an amount C which is borrowed for 1 year at an annual interest rate i. At the end of the year this amount must be repaid together with an amount iC representing the interest. Therefore, the total amount that must be paid back at the end of the year is

$$\text{pay back} = C + iC = C(1+i).$$
(17.9)

If an amount C' is invested at an annual interest rate i to cover an expense C that will be incurred at the end of the year, we have $C = C' + iC' = C'(1+i)$. Thus, the amount that must be invested is

$$C' = \frac{C}{1+i}.$$
(17.10)

This approach may be easily extended to an expense C that will occur in n years in the future to show that the present value of such a future expense is $C/(1+i)^n$.

Example 17.1:

Calculate the present value of the nuclear fuel expenditures that is purchased one year before the fuel is loaded into the reactor and disposed one year after the fuel is discharged from the reactor, where the front-end and the back-end fuel expenses are equal to F and B, respectively. The fuel stays in the core for three years. For convenience, choose the reference point in time the time at which the fuel is loaded into the reactor.

Solution

Since the front-end expense F occurred one year in the past, its present value is found as $F(1+i)$. The back-end expense will occur in four years in the future, and its present value is found as $B/(1+i)^4$. Thus the total present value of the fuel expenditures is

$$\text{Fuel expenditures} = F(1+i) + \frac{B}{(1+i)^4}.$$
(17.11)

Capital Cost

Capital costs are expenses incurred to bring a project to a commercially operable status. They typically include costs needed for purchasing land, construction, and equipment used in the production of goods or in the rendering of services. Additional costs, such as property insurance, property taxes, and replacement costs depend on local conditions and because of this they are not included in general capital cost calculations.

The construction of a power plant occurs over an extended period of time and involves a sequence of non-uniform expenditures. Construction costs contain the costs of the entire construction of the plant and include costs of materials, labor, supervision, supplies, tools, equipment, and transportation. An example of a time distribution of annual construction costs is shown in Fig. 17.3. The complete capital cost is determined by both the magnitude C_k and timing of these expenditures over years $k = 0$ to $N - 1$, if the construction period extends over N years. Applying the present value

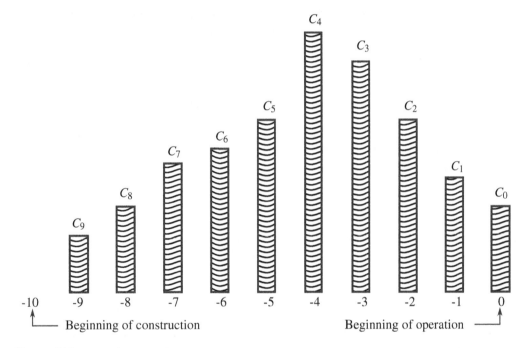

Figure 17.3 Annual construction costs.

concept and assuming that the payment is made at the end of the year, the total capital cost for the construction period extended over N years is,

$$C = C_0 + C_1(1+i) + C_2(1+i)^2 + \ldots + C_{N-1}(1+i)^{N-1} = \sum_{k=0}^{N-1} C_k(1+i)^k. \tag{17.12}$$

Here we followed a convention to take as the reference point in time the time when the plant begins the operation.

After the plant is constructed and ready for operation, it is necessary to recover revenues sufficient to repay the original investment. Two schedules of repayment of capital investment are shown in Fig. 17.4. With equal payments to principal scheme, at the end of year k the payment should be as follows,

$$P_k = \frac{C}{K} + i\left[C - (k-1)\frac{C}{K}\right] = \frac{C}{K}[1 + i(K - k + 1)]. \tag{17.13}$$

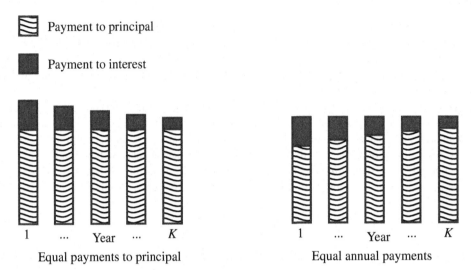

Figure 17.4 Repayment of capital investment with equal payments to principal (left) and equal annual payments (right).

It can be seen that the total annual payment decreases as the outstanding balance decreases. In some cases, a uniform set of payments is preferred. If the capital investment C should be repaid during K years with uniform annual payment C_u, the following equation is valid,

$$C = \left[\frac{1}{(1+i)^K} + \frac{1}{(1+i)^{K-1}} + \dots + \frac{1}{(1+i)^2} + \frac{1}{(1+i)} \right] \times C_u \qquad (17.14)$$

Solving for C_u yields,

$$C_u = \frac{C}{\sum_{k=1}^{K} \frac{1}{(1+i)^k}}. \qquad (17.15)$$

Finding the sum of the series, we get,

$$C_u = C \times \left[\frac{i \times (1+i)^K}{(1+i)^K - 1} \right]. \qquad (17.16)$$

This expression is commonly known as the sinking-fund repayment equation or the amortization equation.

If the sinking-fund repayment scheme is used, the annual payment C_u must be covered by revenues obtained from the sale of electricity. For a given amount of electricity E produced in any year, the levelized charge for electricity L_{cap} sufficient to cover capital expenditures is,

$$L_{cap} = \frac{C_u}{E}. \qquad (17.17)$$

In a similar manner additional levelized charges for electricity to cover taxes associated with capital expenses (L_{tc}) and fixed charges associated with capital investment (L_{fc}) can be calculated. Once these levelized charges are known, the annual revenue which will be sufficient to cover all expenditures associated with capital is as follows,

$$\text{Revenue} = (L_{cap} + L_{tc} + L_{fc})E. \qquad (17.18)$$

During plant operation two additional costs appear. The first one results from the operation and maintenance expenses, and the second one from fuel expenses. In general, the levelized charge for electricity to cover the operation and main expenses can be found as,

$$L_{om} = \frac{C_{om-fixed} + C_{om-variable} \times C_F}{E},$$ (17.19)

where $C_{om-fixed}$ is the fixed annual operation and maintenance expense, $C_{om-variable}$ is the variable (production-dependent) annual operation and maintenance expense, and C_F is the plant capacity factor.

The fuel expenses have to be taken into account for nuclear power plants and fossil fuel power plants. These expenses are region-dependent and prone to time variations. This is especially the case for the price of natural gas, which in some regions can be only about one-third of the price paid in the rest of the world. In addition, the share of fuel expenditures on total costs varies largely between technologies: whereas nuclear plants are characterized by high investment but relatively low fuel costs, this ratio is typically reversed in the case of natural gas. As an example, we describe the calculation principles of the nuclear fuel cycle expenses.

Nuclear Fuel Cycle Expenses

For simplicity, we consider here the direct expenses for a single batch of nuclear fuel. A thorough evaluation of a nuclear fuel cost is complex since the nuclear fuel cycle consists of many processes that occur at different points in time. The front-end processes that should be considered include mining and milling of uranium, conversion to UF_6 for use in enrichment plant, uranium enrichment, and fabrication into fuel elements. The back-end of fuel cycle is composed of cooling fuel in a storage basin at the reactor site, storage at a site away from the reactor, reprocessing and recovering of fissile material from used fuel, and disposal of reprocessing wastes.

Assuming that the fuel total cost for a single equilibrium batch is F (in monetary units per year) and the total cost of spent fuel shipping and disposal is B (in monetary units per year), the levelized direct fuel cycle expenses are found as,

$$L_b = \frac{F(1+i)^{ld} + \frac{B}{(1+i)^{lg+N}}}{\frac{E}{N}\left[\frac{1}{(1+i)} + \frac{1}{(1+i)^2} + ... + \frac{1}{(1+i)^N}\right]},$$ (17.20)

where ld is the lead time (time required for fuel preparation), lg is the lag time (time between discharge from the reactor and either reprocessing or ultimate disposal), N is the residence time in the reactor, i is the interest rate and E is the electric energy production per year (in kWh per year).

Example 17.2:

Calculate the levelized charge for electricity necessary to cover the direct expenses associated with a batch of fuel that is purchased one year before the fuel is loaded into the reactor and disposed one year after the fuel is discharged from the reactor, where the front-end and the back-end fuel expenses are equal to $F = 40$ M€/y and $B = 5$ M€/y, respectively. The fuel stays in the core for three years and the plant produces $E = 8 \times 10^9$ kWh of electric power per year. The effective interest rate is $i = 0.04$. For convenience, choose the reference point in time the time at which the fuel is loaded into the reactor.

Solution

Taking the lead time and the lag time equal to one year and substituting given data into Eq. (17.2) gives,

$$L_b = \frac{4 \times 10^7 \times 1.04 + \frac{5 \times 10^6}{(1.04)^4}}{\frac{8 \times 10^9}{3}\left[\frac{1}{1.04} + \frac{1}{(1.04)^2} + \frac{1}{(1.04)^3}\right]} \frac{€/y}{kWh/y} \approx 0.0062 \frac{€}{kWh}.$$ (17.21)

17.3.2 LEVELIZED COST OF ENERGY

The levelized cost of energy is a measure of the average net present cost of energy transformation for a given power plant. It can also be considered as the minimum constant price at which energy must be sold in order to break even over the lifetime of the plant. Most frequently the generation of electricity is considered, but the concept is also used for energy storage and other energy transformation applications.

We will first consider methods to determine the **levelized cost of electricity** (LCOE), since this is the most commonly used concept. Before making a decision to construct a power plant, it is important to know the average amount of charge for electricity since it determines revenues, and revenues must be sufficient to cover costs. Estimating the charge for electricity is particularly difficult in the case of power plants because costs associated with the plant are incurred over several decades. The main costs that are incurred are as follows:

- the capital cost incurred while constructing the plant,
- the operation and maintenance cost incurred over lifetime of the plant,
- the fuel cost (for fuel-driven plants) incurred during the plant operation,
- income taxes incurred as a consequence of engaging in a commercial enterprise,
- other costs.

Usually the capital and the fuel cost contributions are the most significant ones.

With annual discounting, the levelized cost of electricity L_{coe} can be obtained with the following equation, expressing the equality between the present value of the sum of discounted revenues and the present value of the sum of discounted costs,

$$\sum L_{coe} \times E \times (1+i)^{-k} = \sum \left(C_{cap,k} + C_{om,k} + C_{fuel,k} + C_{carbon,k} + C_{decom,k} \right) \times (1+i)^{-k}, \quad (17.22)$$

where E is the amount of electricity produced annually, i—the effective discount rate corresponding to the cost of capital, $C_{cap,k}$—total capital construction cost in year k, $C_{om,k}$—operation and maintenance cost in year k, $C_{fuel,k}$—fuel cost in year k, $C_{carbon,k}$—carbon cost in year k, $C_{decom,k}$—decommissioning and waste management cost in year k. Since L_{coe} is constant over time, it can be found as,

$$L_{coe} = \frac{\sum \left(C_{cap,k} + C_{om,k} + C_{fuel,k} + C_{carbon,k} + C_{decom,k} \right) \times (1+i)^{-k}}{\sum E \times (1+i)^{-k}}. \quad (17.23)$$

In most cases, when E is constant over the time, it can be brought out of the summation.

The metric presented by Eq. (17.23) is commonly used to assess cost competitiveness of power generation technologies. The main strength of this metric is that it is simple and combines all the direct technology costs into a single expression. Thanks to this, it can be applied equally well to technologies with a wide range of technical solutions and lifetimes.

The LCOE data for selected plants are shown in Table 17.1. The table is based on data provided for 243 plants in 24 countries in the ninth report in the series on the levelized costs of generating electricity produced jointly every five years by the International Energy Agency (IEA) and the OECD[3] Nuclear Energy Agency (NEA) [48]. The key finding of the report is that the costs of low-carbon generation technologies are falling and are increasingly below the costs of conventional fossil fuel generation. With the assumed moderate emission costs of USD 30/tCO$_2$ their costs are now competitive with dispatchable fossil fuel-based electricity generation in many countries. Electricity from new nuclear power plants has decreasing expected costs due to learning from first-of-a-kind projects in several countries. Nuclear thus remains the dispatchable low-carbon technology

---

[3]Organization for Economic Co-operation and Development.

Table 17.1
Levelized Cost of Electricity for Selected Plants at 7% Discount Rate

Country	Net Capacity [MWe]	Capacity Factor [%]	Invest-ment [USD/MWh]	Decommi-ssioning [USD/MWh]	Fuel/O&M [USD/MWh]	Carbon [USD/MWh]	LCOE [USD/MWh]
Combined-Cycle Gas Turbine							
China	475	85	6.49	0.03	53.54/13.49	10.45	84.00
Italy	790	85	6.85	0.04	45.50/6.99	10.10	69.47
Japan	1372	85	12.86	0.07	81.05/7.7	10.84	97.06
Korea	982	85	9.72	0.05	58.23/8.41	10.34	90.19
United States	727	85	11.04	0.06	18.38/5.30	10.20	48.88
Open-Cycle Gas Turbine							
Brazil	980	30	23.45	0.13	24.82/16.71	13.77	78.88
Canada	243	30	16.44	0.09	27.30/18.53	15.15	77.50
Italy	130	30	10.30	0.06	73.98/11.11	16.42	111.87
Coal Plants							
Australia[a]	722	85	27.21	0.07	29.65/8.69	25.47	91.10
China[b]	347	85	8.96	0.02	28.02/14.97	22.70	74.66
Japan[c]	749	85	27.05	0.07	28.73/19.31	24.68	99.84
United States[d]	641	85	46.48	0.13	17.16/29.50	23.99	117.27
Nuclear Plants							
France[e]	1650	85	47.46	0.05	9.33/14.26	-	71.10
France[f]	1000	85	12.45	0.46	9.33/12.92	-	35.15
France[g]	1000	85	8.25	0.15	9.33/12.92	-	30.65
Japan[h]	1152	85	46.87	0.05	13.92/25.84	-	86.67
Sweden[f]	1000	85	8.79	0.32	9.33/12.92	-	31.35
Sweden[g]	1000	85	5.83	0.10	9.33/12.92	-	28.17
Hydropower							
Germany[i]	24.5	65	42.61	5.89	-/8.50	-	57.00
United States[i]	94	66	81.84	0.01	-/5.35	-	82.70
Onshore Wind							
China	50	26	44.87	0.32	-/13.18	-	58.37
United States	100	41	36.46	0.17	-/11.36	-	47.99
Offshore Wind							
China	50	35	62.66	0.48	-/18.68	-	81.82
United States	600	45	47.31	0.33	-/22.92	-	70.56
Solar							
China	20	18	42.40	0.26	-/8.02	-	50.68
Japan	2	14	146.50	1.22	-/24.33	-	172.05
United States	100	23	47.88	0.11	-/6.68	-	54.66

Source: Tables 3.11 to 3.17 in [48].
[a] Supercritical pulverized, with electric conversion efficiency 40%.
[b] Ultra-supercritical, with electric conversion efficiency 45%.
[c] Ultra-supercritical, with electric conversion efficiency 41%.
[d] Ultra-supercritical, with electric conversion efficiency 43%.
[e] New build EPR, with electric conversion efficiency 33%.
[f] Long-term operation 10 years, with electric conversion efficiency 33%.
[g] Long-term operation 20 years, with electric conversion efficiency 33%.
[h] New build ALWR, with electric conversion efficiency 33%.
[i] Run-of-river.

with the lowest expected costs in 2025. Only large hydro reservoirs can provide a similar contribution at comparable costs but remain highly dependent on the natural endowments of individual countries [48].

The simplicity of LCOE stems from many assumptions that need to be made, such as the assumption of the capacity factors, the lifetime length, the discount rate, and the equal value of the produced electricity. In this way this approach neglects the differences in individual systems and markets that considerably influence the competitive position of technologies. These system-specific characteristics interact with the technical and economic characteristics of different technologies, i.e. their variability, dispatchability, response time, and cost structure. This also includes the fact that not all units are dispatched to the same extend across technologies and markets, or that revenues in many markets are determined by fluctuating price and not, as assumed in the LCOE analysis, by a stable price over a technology's lifetime [48].

For example, in a system with a significant share of variable technologies such as wind and solar, the electricity price can become negative if the electricity production exceeds the consumption. This is particularly true if the system does not have enough capacity to store or export the excessive electricity. At the same time, the dispatchable base-load plants have to reduce their electricity production below their capacity, negatively affecting their over-all capacity factor, and possibly, the efficiency. Clearly, the value of electricity generation is reduced if it is not correlated with the electricity demand. It means that LCOE increasingly needs to include the varying value of electricity in order to obtain a meaningful picture of the relative competitiveness of different electricity generating technologies.

In order to complement the LCOE approach and enable a more system specific cost comparison, the IEA has developed a methodology that adjusts the costs by a system value component known as the value adjusted LCOE (VALCOE). It modifies the LCOE of an individual technology in a particular electricity system according to its contribution to enabling all aspects of securely operation the system. A more complete picture of economic costs and full costs to society is still possible by inclusion of the impacts on human health (both through air pollution and through major accidents), the environment, employment, the availability of natural resources, and the security of supply [48].

PROBLEMS

PROBLEM 17.1

Derive Eq. (17.7) for the probability of loss of electric power.

PROBLEM 17.2

Explain the difference between risk, hazard, and danger. Give some examples from everyday life.

PROBLEM 17.3

Applying the present value concept and assuming that payments are made at the end of the year, calculate the change of the construction cost of a project, if the construction period changes from N_1 to N_2 years, and the overnight construction cost remains the same. Assume the beginning of operation as the reference time point and equal annual payments for both cases.

A Notation

A.1 NUMBER NOTATION

In the scientific notation very small and very large numbers are written as values between 1 and less than 10 multiplied by a power of 10. In the engineering notation large and small numbers are converted into a value between 1 and less than 1000, multiplied by a power of 10 in increments of three, such as 10^{-6}, 10^{-3}, 10^3, 10^6, etc. Thus, $3.142 \cdot 10^{11}$ in the scientific notation would be $314.2 \cdot 10^9$ in the engineering notation.

In this book both systems are employed. For microscopic systems the scientific notation is used, whereas for macroscopic and global systems (such as an energy system of a whole country) the engineering notation is preferred.

A.2 NOMENCLATURE AND SYMBOLS

For the reader's convenience, the meanings of variables used in an equation are explained immediately after the equation. When a series of equations occurs and the same set of variables is used, any new variable is explained at the first appearance.

Since this book is cross-disciplinary, it is impossible to avoid that sometimes the same notation is used for different physical quantities. Nevertheless in most cases international notation standards are followed, as summarized below.

LIST OF ROMAN SYMBOLS

Roman	Unit	Description
a	m s^{-2}	acceleration
A	-	atomic mass number
A	m^2	area
$\mathscr{A}$	-	atomic weight
B	m^{-1}	buckling
c	m s^{-1}	speed of light
c_p	$\text{J kg}^{-1}\,\text{K}^{-1}$	specific heat at constant pressure
c_v	$\text{J kg}^{-1}\,\text{K}^{-1}$	specific heat at constant volume
C_D	-	drag coefficient
C_f	-	Fanning friction factor
d	m	particle diameter
D	m	pipe diameter
D_h	m	hydraulic diameter
D_H	m	heated diameter
e	J kg^{-1}	specific energy
e_I	J kg^{-1}	specific internal energy
e_K	J kg^{-1}	specific kinetic energy
e_P	J kg^{-1}	specific potential energy
e_T	J kg^{-1}	specific total energy

DOI: 10.1201/9781003036982-A

LIST OF ROMAN SYMBOLS (CONT.)

Roman	Unit	Description
E	J	energy
E_I	J	internal energy
E_K	J	kinetic energy
E_P	J	potential energy
E_T	J	total energy
g	$\mathrm{m\,s^{-2}}$	acceleration due to gravity
G	$\mathrm{kg\,m^{-2}\,s^{-1}}$	mass flux
G_N	$\mathrm{m^3 \cdot kg^{-1} \cdot s^{-2}}$	Newtonian constant of gravitation ($=6.674\,30(15) \times 10^{-11}$)
H	m	height
i	$\mathrm{J\,kg^{-1}}$	specific enthalpy
I	J	enthalpy
J	$\mathrm{m\,s^{-1}}$	cross-section average superficial velocity
L	m	length
L	J	work
m	kg	mass
M	$\mathrm{kg\,mol^{-1}}$	molar mass
$\mathscr{M}$	-	molecular weight
n	mol	number of moles
$\mathbf{n}$	-	outward-pointing unit vector normal to surface
N	W	power
$\mathscr{N}$	-	number of atoms/molecules
N_0	-	Avogadro's number
N_A	$\mathrm{mol^{-1}}$	Avogadro's constant
P_H	m	heated perimeter
P_w	m	wetted perimeter
q	C	electric charge
q	W	thermal power
q'	$\mathrm{W\,m^{-1}}$	thermal linear power
q''	$\mathrm{W\,m^{-2}}$	heat flux
q'''	$\mathrm{W\,m^{-3}}$	thermal power density
Q	J	thermal energy
r	m	radial coordinate
R	m	radius
s	$\mathrm{J\,kg^{-1}\,K^{-1}}$	specific entropy
S	$\mathrm{J\,K^{-1}}$	entropy
t	s	time
T	K	temperature
$\mathbf{T}$	$\mathrm{N\,m^{-2}}$	total stress
u,v,w	$\mathrm{m\,s^{-1}}$	velocity components
u	$\mathrm{J\,kg^{-1}}$	specific thermal energy
$\mathscr{U}$	J	thermal energy
x,y,z	m	Cartesian coordinates

LIST OF GREEK SYMBOLS

Greek	Unit	Description
α	-	void fraction
κ	-	specific heat ratio
λ	$\mathrm{W\,m^{-1}\,K^{-1}}$	thermal conductivity
μ	Pa s	dynamic viscosity
ν	$\mathrm{m^2\,s^{-1}}$	kinematic viscosity
ρ	$\mathrm{kg\,m^{-3}}$	mass density
σ	$\mathrm{N\,m^{-1}}$	surface tension
τ	$\mathrm{N\,m^{-2}}$	viscous stress

LIST OF SUBSCRIPTS

Subscript	Description
f	saturated liquid phase
g	saturated vapor phase, gas phase
h	hydraulic

LIST OF SUPERSCRIPTS

Superscript	Description
T	transpose

OTHER SYMBOLS

Symbol	Description
$\equiv$	by definition equal to
$\approx$	approximately equal to
$\cong$	equal to a numerical value rounded to four digits
$\langle \rangle$	space averaged

B Constants

B.1 UNIVERSAL CONSTANTS

Table B.1 contains fundamental constants that by definition[1] have exact values. Using these constants, basic SI units, such as meter, second, kilogram, mol and ampere can be defined, as described in Appendix E.1.

Table B.2 contains selected derived or measured constants expressed in the SI units. Some of the constants (e.g. the universal gas constant) are directly derived from the defined constants given in Table B.1, whereas the others are obtained from measurements.

Table B.3 contains constants that are useful in solar energy applications.

B.2 STANDARD CONDITIONS

Standard conditions for temperature and pressure are necessary to allow comparison between different sets of data. Unfortunately there are no universally accepted standards so far. The most used standards are those of the International Union of Pure and Applied Chemistry (IUPAC) and the National Institute of Standards and Technology (NIST).

In chemistry, IUPAC proposed the following definition of **standard temperature and pressure** (STP), valid from 1982:

- temperature of 273.15 K,
- absolute pressure 10^5 Pa (1 bar).

NIST proposed the **normal temperature and pressure** (NTP) standard as follows:

- temperature of 293.15 K,
- absolute pressure 101.325 kPa (1 atm).

The international **standard metric condition** for natural gas and similar fluids are:

- temperature of 288.15 K,
- absolute pressure 101.325 kPa (1 atm).

The volumetric flow rate of such fluids is then expressed either in **standard cubic meter per second**, sm^3/s, or in **normal cubic meter per second**, nm^3/s. It is a good engineering practice to specify the reference conditions of temperature and pressure in any technical publication.

---

[1] According to 2019 redefinition of the SI base units.

Table B.1
Defining Constants of the International System of Units (SI)

Constant	Symbol	Value	Unit
Speed of light in vacuum	c	299 792 458	$m \cdot s^{-1}$
Planck constant	h	$6.626\ 070\ 15 \times 10^{-34}$	$J \cdot s$
Elementary charge	e	$1.602\ 176\ 634 \times 10^{-19}$	C
Boltzmann constant	k_B	$1.380\ 649 \times 10^{-23}$	$J \cdot K^{-1}$
Avogadro constant	N_A	$6.022\ 140\ 76 \times 10^{23}$	mol^{-1}
Luminous efficacy	K_{cd}	638	$lm \cdot W^{-1}$
Hyperfine transition frequency of ^{133}Cs	Δv_{Cs}	9 192 631 770	Hz

Source: CODATA Recommended Values of the Fundamental Constants of Physics and Chemistry, retrieved from
physics.nist.gov/constants on 2020-02-22.

Note: All values are exact.

Table B.2
Other Derived or Measured Universal Constants

Constant	Symbol	Value	Unit
Universal gas constant	R	$8.314\ 462\ 618\ 153\ 24^a$	$J \cdot K^{-1} \cdot mol^{-1}$
Stefan-Boltzmann constant	σ	$5.670\ 374\ 419... \times 10^{-8}$	$W \cdot m^{-2} \cdot K^{-4}$
Electron rest mass	m_e	$9.109\ 383\ 7015(28) \times 10^{-31}$	kg
Proton rest mass	m_p	$1.672\ 621\ 923\ 69(51) \times 10^{-27}$	kg
Neutron rest mass	m_n	$1.674\ 927\ 498\ 04(95) \times 10^{-27}$	kg
Atomic mass unit, or dalton	u	$1.660\ 539\ 066\ 60(50) \times 10^{-27}$	kg
Vacuum magnetic permeability	μ_0	$1.256\ 637\ 062\ 12(19) \times 10^{-6}$	$N\ A^{-2}$
Vacuum electric permittivity	ε_0	$8.854\ 187\ 8128(13) \times 10^{-12}$	$F\ m^{-1}$
Newtonian constant of gravitation	G_N	$6.674\ 30(15) \times 10^{-11}$	$m^3 \cdot kg^{-1} \cdot s^{-2}$
Standard acceleration of gravity	g	$9.806\ 65^a$	$m \cdot s^{-2}$

Source: CODATA Recommended Values of the Fundamental Constants of Physics and Chemistry, retrieved from
physics.nist.gov/constants on 2020-02-22.

[a] Exact value

Table B.3
Useful Constants and Quantities

Constant	Symbol	Value	Unit
Earth mean equatorial radius	$R_\oplus$	$6.378\ 1366(1) \times 10^6$	m
Mass of Earth	$M_\oplus$	$5.972\ 3(9) \times 10^{24}$	kg
Mean radius of Earth's orbit		149.6×10^6	km
Aphelion of Earth's orbit		152.1×10^6	km
Perihelion of Earth's orbit		147.1×10^6	km
Orbital period of Earth		$365.256\ 365\ 004$	day
Solar radius[a]	$R_\odot$	$695\ 700$	km
Sun mean equatorial radius		$696\ 342(65)^{b}$	km
Mass of Sun	$M_\odot$	1.9884×10^{30}	kg
Average solar constant above atmosphere	G_{SC}	1.361^{c}	kW/m^2
Temperature of the Sun's surface	$T_\odot$	$5\ 778$	K
Solar luminosity[d]	$L_{sol}, L_\odot$	3.828×10^{26}	W

[a] Solar radius is a unit of distance defined as the radius of the layer in the Sun's photosphere where the optical depth equals 2/3.

[b] As measured from space during the 2003 and 2006 Mercury transit [28].

[c] Based on satellite measurements [56].

[d] Solar luminosity is a unit of radiant flux, defined by the International Astronomical Union. It corresponds to the total power output of the Sun.

C Data

C.1 ATOMIC DATA OF CHEMICAL ELEMENTS

The table below contains a list of chemical elements and their representative isotopic compositions[1]. The table contains the following columns:

Z—atomic number of the element.

Symbol—short symbol of the element.

Element—full name of the element.

Rel. At. Mass of Element—relative atomic mass of the element, $A_r(X)$, commonly called "standard atomic weight", where X is an element. These values are scaled to $A_r(^{12}C) = 12$, where ^{12}C is a neutral atom of carbon-12 in its nuclear and electronic ground state. The values in parentheses (), following the last significant digit to which they are attributed, are uncertainties. For example $A_r(\text{He}) = 4.002602(2)$, which should be interpreted as $A_r(\text{He}) = 4.002602 \pm 0.000002$. If # is present, the value and uncertainty were derived not from pure experimental data, but at least partly from systematic trends. In many cases the relative atomic mass is given as an atomic-mass interval with the symbol $[a, b]$, which should be interpreted as $a \le A_r(X) \le b$, and denotes the set of atomic mass values in normal materials. Brackets [] enclosing a single value indicate the mass number of the most stable isotope.

N—number of neutrons in the nucleus of the isotope.

A—mass number of the isotope.

Abundance—mole fraction of the isotope in the element. Typically these values represent the isotopic composition of elements most commonly encountered in laboratories.

Rel. At. Mass of Isotope—relative atomic mass of the isotope, $A_r(^AX)$, where X is an element and A is the mass number of the isotope.

Z	Symbol	Element	Rel. At. Mass of Element	N	A	Abundance	Rel. At. Mass of Isotope
1	H	Hydrogen	$[1.00784, 1.00811]^m$	0	1	0.999885(70)	1.00782503223(9)
				1	2	0.000115(70)	2.01410177812(12)
2	He	Helium	$4.002602(2)^{g,r}$	1	3	0.00000134(3)	3.0160293201(25)
				2	4	0.99999866(3)	4.00260325413(6)
3	Li	Lithium	$[6.938, 6.997]^m$	3	6	0.0759(4)	6.0151228874(16)
				4	7	0.9241(4)	7.0160034366(45)
4	Be	Beryllium	9.0121831(5)	5	9	1	9.012183065(82)
5	B	Boron	$[10.806, 10.821]^m$	5	10	0.199(7)	10.01293695(41)
				6	11	0.801(7)	11.00930536(45)
6	C	Carbon	$[12.0096, 12.0116]$	6	12	0.9893(8)	12.0000000(00)
				7	13	0.0107(8)	13.00335483507(23)
7	N	Nitrogen	$[14.00643, 14.00728]$	7	14	0.99636(20)	14.00307400443(20)
				8	15	0.00364(20)	15.00010889888(64)
8	O	Oxygen	$[15.99903, 15.99977]$	8	16	0.99757(16)	15.99491461957(17)
				9	17	0.00038(1)	16.99913175650(69)
				10	18	0.00205(14)	17.99915961286(76)
9	F	Fluorine	18.998403163(6)	10	19	1	18.99840316273(92)
10	Ne	Neon	$20.1797(6)^{g,m}$	10	20	0.9048(3)	19.9924401762(17)
				11	21	0.0027(1)	20.993846685(41)
				12	22	0.0925(3)	21.991385114(18)
11	Na	Sodium	22.98976928(2)	12	23	1	22.9897692820(19)

---

[1] Data retrieved from www.nist.gov on 2020-03-06.

Z	Symbol	Element	Rel. At. Mass of Element	N	A	Abundance	Rel. At. Mass of Isotope
12	Mg	Magnesium	[24.304,24.307]	12	24	0.7899(4)	23.985041697(14)
				13	25	0.1000(1)	24.985836976(50)
				14	26	0.1101(3)	25.982592968(31)
13	Al	Aluminum	26.9815385(7)	14	27	1	26.98153853(11)
14	Si	Silicon	[28.084,28.086]	14	28	0.92223(19)	27.97692653465(44)
				15	29	0.04685(8)	28.97649466490(52)
				16	30	0.03092(11)	29.973770136(23)
15	P	Phosphorus	30.973761998(5)	16	31	1	30.97376199842(70)
16	S	Sulfur	[32.059,32.076]	16	32	0.9499(26)	31.9720711744(14)
				17	33	0.0075(2)	32.9714589098(15)
				18	34	0.0425(24)	33.967867004(47)
				20	36	0.0001(1)	35.96708071(20)
17	Cl	Chlorine	[35.446,35.457]^m	18	35	0.7576(10)	34.968852682(37)
				20	37	0.2424(10)	36.965902602(55)
18	Ar	Argon	39.948(1)^{g,r}	18	36	0.003336(21)	35.967545105(28)
				20	38	0.000629(7)	37.96273211(21)
				22	40	0.996035(25)	39.9623831237(24)
19	K	Potassium	39.0983(1)	20	39	0.932581(44)	38.9637064864(49)
				21	40	0.000117(1)	39.963998166(60)
				22	41	0.067302(44)	40.9618252579(41)
20	Ca	Calcium	40.078(4)^g	20	40	0.96941(156)	39.962590863(22)
				22	42	0.00647(23)	41.95861783(16)
				23	43	0.00135(10)	42.95876644(24)
				24	44	0.02086(110)	43.95548156(35)
				26	46	0.00004(3)	45.9536890(24)
				28	48	0.00187(21)	47.95252276(13)
21	Sc	Scandium	44.955908(5)	24	45	1	44.95590828(77)
22	Ti	Titanium	47.867(1)	24	46	0.0825(3)	45.95262772(35)
				25	47	0.0744(2)	46.95175879(38)
				26	48	0.7372(3)	47.94794198(38)
				27	49	0.0541(2)	48.94786568(39)
				28	50	0.0518(2)	49.94478689(39)
23	V	Vanadium	50.9415(1)	27	50	0.00250(4)	49.94715601(95)
				28	51	0.99750(4)	50.94395704(94)
24	Cr	Chromium	51.9961(6)	26	50	0.04345(13)	49.94604183(94)
				28	52	0.83789(18)	51.94050623(63)
				29	53	0.09501(17)	52.94064815(62)
				30	54	0.02365(7)	53.93887916(61)
25	Mn	Manganese	54.938044(3)	30	55	1	54.93804391(48)
26	Fe	Iron	55.845(2)	28	54	0.05845(35)	53.93960899(53)
				30	56	0.91754(36)	55.93493633(49)
				31	57	0.02119(10)	56.93539284(49)
				32	58	0.00282(4)	57.93327443(53)
27	Co	Cobalt	58.933194(4)	32	59	1	58.93319429(56)
28	Ni	Nickel	58.6934(4)^r	30	58	0.68077(19)	57.93534241(52)
				32	60	0.26223(15)	59.93078588(52)
				33	61	0.011399(13)	60.93105557(52)
				34	62	0.036346(40)	61.92834537(55)
				36	64	0.009255(19)	63.92796682(58)
29	Cu	Copper	63.546(3)^r	34	63	0.6915(15)	62.92959772(56)
				36	65	0.3085(15)	64.92778970(71)
30	Zn	Zinc	65.38(2)^r	34	64	0.4917(75)	63.92914201(71)
				36	66	0.2773(98)	65.92603381(94)
				37	67	0.0404(16)	66.92712775(96)
				38	68	0.1845(63)	67.92484455(98)
				40	70	0.0061(10)	69.9253192(21)
31	Ga	Gallium	69.723(1)	38	69	0.60108(9)	68.9255735(13)
				40	71	0.39892(9)	70.92470258(87)

Z	Symbol	Element	Rel. At. Mass of Element	N	A	Abundance	Rel. At. Mass of Isotope
32	Ge	Germanium	72.630(8)	38	70	0.2057(27)	69.92424875(90)
				40	72	0.2745(32)	71.922075826(81)
				41	73	0.0775(12)	72.923458956(61)
				42	74	0.3650(20)	73.921177761(13)
				44	76	0.0773(12)	75.921402726(19)
33	As	Arsenic	74.921595(6)	42	75	1	74.92159457(95)
34	Se	Selenium	78.971(8)^r	40	74	0.0089(4)	73.922475934(15)
				42	76	0.0937(29)	75.919213704(17)
				43	77	0.0763(16)	76.919914154(67)
				44	78	0.2377(28)	77.91730928(20)
				46	80	0.4961(41)	79.9165218(13)
				48	82	0.0873(22)	81.9166995(15)
35	Br	Bromine	[79.901,79.907]	44	79	0.5069(7)	78.9183376(14)
				46	81	0.4931(7)	80.9162897(14)
36	Kr	Krypton	83.798(2)^{g,m}	42	78	0.00355(3)	77.92036494(76)
				44	80	0.02286(10)	79.91637808(75)
				46	82	0.11593(31)	81.91348273(94)
				47	83	0.11500(19)	82.91412716(32)
				48	84	0.56987(15)	83.9114977282(44)
				50	86	0.17279(41)	85.9106106269(41)
37	Rb	Rubidium	85.4678(3)^g	48	85	0.7217(2)	84.9117897379(54)
				50	87	0.2783(2)	86.9091805310(60)
38	Sr	Strontium	87.62(1)^{g,r}	46	84	0.0056(1)	83.9134191(13)
				48	86	0.0986(1)	85.9092606(12)
				49	87	0.0700(1)	86.9088775(12)
				50	88	0.8258(1)	87.9056125(12)
39	Y	Yttrium	88.90584(2)	50	89	1	88.9058403(24)
40	Zr	Zirconium	91.224(2)^g	50	90	0.5145(40)	89.9046977(20)
				51	91	0.1122(5)	90.9056396(20)
				52	92	0.1715(8)	91.9050347(20)
				54	94	0.1738(28)	93.9063108(20)
				56	96	0.0280(9)	95.9082714(21)
41	Nb	Niobium	92.90637(2)	52	93	1	92.9063730(20)
42	Mo	Molybdenum	95.95(1)^g	50	92	0.1453(30)	91.90680796(84)
				52	94	0.0915(9)	93.90508490(48)
				53	95	0.1584(11)	94.90583877(47)
				54	96	0.1667(15)	95.90467612(47)
				55	97	0.0960(14)	96.90601812(49)
				56	98	0.2439(37)	97.90540482(49)
				58	100	0.0982(31)	99.9074718(11)
43	Tc	Technetium	[98]	55	98	0	97.9072124(36)
44	Ru	Ruthenium	101.07(2)^g	52	96	0.0554(14)	95.90759025(49)
				54	98	0.0187(3)	97.9052868(69)
				55	99	0.1276(14)	98.9059341(11)
				56	100	0.1260(7)	99.9042143(11)
				57	101	0.1706(2)	100.9055769(12)
				58	102	0.3155(14)	101.9043441(12)
				60	104	0.1862(27)	103.9054275(28)
45	Rh	Rhodium	102.90550(2)	58	103	1	102.9054980(26)
46	Pd	Palladium	106.42(1)^g	56	102	0.0102(1)	101.9056022(28)
				58	104	0.1114(8)	103.9040305(14)
				59	105	0.2233(8)	104.9050796(12)
				60	106	0.2733(3)	105.9034804(12)
				62	108	0.2646(9)	107.9038916(12)
				64	110	0.1172(9)	109.90517220(75)
47	Ag	Silver	107.8682(2)^g	60	107	0.51839(8)	106.9050916(26)
				62	109	0.48161(8)	108.9047553(14)

Z	Symbol	Element	Rel. At. Mass of Element	N	A	Abundance	Rel. At. Mass of Isotope
48	Cd	Cadmium	112.414(4)^g	58	106	0.0125(6)	105.9064599(12)
				60	108	0.0089(3)	107.9041834(12)
				62	110	0.1249(18)	109.90300661(61)
				63	111	0.1280(12)	110.90418287(61)
				64	112	0.2413(21)	111.90276287(60)
				65	113	0.1222(12)	112.90440813(45)
				66	114	0.2873(42)	113.90336509(43)
				68	116	0.0749(18)	115.90476315(17)
49	In	Indium	114.818(1)	64	113	0.0429(5)	112.90406184(91)
				66	115	0.9571(5)	114.903878776(12)
50	Sn	Tin	118.710(7)^g	62	112	0.0097(1)	111.90482387(61)
				64	114	0.0066(1)	113.9027827(10)
				65	115	0.0034(1)	114.903344699(16)
				66	116	0.1454(9)	115.90174280(10)
				67	117	0.0768(7)	116.90295398(52)
				68	118	0.2422(9)	117.90160657(54)
				69	119	0.0859(4)	118.90331117(78)
				70	120	0.3258(9)	119.90220163(97)
				72	122	0.0463(3)	121.9034438(26)
				74	124	0.0579(5)	123.9052766(11)
51	Sb	Antimony	121.760(1)^g	70	121	0.5721(5)	120.9038120(30)
				72	123	0.4279(5)	122.9042132(23)
52	Te	Tellurium	127.60(3)^g	68	120	0.0009(1)	119.9040593(33)
				70	122	0.0255(12)	121.9030435(16)
				71	123	0.0089(3)	122.9042698(16)
				72	124	0.0474(14)	123.9028171(16)
				73	125	0.0707(15)	124.9044299(16)
				74	126	0.1884(25)	125.9033109(16)
				76	128	0.3174(8)	127.90446128(93)
				78	130	0.3408(62)	129.906222748(12)
53	I	Iodine	126.90447(3)	74	127	1	126.9044719(39)
54	Xe	Xenon	131.293(6)^{g,m}	70	124	0.000952(3)	123.9058920(19)
				72	126	0.000890(2)	125.9042983(38)
				74	128	0.019102(8)	127.9035310(11)
				75	129	0.264006(82)	128.9047808611(60)
				76	130	0.040710(13)	129.903509349(10)
				77	131	0.212324(30)	130.90508406(24)
				78	132	0.269086(33)	131.9041550856(56)
				80	134	0.104357(21)	133.90539466(90)
				82	136	0.088573(44)	135.907214484(11)
55	Cs	Cesium	132.90545196(6)	78	133	1	132.9054519610(80)
56	Ba	Barium	137.327(7)	74	130	0.00106(1)	129.9063207(28)
				76	132	0.00101(1)	131.9050611(11)
				78	134	0.02417(18)	133.90450818(30)
				79	135	0.06592(12)	134.90568838(29)
				80	136	0.07854(24)	135.90457573(29)
				81	137	0.11232(24)	136.90582714(30)
				82	138	0.71698(42)	137.90524700(31)
57	La	Lanthanum	138.90547(7)^g	81	138	0.0008881(71)	137.9071149(37)
				82	139	0.9991119(71)	138.9063563(24)
58	Ce	Cerium	140.116(1)^g	78	136	0.00185(2)	135.90712921(41)
				80	138	0.00251(2)	137.905991(11)
				82	140	0.88450(51)	139.9054431(23)
				84	142	0.11114(51)	141.9092504(29)
59	Pr	Praseodymium	140.90766(2)	82	141	1	140.9076576(23)

Z	Symbol	Element	Rel. At. Mass of Element	N	A	Abundance	Rel. At. Mass of Isotope
60	Nd	Neodymium	144.242(3)^g	82	142	0.27152(40)	141.9077290(20)
				83	143	0.12174(26)	142.9098200(20)
				84	144	0.23798(19)	143.9100930(20)
				85	145	0.08293(12)	144.9125793(20)
				86	146	0.17189(32)	145.9131226(20)
				88	148	0.05756(21)	147.9168993(26)
				90	150	0.05638(28)	149.9209022(18)
61	Pm	Promethium	[145]	84	145	0	144.9127559(33)
62	Sm	Samarium	150.36(2)^g	82	144	0.0307(7)	143.9120065(21)
				85	147	0.1499(18)	146.9149044(19)
				86	148	0.1124(10)	147.9148292(19)
				87	149	0.1382(7)	148.9171921(18)
				88	150	0.0738(1)	149.9172829(18)
				90	152	0.2675(16)	151.9197397(18)
				92	154	0.2275(29)	153.9222169(20)
63	Eu	Europium	151.964(1)^g	88	151	0.4781(6)	150.9198578(18)
				90	153	0.5219(6)	152.9212380(18)
64	Gd	Gadolinium	157.25(3)^g	88	152	0.0020(1)	151.9197995(18)
				90	154	0.0218(3)	153.9208741(17)
				91	155	0.1480(12)	154.9226305(17)
				92	156	0.2047(9)	155.9221312(17)
				93	157	0.1565(2)	156.9239686(17)
				94	158	0.2484(7)	157.9241123(17)
				96	160	0.2186(19)	159.9270624(18)
65	Tb	Terbium	158.92535(2)	94	159	1	158.9253547(19)
66	Dy	Dysprosium	162.500(1)^g	90	156	0.00056(3)	155.9242847(17)
				92	158	0.00095(3)	157.9244159(31)
				94	160	0.02329(18)	159.9252046(20)
				95	161	0.18889(42)	160.9269405(20)
				96	162	0.25475(36)	161.9268056(20)
				97	163	0.24896(42)	162.9287383(20)
				98	164	0.28260(54)	163.9291819(20)
67	Ho	Holmium	164.93033(2)	98	165	1	164.9303288(21)
68	Er	Erbium	167.259(3)^g	94	162	0.00139(5)	161.9287884(20)
				96	164	0.01601(3)	163.9292088(20)
				98	166	0.33503(36)	165.9302995(22)
				99	167	0.22869(9)	166.9320546(22)
				100	168	0.26978(18)	167.9323767(22)
				102	170	0.14910(36)	169.9354702(26)
69	Tm	Thulium	168.93422(2)	100	169	1	168.9342179(22)
70	Yb	Ytterbium	173.054(5)^g	98	168	0.00123(3)	167.9338896(22)
				100	170	0.02982(39)	169.9347664(22)
				101	171	0.1409(14)	170.9363302(22)
				102	172	0.2168(13)	171.9363859(22)
				103	173	0.16103(63)	172.9382151(22)
				104	174	0.32026(80)	173.9388664(22)
				106	176	0.12996(83)	175.9425764(24)
71	Lu	Lutetium	174.9668(1)^g	104	175	0.97401(13)	174.9407752(20)
				105	176	0.02599(13)	175.9426897(20)
72	Hf	Hafnium	178.49(2)	102	174	0.0016(1)	173.9400461(28)
				104	176	0.0526(7)	175.9414076(22)
				105	177	0.1860(9)	176.9432277(20)
				106	178	0.2728(7)	177.9437058(20)
				107	179	0.1362(2)	178.9458232(20)
				108	180	0.3508(16)	179.9465570(20)
73	Ta	Tantalum	180.94788(2)	107	180	0.0001201(32)	179.9474648(24)
				108	181	0.9998799(32)	180.9479958(20)

Z	Symbol	Element	Rel. At. Mass of Element	N	A	Abundance	Rel. At. Mass of Isotope
74	W	Tungsten	183.84(1)	106	180	0.0012(1)	179.9467108(20)
				108	182	0.2650(16)	181.94820394(91)
				109	183	0.1431(4)	182.95022275(90)
				110	184	0.3064(2)	183.95093092(94)
				112	186	0.2843(19)	185.9543628(17)
75	Re	Rhenium	186.207(1)	110	185	0.3740(2)	184.9529545(13)
				112	187	0.6260(2)	186.9557501(16)
76	Os	Osmium	190.23(3)^g	108	184	0.0002(1)	183.9524885(14)
				110	186	0.0159(3)	185.9538350(16)
				111	187	0.0196(2)	186.9557474(16)
				112	188	0.1324(8)	187.9558352(16)
				113	189	0.1615(5)	188.9581442(17)
				114	190	0.2626(2)	189.9584437(17)
				116	192	0.4078(19)	191.9614770(29)
77	Ir	Iridium	192.217(3)	114	191	0.373(2)	190.9605893(21)
				116	193	0.627(2)	192.9629216(21)
78	Pt	Platinum	195.084(9)	112	190	0.00012(2)	189.9599297(63)
				114	192	0.00782(24)	191.9610387(32)
				116	194	0.3286(40)	193.9626809(10)
				117	195	0.3378(24)	194.9647917(10)
				118	196	0.2521(34)	195.96495209(99)
				120	198	0.07356(130)	197.9678949(23)
79	Au	Gold	196.966569(5)	118	197	1	196.96656879(71)
80	Hg	Mercury	200.592(3)	116	196	0.0015(1)	195.9658326(32)
				118	198	0.0997(20)	197.96676860(52)
				119	199	0.1687(22)	198.96828064(46)
				120	200	0.2310(19)	199.96832659(47)
				121	201	0.1318(9)	200.97030284(69)
				122	202	0.2986(26)	201.97064340(69)
				124	204	0.0687(15)	203.97349398(53)
81	Tl	Thallium	[204.382,204.385]	122	203	0.2952(1)	202.9723446(14)
				124	205	0.7048(1)	204.9744278(14)
82	Pb	Lead	207.2(1)^{g,r}	122	204	0.014(1)	203.9730440(13)
				124	206	0.241(1)	205.9744657(13)
				125	207	0.221(1)	206.9758973(13)
				126	208	0.524(1)	207.9766525(13)
83	Bi	Bismuth	208.98040(1)	126	209	1	208.9803991(16)
84	Po	Polonium	[209]	125	209	0	208.9824308(20)
85	At	Astatine	[210]	125	210	0	209.9871479(83)
86	Rn	Radon	[222]	136	222	0	222.0175782(25)
87	Fr	Francium	[223]	136	223	0	223.0197360(25)
88	Ra	Radium	[226]	138	226	0	226.0254103(25)
89	Ac	Actinium	[227]	138	227	0	227.0277523(25)
90	Th	Thorium	232.0377(4)^g	142	232	1	232.0380558(21)
91	Pa	Protactinium	231.03588(2)	140	231	1	231.0358842(24)
92	U	Uranium	238.02891(3)^{g,m}	142	234	0.000054(5)	234.0409523(19)
				143	235	0.007204(6)	235.0439301(19)
				146	238	0.992742(10)	238.0507884(20)
93	Np	Neptunium	[237]	144	237	0	237.0481736(19)
94	Pu	Plutonium	[244]	150	244	0	244.0642053(56)

^{g} Geological materials are known in which the element has an isotopic composition outside the limits for normal materials. The difference between the relative atomic mass of the element in such materials and that given in the table may exceed the stated uncertainty.

^{m} Modified isotopic compositions may be found in commercially available material because the material has been subjected to isotope fractionation. Significant deviations in atomic mass of the element from that given in the table can occur.

^{r} Range in isotopic composition of normal terrestrial material prevents a more precise relative atomic mass of the element being given. The tabulated atomic-mass value and uncertainty should be applicable to normal materials.

C.2 WATER-STEAM PROPERTY DATA

The properties shown in this section are based on the IAPWS-IF97 standard. Density (ρ), specific enthalpy (i), dynamic viscosity (μ), specific heat (c_p), thermal conductivity (λ), and specific entropy (s) are provided for sub-cooled water and superheated steam. For saturation conditions, the following properties are provided: pressure (p), temperature (T_{sat}), specific enthalpy of water (i_f), specific enthalpy of steam (i_g), latent heat ($i_{fg} = i_g - i_f$), specific entropy of water (s_f), specific entropy of steam (s_g), entropy increase during evaporation ($s_{fg} = s_g - s_f$), and surface tension (σ).

C.2.1 SUB-COOLED AND SUPERHEATED CONDITIONS

p (bar)	T (°C)	ρ (kg/m^3)	i (kJ/kg)	μ (μPa·s)	c_p (kJ/kg K)	λ (W/kg K)	s (kJ/kg K)
1	50	988.047	209.412	546.852	4.17956	0.64051	0.70375
1	70	977.779	293.074	403.900	4.18810	0.65961	0.95495
1	90	965.318	376.992	314.413	4.20502	0.67302	1.19263
1	95	961.894	398.030	297.286	4.21057	0.67555	1.25017
1	99.6059^f	958.637	417.436	282.947	4.21615	0.67759	1.30256
1	99.6059^g	0.59031	2674.95	12.2561	2.07594	0.02475	7.35881
1	105	0.58124	2686.09	12.4568	2.05460	0.02514	7.38847
1	110	0.57313	2696.32	12.6444	2.03992	0.02551	7.41536
1	130	0.54309	2736.72	13.4059	2.00391	0.02710	7.51814
1	150	0.51634	2776.59	14.1830	1.98566	0.02880	7.61467
10	130	935.211	546.882	213.084	4.26285	0.68525	1.63392
10	150	917.304	632.575	182.593	4.30857	0.68421	1.84137
10	170	897.586	719.320	159.605	4.36867	0.67886	2.04166
10	175	892.358	741.207	154.727	4.38647	0.67686	2.09077
10	179.886^f	887.127	762.683	150.248	4.40511	0.67465	2.13843
10	179.886^g	5.14539	2777.12	15.0220	2.71498	0.03540	6.58498
10	185	5.06580	2790.70	15.2439	2.60350	0.03549	6.61479
10	190	4.99205	2803.52	15.4608	2.52852	0.03564	6.64262
10	210	4.72720	2852.20	16.3254	2.36144	0.03662	6.74555
10	230	4.49848	2898.45	17.1867	2.27017	0.03803	6.83935
20	165	903.304	698.085	165.047	4.34830	0.68132	1.99095
20	185	882.210	785.746	146.059	4.42128	0.67287	2.18657
20	205	858.983	875.096	130.947	4.51844	0.66021	2.37744
20	210	852.803	897.760	127.628	4.54761	0.65637	2.42460
20	212.385^f	849.798	908.622	126.107	4.56234	0.65444	2.44702
20	212.385^g	10.0421	2798.38	16.1449	3.19036	0.04165	6.33916
20	215	9.95172	2806.59	16.2653	3.09115	0.04156	6.35603
20	220	9.78788	2821.67	16.4955	2.94874	0.04146	6.38676
20	240	9.21758	2877.21	17.4060	2.64811	0.04178	6.49719
20	260	8.74124	2928.47	18.3042	2.49094	0.04280	6.59522
30	185	882.895	786.231	146.307	4.41656	0.67368	2.18516
30	205	859.771	875.472	131.199	4.51220	0.66112	2.37580
30	225	834.169	966.942	118.797	4.64161	0.64424	2.56319
30	230	827.319	990.247	116.016	4.68102	0.63931	2.60974
30	233.858^f	821.895	1008.37	113.947	4.71380	0.63530	2.64562
30	233.858^g	15.0006	2803.26	16.9033	3.61228	0.04670	6.18579
30	235	14.9339	2807.35	16.9578	3.55258	0.04660	6.19384
30	240	14.6569	2824.56	17.1974	3.34354	0.04628	6.22755
30	260	13.7196	2886.42	18.1401	2.90697	0.04594	6.34585
30	280	12.9607	2942.16	19.0637	2.68542	0.04660	6.44851
50	215	848.994	921.516	125.226	4.55759	0.65517	2.46625
50	235	822.272	1014.05	113.872	4.70373	0.63636	2.65202
50	255	792.192	1110.06	104.056	4.91017	0.61276	2.83731

p (bar)	T (°C)	ρ (kg/m^3)	i (kJ/kg)	μ (μPa·s)	c_p (kJ/kg K)	λ (W/kg K)	s (kJ/kg K)
50	260	784.020	1134.77	101.771	4.97552	0.60602	2.88388
50	263.943^f	777.360	1154.50	100.008	5.03218	0.60046	2.92075
50	263.943^g	25.3509	2794.23	18.0327	4.43784	0.05564	5.97370
50	265	25.2219	2798.88	18.0857	4.35703	0.05545	5.98235
50	270	24.6503	2819.84	18.3371	4.04602	0.05469	6.02113
50	290	22.8018	2893.00	19.3191	3.36622	0.05318	6.15348
50	310	21.3827	2956.58	20.2736	3.02176	0.05315	6.26445
70	235	824.240	1014.38	114.398	4.68438	0.63860	2.64788
70	255	794.608	1109.91	104.626	4.88106	0.61545	2.83226
70	275	760.633	1210.24	95.8194	5.17298	0.58690	3.01868
70	280	751.244	1236.34	93.7016	5.26978	0.57878	3.06608
70	285.83^f	739.724	1267.44	91.2529	5.40039	0.56873	3.12199
70	285.83^g	36.5236	2772.57	18.9606	5.35404	0.06437	5.81463
70	290	35.6584	2793.98	19.1739	4.93602	0.06321	5.85279
70	295	34.7341	2817.70	19.4277	4.57120	0.06211	5.89473
70	315	31.8273	2899.57	20.4198	3.72521	0.05968	6.03644
70	335	29.6751	2969.28	21.3836	3.28458	0.05912	6.15302
90	255	796.965	1109.82	105.187	4.85349	0.61808	2.82731
90	275	763.650	1209.44	96.4513	5.12787	0.59017	3.01243
90	295	724.282	1316.03	88.1505	5.57120	0.55578	3.20336
90	300	713.071	1344.27	86.0630	5.73047	0.54590	3.25286
90	303.347^f	705.158	1363.65	84.6519	5.85416	0.53893	3.28657
90	303.347^g	48.7973	2742.88	19.8302	6.47619	0.07378	5.67901
90	305	48.2101	2753.35	19.9126	6.19242	0.07302	5.69714
90	310	46.6216	2782.61	20.1625	5.55785	0.07106	5.74754
90	330	41.9577	2878.87	21.1488	4.25988	0.06661	5.90996
90	350	38.7321	2957.22	22.1128	3.63702	0.06497	6.03781
110	270	775.183	1183.49	99.1686	5.01117	0.60069	2.96011
110	290	738.514	1287.02	90.9210	5.37032	0.56894	3.14725
110	310	693.719	1400.08	82.6809	6.00704	0.52945	3.34447
110	315	680.510	1430.72	80.5130	6.25681	0.51793	3.39679
110	318.081^f	671.796	1450.28	79.1380	6.44269	0.51042	3.42995
110	318.081^g	62.5239	2706.39	20.7156	7.91681	0.08464	5.55453
110	320	61.4344	2721.07	20.8020	7.40624	0.08328	5.57932
110	325	58.9813	2755.61	21.0326	6.48019	0.08036	5.63730
110	345	52.2055	2864.80	21.9762	4.72521	0.07375	5.81702
110	365	47.7809	2950.60	22.9199	3.93839	0.07069	5.95368
130	280	760.638	1233.53	95.6254	5.12483	0.58889	3.04664
130	300	721.600	1339.92	87.6056	5.55148	0.55462	3.23555
130	320	672.705	1458.02	79.2983	6.36059	0.51168	3.43800
130	325	657.886	1490.64	77.0374	6.70089	0.49904	3.49276
130	330.857^f	638.371	1531.40	74.2013	7.25793	0.48288	3.56058
130	330.857^g	78.2159	2662.89	21.6783	9.90715	0.09803	5.43388
130	335	74.6713	2700.35	21.8236	8.32392	0.09370	5.49568
130	340	71.2780	2738.92	22.0193	7.19537	0.08981	5.55886
130	360	62.2874	2858.09	22.8859	5.08166	0.08091	5.75029
130	380	56.6452	2949.64	23.7914	4.17422	0.07548	5.89272
150	295	735.696	1310.98	90.2953	5.35874	0.56769	3.18002
150	315	691.612	1423.51	82.3501	5.95672	0.52868	3.37462
150	335	632.628	1553.95	73.4367	7.32307	0.47855	3.59258
150	340	613.094	1592.27	70.7684	8.06472	0.46328	3.65534
150	342.158^f	603.514	1610.15	69.5045	8.52522	0.45616	3.68445
150	342.158^g	96.7109	2610.86	22.7932	12.9821	0.11579	5.31080
150	345	92.5891	2644.47	22.8229	10.8504	0.11062	5.36530
150	350	87.1027	2693.00	22.9353	8.78851	0.10406	5.44350

p (bar)	T (°C)	ρ (kg/m³)	i (kJ/kg)	μ (μPa·s)	c_p (kJ/kg K)	λ (W/kg K)	s (kJ/kg K)
150	370	74.1123	2831.40	23.6517	5.68675	0.08878	5.66236
150	390	66.6294	2932.11	24.4949	4.52636	0.08154	5.81664
175	305	719.996	1363.23	87.2869	5.51155	0.55453	3.26519
175	325	672.722	1479.84	79.3640	6.23114	0.51278	3.46340
175	345	606.735	1619.24	70.0094	8.09451	0.45862	3.69248
175	350	583.284	1662.45	66.9931	9.31471	0.44189	3.76210
175	354.671^f	554.671	1710.76	24.5980	11.7161	0.42460	3.83933
175	354.671^g	126.154	2529.11	24.5959	20.3861	0.15041	5.14280
175	360	112.735	2612.11	24.3426	12.5541	0.13220	5.27448
175	365	104.987	2667.41	24.3260	9.88367	0.12202	5.36149
175	385	87.7186	2819.31	24.8116	6.12497	0.09637	5.59616
175	405	78.2387	2926.89	25.5561	4.80232	0.08825	5.75727
200	315	703.574	1416.48	84.3906	5.68601	0.54063	3.35047
200	335	652.646	1537.85	76.4101	6.56010	0.49605	3.55332
200	355	577.560	1689.10	66.3874	9.24850	0.43768	3.79782
200	360	548.088	1739.97	62.8131	11.4527	0.41979	3.87860
200	365.746^f	490.524	1827.10	27.5020	23.1986	0.40374	4.01538
200	365.746^g	170.699	2411.39	27.4892	45.6779	0.22650	4.92990
200	370	144.458	2526.33	26.3168	18.6702	0.17938	5.10937
200	375	130.248	2602.59	25.9146	12.7497	0.14516	5.22745
200	395	104.325	2783.66	25.9134	6.93650	0.10712	5.50299
200	415	91.6515	2902.98	26.4952	5.23792	0.09618	5.67907
220.64	325	685.279	1471.51	81.4013	5.91075	0.52493	3.43824
220.64	345	629.437	1599.29	73.2031	7.02522	0.47699	3.64828
220.64	365	538.987	1770.57	61.8605	11.5669	0.41312	3.92076
220.64	370	494.900	1840.38	56.8630	17.8178	0.39475	4.02962
220.64	373.946^c	333.590	2068.59	40.4857	15508.6	0.80297	4.38272
220.64	375	210.605	2335.69	30.5518	66.7832	0.33572	4.79535
220.64	380	165.350	2497.57	27.9948	20.0991	0.18704	5.04437
220.64	400	121.890	2732.92	26.9019	8.10092	0.11983	5.40004
220.64	420	104.816	2867.92	27.2430	5.77696	0.10433	5.59782
250	335	668.367	1526.35	78.8446	6.11310	0.51075	3.52197
250	355	608.436	1659.94	70.5147	7.44975	0.46032	3.73786
250	375	505.649	1849.18	58.2685	13.6039	0.38691	4.03418
250	380	451.047	1935.30	52.4156	23.1002	0.38914	4.16646
250	384.863^p	317.491	2150.71	39.5717	71.1190	0.36068	4.49474
250	390	215.205	2395.46	31.7048	28.4665	0.22626	4.86555
250	395	184.181	2503.96	29.9468	17.2634	0.18248	5.02857
250	415	137.975	2730.55	28.4612	8.19086	0.12882	5.36331
250	435	118.460	2868.26	28.5967	5.92912	0.11239	5.56073

^{f} Saturated liquid phase.
^{g} Saturated vapor phase.
^{c} Critical point temperature.
^{p} Pseudo-critical point temperature.

C.2.2 SATURATED CONDITIONS

p (bar)	T_{sat} (°C)	i_f (kJ/kg)	i_g (kJ/kg)	i_{fg} (kJ/kg)	s_f (J/kg K)	s_g (J/kg K)	s_{fg} (J/kg K)	σ (10^{-3}N/m)
1	99.61	417.436	2674.95	2257.51	1302.56	7358.81	6056.25	58.99
2	120.2	504.684	2706.24	2201.56	1530.10	7126.86	5596.76	54.93
3	133.5	561.455	2724.89	2163.44	1671.76	6991.57	5319.80	52.20
4	143.6	604.723	2738.06	2133.33	1776.60	6895.42	5118.82	50.10
5	151.8	640.185	2748.11	2107.92	1860.60	6820.58	4959.98	48.35

p (bar)	T_{sat} (°C)	i_f (kJ/kg)	i_g (kJ/kg)	i_{fg} (kJ/kg)	s_f (J/kg K)	s_g (J/kg K)	s_{fg} (J/kg K)	σ (10^{-3}N/m)
6	158.8	670.501	2756.14	2085.64	1931.10	6759.17	4828.07	46.84
7	165.0	697.143	2762.75	2065.61	1992.08	6706.98	4714.90	45.51
8	170.4	721.018	2768.30	2047.28	2045.99	6661.54	4615.55	44.32
8	170.4	721.018	2768.30	2047.28	2045.99	6661.54	4615.55	44.32
9	175.4	742.725	2773.04	2030.31	2094.40	6621.24	4526.83	43.22
10	179.9	762.683	2777.12	2014.44	2138.43	6584.98	4446.55	42.22
11	184.1	781.198	2780.67	1999.47	2178.86	6551.99	4373.12	41.28
12	188.0	798.499	2783.77	1985.27	2216.30	6521.69	4305.39	40.40
13	191.6	814.764	2786.49	1971.73	2251.18	6493.65	4242.46	39.58
14	195.0	830.132	2788.89	1958.76	2283.88	6467.52	4183.64	38.80
15	198.3	844.717	2791.01	1946.29	2314.68	6443.05	4128.37	38.06
16	201.4	858.610	2792.88	1934.27	2343.81	6420.02	4076.21	37.36
17	204.3	871.888	2794.53	1922.64	2371.46	6398.25	4026.79	36.69
18	207.1	884.614	2795.99	1911.37	2397.79	6377.60	3979.80	36.04
19	209.8	896.844	2797.26	1900.42	2422.94	6357.94	3934.99	35.42
20	212.4	908.622	2798.38	1889.76	2447.02	6339.16	3892.14	34.83
21	214.9	919.989	2799.36	1879.37	2470.13	6321.20	3851.06	34.26
22	217.3	930.981	2800.20	1869.22	2492.36	6303.95	3811.59	33.70
23	219.6	941.626	2800.92	1859.30	2513.77	6287.37	3773.60	33.17
24	221.8	951.952	2801.54	1849.58	2534.44	6271.40	3736.95	32.65
25	224.0	961.983	2802.04	1840.06	2554.43	6255.97	3701.55	32.15
26	226.1	971.740	2802.45	1830.71	2573.77	6241.06	3667.29	31.66
27	228.1	981.241	2802.78	1821.54	2592.52	6226.62	3634.10	31.18
28	230.1	990.503	2803.02	1812.51	2610.73	6212.61	3601.89	30.72
29	232.0	999.542	2803.18	1803.63	2628.41	6199.01	3570.60	30.27
30	233.9	1008.37	2803.26	1794.89	2645.62	6185.79	3540.17	29.83
31	235.7	1017.00	2803.28	1786.28	2662.38	6172.92	3510.54	29.41
32	237.5	1025.45	2803.24	1777.79	2678.71	6160.37	3481.66	28.99
33	239.2	1033.72	2803.13	1769.41	2694.64	6148.14	3453.50	28.58
34	240.9	1041.83	2802.96	1761.14	2710.19	6136.19	3426.00	28.18
35	242.6	1049.78	2802.74	1752.97	2725.39	6124.51	3399.12	27.79
36	244.2	1057.57	2802.47	1744.90	2740.25	6113.09	3372.84	27.41
37	245.8	1065.23	2802.15	1736.91	2754.79	6101.92	3347.13	27.04
38	247.3	1072.76	2801.78	1729.02	2769.03	6090.97	3321.94	26.67
39	248.9	1080.15	2801.36	1721.21	2782.98	6080.24	3297.26	26.31
40	250.4	1087.43	2800.90	1713.47	2796.65	6069.71	3273.06	25.96
41	251.8	1094.58	2800.39	1705.81	2810.07	6059.38	3249.31	25.61
42	253.3	1101.63	2799.85	1698.22	2823.23	6049.23	3226.00	25.27
43	254.7	1108.57	2799.27	1690.70	2836.15	6039.25	3203.10	24.94
44	256.1	1115.40	2798.65	1683.25	2848.85	6029.45	3180.60	24.61
45	257.4	1122.14	2798.00	1675.85	2861.33	6019.80	3158.47	24.29
46	258.8	1128.79	2797.31	1668.52	2873.60	6010.30	3136.71	23.98
47	260.1	1135.34	2796.59	1661.24	2885.66	6000.95	3115.29	23.66
48	261.4	1141.81	2795.83	1654.02	2897.54	5991.74	3094.20	23.36
49	262.7	1148.20	2795.04	1646.85	2909.23	5982.66	3073.43	23.06
50	263.9	1154.50	2794.23	1639.73	2920.75	5973.70	3052.96	22.76
51	265.2	1160.73	2793.38	1632.65	2932.09	5964.87	3032.78	22.47
52	266.4	1166.88	2792.51	1625.62	2943.27	5956.15	3012.89	22.18
53	267.6	1172.96	2791.60	1618.64	2954.29	5947.55	2993.26	21.90
54	268.8	1178.98	2790.67	1611.69	2965.15	5939.05	2973.89	21.62
55	270.0	1184.92	2789.72	1604.79	2975.88	5930.65	2954.78	21.34
56	271.1	1190.81	2788.74	1597.93	2986.45	5922.35	2935.90	21.07
57	272.3	1196.63	2787.73	1591.10	2996.89	5914.15	2917.26	20.81
58	273.4	1202.39	2786.70	1584.31	3007.20	5906.04	2898.83	20.54
59	274.5	1208.09	2785.64	1577.55	3017.38	5898.01	2880.63	20.28
60	275.6	1213.73	2784.56	1570.83	3027.44	5890.07	2862.63	20.02
61	276.7	1219.32	2783.46	1564.14	3037.38	5882.21	2844.83	19.77
62	277.7	1224.86	2782.33	1557.48	3047.20	5874.42	2827.23	19.52

p (bar)	T_{sat} (°C)	i_f (kJ/kg)	i_g (kJ/kg)	i_{fg} (kJ/kg)	s_f (J/kg K)	s_g (J/kg K)	s_{fg} (J/kg K)	σ (10^{-3}N/m)
63	278.8	1230.34	2781.19	1550.84	3056.91	5866.71	2809.81	19.28
64	279.8	1235.78	2780.02	1544.24	3066.50	5859.08	2792.57	19.03
65	280.9	1241.17	2778.83	1537.66	3076.00	5851.51	2775.51	18.79
66	281.9	1246.51	2777.62	1531.11	3085.39	5844.01	2758.62	18.55
67	282.9	1251.81	2776.39	1524.58	3094.68	5836.58	2741.89	18.32
68	283.9	1257.06	2775.13	1518.07	3103.88	5829.20	2725.32	18.09
69	284.9	1262.27	2773.86	1511.59	3112.98	5821.89	2708.91	17.86
70	285.8	1267.44	2772.57	1505.13	3121.99	5814.63	2692.64	17.63
71	286.8	1272.57	2771.26	1498.69	3130.92	5807.43	2676.52	17.41
72	287.7	1277.65	2769.93	1492.27	3139.76	5800.29	2660.53	17.19
73	288.7	1282.70	2768.58	1485.87	3148.51	5793.19	2644.68	16.97
74	289.6	1287.72	2767.21	1479.49	3157.19	5786.15	2628.96	16.75
75	290.5	1292.70	2765.82	1473.12	3165.78	5779.16	2613.37	16.54
76	291.4	1297.64	2764.41	1466.78	3174.30	5772.21	2597.91	16.33
77	292.4	1302.55	2762.99	1460.44	3182.74	5765.30	2582.56	16.12
78	293.2	1307.42	2761.55	1454.12	3191.12	5758.44	2567.33	15.91
79	294.1	1312.27	2760.09	1447.82	3199.42	5751.63	2552.21	15.71
80	295.0	1317.08	2758.61	1441.53	3207.65	5744.85	2537.20	15.51
81	295.9	1321.86	2757.12	1435.25	3215.82	5738.11	2522.29	15.31
82	296.7	1326.61	2755.60	1428.99	3223.92	5731.41	2507.49	15.11
83	297.6	1331.34	2754.07	1422.74	3231.96	5724.75	2492.79	14.91
84	298.4	1336.03	2752.52	1416.49	3239.93	5718.12	2478.19	14.72
85	299.3	1340.70	2750.96	1410.26	3247.85	5711.52	2463.67	14.53
86	300.1	1345.34	2749.38	1404.04	3255.70	5704.96	2449.26	14.34
87	300.9	1349.96	2747.78	1397.82	3263.50	5698.43	2434.92	14.15
88	301.7	1354.54	2746.16	1391.62	3271.25	5691.93	2420.68	13.96
89	302.5	1359.11	2744.53	1385.42	3278.94	5685.45	2406.52	13.78
90	303.3	1363.65	2742.88	1379.23	3286.57	5679.01	2392.44	13.60
91	304.1	1368.17	2741.22	1373.05	3294.16	5672.59	2378.44	13.41
92	304.9	1372.66	2739.53	1366.87	3301.69	5666.20	2364.51	13.24
93	305.7	1377.14	2737.83	1360.70	3309.18	5659.84	2350.66	13.06
94	306.5	1381.59	2736.12	1354.53	3316.61	5653.49	2336.88	12.88
95	307.3	1386.02	2734.38	1348.37	3324.00	5647.17	2323.17	12.71
96	308.0	1390.43	2732.64	1342.21	3331.35	5640.88	2309.53	12.54
97	308.8	1394.81	2730.87	1336.06	3338.65	5634.60	2295.95	12.37
98	309.5	1399.18	2729.09	1329.90	3345.90	5628.35	2282.44	12.20
99	310.3	1403.54	2727.29	1323.75	3353.12	5622.11	2268.99	12.03
100	311.0	1407.87	2725.47	1317.61	3360.29	5615.89	2255.60	11.86
101	311.7	1412.18	2723.64	1311.46	3367.42	5609.69	2242.27	11.70
102	312.5	1416.48	2721.79	1305.31	3374.52	5603.50	2228.99	11.54
103	313.2	1420.76	2719.93	1299.17	3381.57	5597.34	2215.77	11.38
104	313.9	1425.02	2718.04	1293.02	3388.59	5591.18	2202.59	11.22
105	314.6	1429.27	2716.14	1286.88	3395.57	5585.04	2189.48	11.06
106	315.3	1433.50	2714.23	1280.73	3402.51	5578.92	2176.40	10.90
107	316.0	1437.72	2712.30	1274.58	3409.42	5572.80	2163.38	10.75
108	316.7	1441.92	2710.35	1268.43	3416.30	5566.70	2150.40	10.59
109	317.4	1446.11	2708.38	1262.27	3423.14	5560.61	2137.47	10.44
110	318.1	1450.28	2706.39	1256.12	3429.95	5554.53	2124.58	10.29
111	318.8	1454.44	2704.39	1249.96	3436.73	5548.46	2111.73	10.14
112	319.4	1458.58	2702.37	1243.79	3443.48	5542.40	2098.92	9.988
113	320.1	1462.72	2700.34	1237.62	3450.20	5536.34	2086.14	9.841
114	320.8	1466.84	2698.28	1231.45	3456.89	5530.29	2073.41	9.695
115	321.4	1470.95	2696.21	1225.26	3463.55	5524.25	2060.70	9.550
116	322.1	1475.05	2694.12	1219.08	3470.18	5518.22	2048.03	9.406
117	322.7	1479.13	2692.02	1212.88	3476.79	5512.19	2035.40	9.264
118	323.4	1483.21	2689.89	1206.68	3483.37	5506.16	2022.79	9.123
119	324.0	1487.27	2687.75	1200.47	3489.93	5500.14	2010.21	8.983
120	324.7	1491.33	2685.58	1194.26	3496.46	5494.12	1997.66	8.844

p (bar)	T_{sat} (°C)	i_f (kJ/kg)	i_g (kJ/kg)	i_{fg} (kJ/kg)	s_f (J/kg K)	s_g (J/kg K)	s_{fg} (J/kg K)	σ (10^{-3}N/m)
121	325.3	1495.37	2683.40	1188.03	3502.97	5488.10	1985.13	8.707
122	325.9	1499.41	2681.20	1181.79	3509.45	5482.08	1972.63	8.570
123	326.6	1503.43	2678.98	1175.55	3515.91	5476.06	1960.15	8.435
124	327.2	1507.45	2676.74	1169.29	3522.35	5470.04	1947.69	8.301
125	327.8	1511.46	2674.49	1163.02	3528.77	5464.02	1935.25	8.168
126	328.4	1515.47	2672.21	1156.74	3535.17	5458.00	1922.83	8.037
127	329.0	1519.46	2669.91	1150.45	3541.55	5451.98	1910.43	7.906
128	329.7	1523.45	2667.59	1144.14	3547.91	5445.95	1898.04	7.777
129	330.3	1527.43	2665.25	1137.82	3554.25	5439.92	1885.67	7.648
130	330.9	1531.40	2662.89	1131.49	3560.58	5433.88	1873.30	7.521
131	331.5	1535.37	2660.51	1125.14	3566.88	5427.84	1860.95	7.395
132	332.0	1539.33	2658.11	1118.78	3573.17	5421.79	1848.61	7.270
133	332.6	1543.29	2655.69	1112.40	3579.45	5415.73	1836.28	7.146
134	333.2	1547.24	2653.24	1106.00	3585.71	5409.66	1823.95	7.023
135	333.8	1551.19	2650.77	1099.58	3591.96	5403.59	1811.63	6.901
136	334.4	1555.14	2648.28	1093.15	3598.19	5397.50	1799.31	6.780
137	335.0	1559.08	2645.77	1086.70	3604.41	5391.41	1787.00	6.660
138	335.5	1563.01	2643.24	1080.22	3610.62	5385.30	1774.68	6.541
139	336.1	1566.95	2640.68	1073.73	3616.81	5379.18	1762.37	6.423
140	336.7	1570.88	2638.09	1067.21	3623.00	5373.05	1750.05	6.306
141	337.2	1574.81	2635.49	1060.68	3629.18	5366.90	1737.72	6.190
142	337.8	1578.74	2632.85	1054.12	3635.34	5360.74	1725.39	6.075
143	338.3	1582.66	2630.20	1047.53	3641.50	5354.56	1713.06	5.962
144	338.9	1586.59	2627.51	1040.93	3647.66	5348.37	1700.71	5.849
145	339.5	1590.51	2624.81	1034.29	3653.80	5342.15	1688.35	5.737
146	340.0	1594.44	2622.07	1027.63	3659.94	5335.92	1675.98	5.626
147	340.5	1598.37	2619.31	1020.95	3666.07	5329.67	1663.60	5.516
148	341.1	1602.29	2616.52	1014.23	3672.20	5323.40	1651.20	5.407
149	341.6	1606.22	2613.71	1007.49	3678.32	5317.11	1638.79	5.298
150	342.2	1610.15	2610.86	1000.71	3684.45	5310.80	1626.35	5.191
151	342.7	1614.08	2607.99	993.909	3690.57	5304.46	1613.90	5.085
152	343.2	1618.02	2605.09	987.073	3696.68	5298.11	1601.42	4.980
153	343.7	1621.96	2602.16	980.205	3702.80	5291.72	1588.92	4.875
154	344.3	1625.90	2599.21	973.303	3708.92	5285.31	1576.39	4.772
155	344.8	1629.85	2596.22	966.366	3715.04	5278.88	1563.84	4.669
156	345.3	1633.80	2593.20	959.395	3721.16	5272.41	1551.26	4.567
157	345.8	1637.76	2590.15	952.386	3727.28	5265.92	1538.64	4.467
158	346.3	1641.72	2587.06	945.341	3733.41	5259.40	1525.99	4.367
159	346.8	1645.69	2583.95	938.256	3739.54	5252.85	1513.31	4.268
160	347.4	1649.67	2580.80	931.132	3745.68	5246.27	1500.59	4.170
161	347.9	1653.66	2577.62	923.968	3751.82	5239.66	1487.84	4.072
162	348.4	1657.65	2574.41	916.762	3757.97	5233.01	1475.04	3.976
163	348.9	1661.65	2571.16	909.513	3764.13	5226.33	1462.20	3.881
164	349.4	1665.66	2567.88	902.220	3770.30	5219.61	1449.32	3.786
165	349.9	1669.68	2564.57	894.882	3776.48	5212.86	1436.39	3.693
166	350.3	1673.75	2561.25	887.498	3782.72	5206.13	1423.42	3.600
167	350.8	1677.80	2557.85	880.052	3788.93	5199.29	1410.36	3.508
168	351.3	1681.86	2554.41	872.551	3795.16	5192.40	1397.25	3.417
169	351.8	1685.94	2550.93	864.993	3801.40	5185.47	1384.07	3.327
170	352.3	1690.04	2547.41	857.377	3807.67	5178.49	1370.82	3.237
171	352.8	1694.15	2543.85	849.701	3813.95	5171.46	1357.51	3.149
172	353.3	1698.27	2540.23	841.963	3820.26	5164.38	1344.12	3.061
173	353.7	1702.42	2536.58	834.160	3826.59	5157.24	1330.65	2.975
174	354.2	1706.58	2532.87	826.291	3832.95	5150.05	1317.11	2.889
175	354.7	1710.76	2529.11	818.351	3839.33	5142.80	1303.47	2.804
176	355.1	1714.97	2525.31	810.339	3845.73	5135.48	1289.75	2.720
177	355.6	1719.19	2521.45	802.252	3852.17	5128.10	1275.93	2.637
178	356.1	1723.45	2517.53	794.087	3858.64	5120.65	1262.01	2.554

p (bar)	T_{sat} (°C)	i_f (kJ/kg)	i_g (kJ/kg)	i_{fg} (kJ/kg)	s_f (J/kg K)	s_g (J/kg K)	s_{fg} (J/kg K)	σ (10^{-3}N/m)
179	356.5	1727.72	2513.56	785.839	3865.14	5113.13	1247.99	2.473
180	357.0	1732.02	2509.53	777.507	3871.68	5105.54	1233.86	2.392
181	357.4	1736.35	2505.44	769.085	3878.26	5097.87	1219.61	2.312
182	357.9	1740.71	2501.28	760.570	3884.88	5090.11	1205.23	2.233
183	358.4	1745.10	2497.06	751.958	3891.54	5082.27	1190.73	2.155
184	358.8	1749.53	2492.77	743.245	3898.24	5074.34	1176.10	2.078
185	359.3	1753.99	2488.41	734.424	3905.00	5066.31	1161.31	2.002
186	359.7	1758.48	2483.98	725.491	3911.81	5058.19	1146.38	1.927
187	360.1	1763.02	2479.46	716.441	3918.67	5049.96	1131.28	1.852
188	360.6	1767.60	2474.86	707.266	3925.60	5041.61	1116.02	1.779
189	361.0	1772.22	2470.18	697.961	3932.59	5033.16	1100.57	1.706
190	361.5	1776.89	2465.41	688.518	3939.64	5024.57	1084.93	1.634
191	361.9	1781.61	2460.54	678.928	3946.78	5015.86	1069.08	1.563
192	362.3	1786.39	2455.57	669.185	3953.99	5007.01	1053.02	1.494
193	362.8	1791.22	2450.50	659.277	3961.28	4998.01	1036.72	1.425
194	363.2	1796.12	2445.31	649.194	3968.67	4988.85	1020.18	1.357
195	363.6	1801.08	2440.01	638.925	3976.16	4979.52	1003.36	1.290
196	364.1	1806.12	2434.57	628.457	3983.75	4970.01	986.265	1.223
197	364.5	1811.23	2429.00	617.776	3991.46	4960.31	968.857	1.158
198	364.9	1816.43	2423.29	606.866	3999.29	4950.41	951.116	1.094
199	365.3	1821.71	2417.42	595.710	4007.26	4940.27	933.015	1.031
200	365.7	1827.10	2411.39	584.287	4015.38	4929.90	914.523	0.969
201	366.2	1832.60	2405.17	572.575	4023.66	4919.26	895.606	0.908
202	366.6	1838.21	2398.76	560.548	4032.12	4908.34	876.223	0.848
203	367.0	1843.96	2392.13	548.178	4040.77	4897.10	856.332	0.789
204	367.4	1849.85	2385.28	535.429	4049.64	4885.52	835.878	0.731
205	367.8	1855.90	2378.16	522.261	4058.76	4873.56	814.800	0.674
206	368.2	1862.13	2370.76	508.630	4068.15	4861.18	793.027	0.619
207	368.6	1868.56	2363.04	494.478	4077.85	4848.32	770.473	0.565
208	369.0	1875.23	2354.97	479.737	4087.89	4834.93	747.035	0.512
209	369.4	1882.16	2346.49	464.327	4098.35	4820.93	722.586	0.460
210	369.8	1889.40	2337.54	448.147	4109.26	4806.23	696.973	0.410
211	370.2	1896.99	2328.06	431.066	4120.73	4790.72	669.994	0.361
212	370.6	1905.02	2317.94	412.917	4132.85	4774.24	641.393	0.313
213	371.0	1913.57	2307.05	393.479	4145.77	4756.60	610.829	0.268
214	371.4	1922.77	2295.22	372.446	4159.69	4737.52	577.829	0.224
215	371.8	1932.81	2282.19	349.378	4174.89	4716.61	541.718	0.181
216	372.2	1943.96	2267.57	323.610	4191.81	4693.28	501.468	0.141
217	372.6	1956.70	2250.75	294.048	4211.16	4666.55	455.392	0.104
218	372.9	1971.88	2230.56	258.683	4234.27	4634.66	400.388	0.069
219	373.3	1991.44	2204.48	213.041	4264.15	4593.70	329.550	0.038
220	373.7	2021.91	2164.20	142.293	4310.86	4530.84	219.977	0.012

D Mathematical Tools

This appendix contains a brief summary of selected mathematical tools used in the text. These tools can be particularly useful in chapters devoted to fluid flow and heat transfer in energy systems. For more rigorous proofs or detailed explanations the reader should consult any of the many introductory mathematics or mathematical physics texts available.

D.1 COORDINATE SYSTEMS

This section provides basic information on coordinate systems. A coordinate system in the three-dimensional space is defined by choosing a set of three linearly independent vectors, $\{\mathbf{e}_1, \mathbf{e}_2, \mathbf{e}_3\}$, representing the three directions of the space.

A velocity vector is denoted by $\mathbf{v}$. In three-dimensional space it has three components denoted in general by (u, v, w) or (v_1, v_2, v_3). The space coordinates are denoted accordingly (x, y, z) or (x_1, x_2, x_3). More specific notations are used for various coordinate systems as described in the following subsections.

D.1.1 CARTESIAN COORDINATES

The Cartesian coordinate system is shown in Fig. D.1. As shown in the figure, the unit vector basis of the Cartesian coordinate system is given as $\{\mathbf{e}_1, \mathbf{e}_2, \mathbf{e}_3\} = \{\mathbf{i}, \mathbf{j}, \mathbf{k}\}$, and the velocity vector can be represented as $\mathbf{v} = v_x\mathbf{i} + v_y\mathbf{j} + v_z\mathbf{k} = u\mathbf{i} + v\mathbf{j} + w\mathbf{k} = v_1\mathbf{e}_1 + v_2\mathbf{e}_2 + v_3\mathbf{e}_3$, where $\mathbf{i}, \mathbf{j}, \mathbf{k}$ are unit vectors along axes x, y, z, and v_x, v_y, v_z are the corresponding velocity components. The last notation is very useful when index summation convention is used. In that case, the velocity vector is given as,

$$\mathbf{v} = \sum_{i=1}^{3} v_i\mathbf{e}_i \equiv v_i\mathbf{e}_i. \tag{D.1}$$

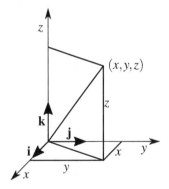
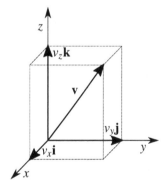

Figure D.1 Cartesian coordinates (x, y, z).

D.1.2 CYLINDRICAL POLAR COORDINATES

The cylindrical polar coordinate system is shown in Fig. D.2. In this coordinate system the unit vector basis is denoted as $\{\mathbf{e}_1, \mathbf{e}_2, \mathbf{e}_3\} = \{\mathbf{e}_r, \mathbf{e}_\theta, \mathbf{e}_z\}$, and the velocity vector can be represented as $\mathbf{v} = v_r\mathbf{e}_r + v_\theta\mathbf{e}_\theta + v_z\mathbf{e}_z = u\mathbf{e}_r + v\mathbf{e}_\theta + w\mathbf{e}_z$.

DOI: 10.1201/9781003036982-D

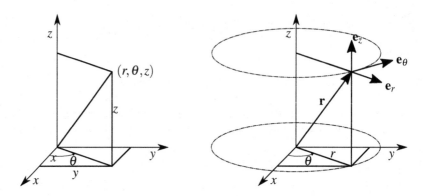

Figure D.2 Cylindrical polar coordinates (r,θ,z).

The following transformation can be used to express the Cartesian coordinates (x,y,z) in terms of the cylindrical polar coordinates (r,θ,z),

$$x = r\cos\theta, \qquad y = r\sin\theta, \qquad z = z. \tag{D.2}$$

The corresponding transformation of the base unit vectors is as follows,

$$\mathbf{i} = \cos\theta\,\mathbf{e}_r - \sin\theta\,\mathbf{e}_\theta, \qquad \mathbf{j} = \sin\theta\,\mathbf{e}_r + \cos\theta\,\mathbf{e}_\theta, \qquad \mathbf{k} = \mathbf{e}_z \tag{D.3}$$

D.1.3 SPHERICAL POLAR COORDINATES

The spherical polar coordinate system is shown in Fig. D.3. The unit basis vectors are $\{\mathbf{e}_1,\mathbf{e}_2,\mathbf{e}_3\} = \{\mathbf{e}_r,\mathbf{e}_\theta,\mathbf{e}_\phi\}$, and the velocity vector can be represented as $\mathbf{v} = v_r\mathbf{e}_r + v_\theta\mathbf{e}_\theta + v_\phi\mathbf{e}_\phi = u\mathbf{e}_r + v\mathbf{e}_\theta + w\mathbf{e}_\phi$.

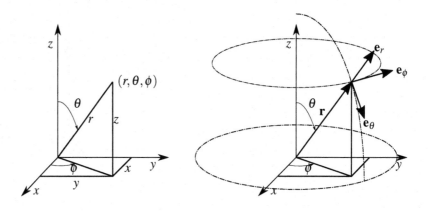

Figure D.3 Spherical coordinates (r,θ,ϕ).

The following transformation can be used to express the Cartesian coordinates (x,y,z) in terms of the spherical polar coordinates (r,θ,ϕ),

$$x = r\sin\theta\cos\phi, \qquad y = r\sin\theta\cos\phi, \qquad z = r\cos\theta. \tag{D.4}$$

The corresponding transformations of the base unit vectors are as follows,

$$\mathbf{i} = \sin\theta\cos\phi\,\mathbf{e}_r + \cos\theta\cos\phi\,\mathbf{e}_\theta - \sin\phi\,\mathbf{e}_\phi, \tag{D.5}$$

$$\mathbf{j} = \sin\theta\sin\phi\,\mathbf{e}_r + \cos\theta\sin\phi\,\mathbf{e}_\theta + \cos\phi\,\mathbf{e}_\phi, \tag{D.6}$$

$$\mathbf{k} = \cos\theta\,\mathbf{e}_r - \sin\theta\,\mathbf{e}_\theta. \tag{D.7}$$

D.2 SCALAR, VECTOR, AND TENSOR FIELDS

A scalar field or scalar-valued function associates a scalar value to every point in a space. Examples include the temperature or pressure distribution in a fluid.

A vector field is a vector each of whose components is a scalar field

$$\mathbf{v}(x,y,z) = v_1(x,y,z)\mathbf{e}_1 + v_2(x,y,z)\mathbf{e}_2 + v_3(x,y,z)\mathbf{e}_3. \tag{D.8}$$

The dot product of two vector fields is a scalar field

$$\mathbf{v}(x,y,z) \cdot \mathbf{u}(x,y,z) = v_1(x,y,z)u_1(x,y,z) + v_2(x,y,z)u_2(x,y,z) + v_3(x,y,z)u_3(x,y,z). \tag{D.9}$$

The cross product of two vector fields is a vector field

$$\mathbf{v} \times \mathbf{u} = (v_2 u_3 - v_3 u_2)\mathbf{e}_1 - (v_1 u_3 - v_3 u_1)\mathbf{e}_2 + (v_1 u_2 - v_2 u_1)\mathbf{e}_3, \tag{D.10}$$

where dependence of vectors and their components on coordinates x, y, z is dropped for clarity.

The dyadic product of any two unit basis vectors $\mathbf{e}_i$ and $\mathbf{e}_j$, also referred to as a unit dyad, is denoted $\mathbf{e}_i\mathbf{e}_j$. A second-order tensor τ can be written as a linear combination of the unit dyads

$$\tau = \sum_{i=1}^{3}\sum_{j=1}^{3} \tau_{ij}\mathbf{e}_i\mathbf{e}_j, \tag{D.11}$$

where the scalars τ_{ij} are referred to as the components of the tensor τ. A tensor field is a tensor each of whose components is a scalar field.

If σ and τ are tensors, the single-dot product of the tensors is

$$\sigma \cdot \tau = \sum_{i=1}^{3}\sum_{l=1}^{3}\left(\sum_{j=1}^{3} \sigma_{ij}\tau_{jl}\right)\mathbf{e}_i\mathbf{e}_l. \tag{D.12}$$

The double-dot product of two tensors

$$\sigma : \tau = \sum_{i=1}^{3}\sum_{j=1}^{3} \sigma_{ij}\tau_{ji}\mathbf{e}_i\mathbf{e}_j. \tag{D.13}$$

If $\mathbf{v}$ is a vector, the dot product of a tensor τ and the vector is

$$\tau \cdot \mathbf{v} = \sum_{i=1}^{3}\left(\sum_{j=1}^{3} \tau_{ij}v_j\right)\mathbf{e}_i, \tag{D.14}$$

and

$$\mathbf{v} \cdot \tau = \sum_{i=1}^{3}\left(\sum_{j=1}^{3} \tau_{ji}v_j\right)\mathbf{e}_i, \tag{D.15}$$

D.3 DIFFERENTIAL OPERATORS

Differential operators represent differentiation of functions and return another function. They are useful to write equations in a compact form.

D.3.1 NABLA

The nabla operator ∇ is defined in a Cartesian system of coordinates (x_1, x_2, x_3) by the orthonormal basis $(\mathbf{e}_1, \mathbf{e}_2, \mathbf{e}_3)$ as,

$$\nabla \equiv \mathbf{e}_1 \frac{\partial}{\partial x_1} + \mathbf{e}_2 \frac{\partial}{\partial x_2} + \mathbf{e}_3 \frac{\partial}{\partial x_3}. \tag{D.16}$$

D.3.2 GRADIENT

The gradient of a differentiable scalar field f is a vector field

$$\nabla f = \left(\sum_{i=1}^{3} \mathbf{e}_i \right) f = \mathbf{e}_1 \frac{\partial f}{\partial x_1} + \mathbf{e}_2 \frac{\partial f}{\partial x_2} + \mathbf{e}_3 \frac{\partial f}{\partial x_3}. \tag{D.17}$$

The gradient of a differentiable vector field $\mathbf{v}$ is a dyadic tensor field

$$\nabla \mathbf{v} = \sum_{i=1}^{3} \sum_{j=1}^{3} \frac{\partial v_j}{\partial x_i} \mathbf{e}_i \mathbf{e}_j. \tag{D.18}$$

D.3.3 DIVERGENCE AND CURL

The divergence of a differentiable vector field $\mathbf{v}$ is a scalar field

$$\nabla \cdot \mathbf{v} = \frac{\partial v_1}{\partial x_1} + \frac{\partial v_2}{\partial x_2} + \frac{\partial v_3}{\partial x_3}. \tag{D.19}$$

The curl of a differentiable vector field $\mathbf{v}$ is a vector field

$$\nabla \times \mathbf{v} = \left(\frac{\partial v_3}{\partial x_2} - \frac{\partial v_2}{\partial x_3} \right) \mathbf{e}_1 + \left(\frac{\partial v_1}{\partial x_3} - \frac{\partial v_3}{\partial x_1} \right) \mathbf{e}_2 + \left(\frac{\partial v_2}{\partial x_1} - \frac{\partial v_1}{\partial x_2} \right) \mathbf{e}_3. \tag{D.20}$$

D.3.4 LAPLACIAN

The Laplacian ∇^2 of a scalar function f with continuous second partial derivatives is defined as the divergence of the gradient

$$\nabla^2 f = \frac{\partial^2 v_1}{\partial x_1^2} + \frac{\partial^2 v_2}{\partial x_2^2} + \frac{\partial^2 v_3}{\partial x_3^2}. \tag{D.21}$$

D.4 INTEGRAL THEOREMS

There are several integral theorems that are particularly useful in fluid mechanics. They are shortly presented below.

D.4.1 DIVERGENCE THEOREM

The *divergence theorem* can be used to replace an integral over a volume V with an integral over a surface S that is enclosing the volume. For any vector field $\mathbf{v}$, the following relationship is valid,

$$\iiint_V (\nabla \cdot \mathbf{v}) dV = \iint_S (\mathbf{n} \cdot \mathbf{v}) dS, \tag{D.22}$$

in which $\mathbf{n}$ is the outwardly directed unit normal vector.

D.4.2 LEIBNIZ'S RULES

Leibniz's rules are useful to find a derivative with respect to time of a time-dependent quantity integrated over a time dependent region.

For a one-dimensional case, assuming that the region is a segment of x-axis, the rule is as follows,

$$\frac{d}{dt}\left[\int_{a(t)}^{b(t)} f(x,t)dx\right] = \int_{a(t)}^{b(t)} \frac{\partial f(x,t)}{\partial t}dx + f(b,t)\frac{db(t)}{dt} - f(a,t)\frac{da(t)}{dt}. \tag{D.23}$$

To facilitate physical interpretation of the rule, it can be written as follows,

$$\frac{d}{dt}\left[\int_{a(t)}^{b(t)} f(x,t)dx\right] = \int_{a(t)}^{b(t)} \frac{\partial f(x,t)}{\partial t}dx + f(b,t)\mathbf{v}_b \cdot \mathbf{n}_b + f(a,t)\mathbf{v}_a \cdot \mathbf{n}_a, \tag{D.24}$$

where

$$\mathbf{v}_a = \frac{da(t)}{dt}, \quad \mathbf{v}_b = \frac{db(t)}{dt}, \quad \mathbf{n}_a = -\mathbf{i}, \quad \mathbf{n}_b = \mathbf{i}. \tag{D.25}$$

Here $\mathbf{i}$ is a unit vector in the x-direction. It can be seen that the two last terms in Eq. (D.24) contain the velocity of displacement of integration boundaries and they represent the effect of the boundary movement on the over-all value of the time derivative.

For a two-dimensional case, assuming that the region is an area $A(t)$ located on the (x,y) plane and surrounded with a contour $C(t)$, the corresponding formulation of Leibniz's rule is as follows,

$$\frac{d}{dt}\left[\iint_{A(t)} f(x,y,t)dA\right] = \iint_{A(t)} \frac{\partial f(x,y,t)}{\partial t}dA + \int_{C(t)} f(x,y,t)\mathbf{v} \cdot \mathbf{n}dC. \tag{D.26}$$

For three-dimensional case, when integration is over a volume $V(t)$, surrounded with a surface $S(t)$, Leibniz's rule becomes,

$$\frac{d}{dt}\left[\iiint_{V(t)} f(x,y,z,t)dV\right] =$$
$$\iiint_{V(t)} \frac{\partial f(x,y,z,t)}{\partial t}dV + \iint_{S(t)} f(x,y,z,t)\mathbf{v} \cdot \mathbf{n}dS \tag{D.27}$$

Vector $\mathbf{n}$ in Eqs. (D.26) and (D.27), similarly as for the one-dimensional case, is a unit normal vector pointing outward from the region of integration. It should be remembered here that vector $\mathbf{v}$ on the right-hand-side of Eq. (D.27) represents the boundary $S(t)$ velocity.

Leibniz's rule is also applicable when integration is taken over a time interval $[t_1; t_2]$ as follows,

$$\frac{d}{dt}\left[\int_{t_1(t)}^{t_2(t)} f(x,\tau)d\tau\right] = \int_{t_1(t)}^{t_2(t)} \frac{\partial f(x,\tau)}{\partial \tau}d\tau + f(t_2,t)\frac{dt_2(t)}{dt} - f(t_1,t)\frac{dt_1(t)}{dt}. \tag{D.28}$$

D.4.3 REYNOLDS TRANSPORT THEOREM

The **Reynolds transport theorem** is given as follows

$$\frac{D}{Dt}\left(\iiint_{V_m(t)} f(\mathbf{r},t)dV\right) = \iiint_{V_m(t)} \frac{\partial f(\mathbf{r},t)}{\partial t}dV + \iint_{S_m(t)} f(\mathbf{r},t)\mathbf{v} \cdot \mathbf{n}dS. \tag{D.29}$$

This equation is useful to transform the derivatives of integrals from material based coordinates (Lagrangian frame of reference) on the left hand side into spatial coordinates (Eulerian frame of reference) on the right hand side. Note that the integration is performed over a material volume V_m bounded by the material surface S_m.

D.4.4 DIVERGENCE THEOREM

The **divergence theorem**, also referred to as **Gauss's integral theorem**, states an equivalence of the volume integral over the volume V and the surface integral over the boundary S of the volume V. For any vector or tensor quantity Ψ the divergence theorem is as follows,

$$\iiint_{V(t)} \nabla \cdot \Psi \, dV = \iint_{S(t)} \Psi \cdot \mathbf{n} \, dS, \tag{D.30}$$

where $\mathbf{n}$ is a unit vector directed normally outward from S. Closely related the **theorem of the rotational** is as follows,

$$\iiint_{V(t)} \nabla \times \Psi \, dV = \iint_{S(t)} \mathbf{n} \times \Psi \, dS. \tag{D.31}$$

For any scalar function Φ, the following **theorem of the gradient** is valid,

$$\iiint_{V(t)} \nabla \Phi \, dV = \iint_{S(t)} \Phi \mathbf{n} \, dS. \tag{D.32}$$

D.5 CONSERVATION EQUATIONS IN FLUID MECHANICS

Differential conservation equations of mass, momentum, and energy are given in this section.

D.5.1 MASS CONSERVATION EQUATION

Vector form:

$$\frac{\partial \rho}{\partial t} + \nabla \cdot (\rho \mathbf{v}) = 0. \tag{D.33}$$

Cartesian coordinates (x, y, z):

$$\frac{\partial \rho}{\partial t} + \frac{\partial (\rho u)}{\partial x} + \frac{\partial (\rho v)}{\partial y} + \frac{\partial (\rho w)}{\partial z} = 0. \tag{D.34}$$

Cylindrical coordinates (r, θ, z):

$$\frac{\partial \rho}{\partial t} + \frac{1}{r} \frac{\partial (\rho r u)}{\partial r} + \frac{1}{r} \frac{\partial (\rho v)}{\partial \theta} + \frac{\partial (\rho w)}{\partial z} = 0. \tag{D.35}$$

D.5.2 MOMENTUM CONSERVATION EQUATIONS

In terms of the total shear stress σ:

$$\rho \left(\frac{\partial \mathbf{v}}{\partial t} + \mathbf{v} \cdot \nabla \mathbf{v} \right) = \nabla \cdot \sigma + \rho \mathbf{g}. \tag{D.36}$$

In terms of the pressure gradient ∇p and the viscous shear stress τ:

$$\rho \left(\frac{\partial \mathbf{v}}{\partial t} + \mathbf{v} \cdot \nabla \mathbf{v} \right) = -\nabla p + \nabla \cdot \tau + \rho \mathbf{g}. \tag{D.37}$$

Cartesian coordinates (x, y, z):
x-component:

$$\rho \left(\frac{\partial u}{\partial t} + u \frac{\partial u}{\partial x} + v \frac{\partial u}{\partial y} + w \frac{\partial u}{\partial z} \right) = -\frac{\partial p}{\partial x} + \frac{\partial \tau_{xx}}{\partial x} + \frac{\partial \tau_{xy}}{\partial y} + \frac{\partial \tau_{xz}}{\partial z} + \rho g_x, \tag{D.38}$$

y-component:

$$\rho \left(\frac{\partial v}{\partial t} + u\frac{\partial v}{\partial x} + v\frac{\partial v}{\partial y} + w\frac{\partial v}{\partial z} \right) = -\frac{\partial p}{\partial y} + \frac{\partial \tau_{yx}}{\partial x} + \frac{\partial \tau_{yy}}{\partial y} + \frac{\partial \tau_{yz}}{\partial z} + \rho g_y, \tag{D.39}$$

z-component:

$$\rho \left(\frac{\partial w}{\partial t} + u\frac{\partial w}{\partial x} + v\frac{\partial w}{\partial y} + w\frac{\partial w}{\partial z} \right) = -\frac{\partial p}{\partial z} + \frac{\partial \tau_{zx}}{\partial x} + \frac{\partial \tau_{zy}}{\partial y} + \frac{\partial \tau_{zz}}{\partial z} + \rho g_z. \tag{D.40}$$

Cylindrical coordinates (r, θ, z):
r-component:

$$\rho \left(\frac{\partial u}{\partial t} + u\frac{\partial u}{\partial r} + \frac{v}{r}\frac{\partial u}{\partial \theta} + w\frac{\partial u}{\partial z} - \frac{v^2}{r} \right) = -\frac{\partial p}{\partial r} +$$
$$\frac{1}{r}\frac{\partial (r\tau_{rr})}{\partial r} + \frac{1}{r}\frac{\partial \tau_{r\theta}}{\partial \theta} + \frac{\partial \tau_{rz}}{\partial z} - \frac{\tau_{\theta\theta}}{r} + \rho g_r \tag{D.41}$$

θ-component:

$$\rho \left(\frac{\partial v}{\partial t} + u\frac{\partial v}{\partial r} + \frac{v}{r}\frac{\partial v}{\partial \theta} + w\frac{\partial v}{\partial z} + \frac{uv}{r} \right) = -\frac{1}{r}\frac{\partial p}{\partial \theta} +$$
$$\frac{1}{r^2}\frac{\partial (r^2\tau_{\theta r})}{\partial r} + \frac{1}{r}\frac{\partial \tau_{\theta\theta}}{\partial \theta} + \frac{\partial \tau_{\theta z}}{\partial z} + \rho g_\theta \tag{D.42}$$

z-component:

$$\rho \left(\frac{\partial w}{\partial t} + u\frac{\partial w}{\partial r} + \frac{v}{r}\frac{\partial w}{\partial \theta} + w\frac{\partial w}{\partial z} \right) = -\frac{\partial p}{\partial z} +$$
$$\frac{1}{r}\frac{\partial (r\tau_{zr})}{\partial r} + \frac{1}{r}\frac{\partial \tau_{z\theta}}{\partial \theta} + \frac{\partial \tau_{zz}}{\partial z} + \rho g_z \tag{D.43}$$

Newtonian fluids with constant ρ and μ:

$$\rho \left(\frac{\partial \mathbf{v}}{\partial t} + \mathbf{v} \cdot \nabla \mathbf{v} \right) = -\nabla p + \mu \nabla^2 \mathbf{v} + \rho \mathbf{g}. \tag{D.44}$$

Cartesian coordinates (x, y, z):
x-component:

$$\rho \left(\frac{\partial u}{\partial t} + u\frac{\partial u}{\partial x} + v\frac{\partial u}{\partial y} + w\frac{\partial u}{\partial z} \right) = -\frac{\partial p}{\partial x} + \mu \left[\frac{\partial^2 u}{\partial x^2} + \frac{\partial^2 u}{\partial y^2} + \frac{\partial^2 u}{\partial z^2} \right] + \rho g_x, \tag{D.45}$$

y-component:

$$\rho \left(\frac{\partial v}{\partial t} + u\frac{\partial v}{\partial x} + v\frac{\partial v}{\partial y} + w\frac{\partial v}{\partial z} \right) = -\frac{\partial p}{\partial y} + \mu \left[\frac{\partial^2 v}{\partial x^2} + \frac{\partial^2 v}{\partial y^2} + \frac{\partial^2 v}{\partial z^2} \right] + \rho g_y, \tag{D.46}$$

z-component:

$$\rho \left(\frac{\partial w}{\partial t} + u\frac{\partial w}{\partial x} + v\frac{\partial w}{\partial y} + w\frac{\partial w}{\partial z} \right) = -\frac{\partial p}{\partial z} + \mu \left[\frac{\partial^2 w}{\partial x^2} + \frac{\partial^2 w}{\partial y^2} + \frac{\partial^2 w}{\partial z^2} \right] + \rho g_z. \tag{D.47}$$

Cylindrical coordinates (r, θ, z):

r-component:

$$\rho \left(\frac{\partial u}{\partial t} + u \frac{\partial u}{\partial r} + \frac{v}{r} \frac{\partial u}{\partial \theta} + w \frac{\partial u}{\partial z} - \frac{v^2}{r} \right) = -\frac{\partial p}{\partial r} +$$
$$\mu \left[\frac{\partial}{\partial r} \left(\frac{1}{r} \frac{\partial (ru)}{\partial r} \right) + \frac{1}{r^2} \frac{\partial^2 u}{\partial \theta^2} + \frac{\partial^2 u}{\partial z^2} - \frac{2}{r^2} \frac{\partial v}{\partial \theta} \right] + \rho g_r \tag{D.48}$$

θ-component:

$$\rho \left(\frac{\partial v}{\partial t} + u \frac{\partial v}{\partial r} + \frac{v}{r} \frac{\partial v}{\partial \theta} + w \frac{\partial v}{\partial z} + \frac{uv}{r} \right) = -\frac{1}{r} \frac{\partial p}{\partial \theta} +$$
$$\mu \left[\frac{\partial}{\partial r} \left(\frac{1}{r} \frac{\partial (rv)}{\partial r} \right) + \frac{1}{r^2} \frac{\partial^2 v}{\partial \theta^2} + \frac{\partial^2 v}{\partial z^2} + \frac{2}{r^2} \frac{\partial u}{\partial \theta} \right] + \rho g_\theta \tag{D.49}$$

z-component:

$$\rho \left(\frac{\partial w}{\partial t} + u \frac{\partial w}{\partial r} + \frac{v}{r} \frac{\partial w}{\partial \theta} + w \frac{\partial w}{\partial z} \right) = -\frac{\partial p}{\partial z} +$$
$$\mu \left[\frac{1}{r} \frac{\partial}{\partial r} \left(r \frac{\partial w}{\partial r} \right) + \frac{1}{r^2} \frac{\partial^2 w}{\partial \theta^2} + \frac{\partial^2 w}{\partial z^2} \right] + \rho g_z \tag{D.50}$$

D.5.3 ENERGY CONSERVATION EQUATIONS

Multiplying both sides of Eq. (D.37) by $\mathbf{v}$, the following equation for the rate of change of the kinetic energy is obtained:

$$\rho \left(\frac{\partial e_K}{\partial t} + \mathbf{v} \cdot \nabla e_K \right) = \nabla \cdot [(\tau - p\mathbf{I}) \cdot \mathbf{v}] + p \nabla \cdot \mathbf{v} + \mathbf{v} \cdot \rho \mathbf{g} - \tau : \nabla \mathbf{v}, \tag{D.51}$$

where $e_K = \frac{1}{2} v^2 = \frac{1}{2} |\mathbf{v}|^2$ is the fluid kinetic energy per unit mass and $\mathbf{I}$ is the unity tensor. For heat transfer problems, the following energy conservation equation is valid:

$$\rho c_p \left(\frac{\partial T}{\partial t} + \mathbf{v} \cdot \nabla T \right) = \nabla \cdot \lambda \nabla T - \nabla \cdot \mathbf{q}_r'' + q''' + \beta T \left(\frac{\partial p}{\partial t} + \mathbf{v} \cdot \nabla p \right) + \phi, \tag{D.52}$$

where $\mathbf{q}_r''$ is the heat flux vector due to radiation, q''' is the heat source, β is the volumetric thermal expansion coefficient of the fluid, and $\phi \equiv \tau : \nabla \mathbf{v} = \nabla \cdot (\tau \cdot \mathbf{v}) - \mathbf{v} \cdot (\nabla \cdot \tau)$ is the energy dissipation term. For incompressible materials $c_p = c_v$ is valid.

Cartesian coordinates (x, y, z):

$$\rho c_p \left(\frac{\partial T}{\partial t} + u \frac{\partial T}{\partial x} + v \frac{\partial T}{\partial y} + w \frac{\partial T}{\partial z} \right) = \frac{\partial}{\partial x} \left(\lambda \frac{\partial T}{\partial x} \right) + \frac{\partial}{\partial y} \left(\lambda \frac{\partial T}{\partial y} \right) + \frac{\partial}{\partial z} \left(\lambda \frac{\partial T}{\partial z} \right)$$
$$- \left(\frac{\partial q_{rx}''}{\partial x} + \frac{\partial q_{ry}''}{\partial y} + \frac{\partial q_{rz}''}{\partial z} \right) + q''' + \beta T \left(\frac{\partial p}{\partial t} + u \frac{\partial p}{\partial x} + v \frac{\partial p}{\partial y} + w \frac{\partial p}{\partial z} \right) + \phi \tag{D.53}$$

where q_{rx}'', q_{ry}'', and q_{rz}'' are components of the radiation heat flux vector $\mathbf{q}_r''$.
Cylindrical coordinates (r, θ, z):

$$\rho c_p \left(\frac{\partial T}{\partial t} + u \frac{\partial T}{\partial r} + \frac{v}{r} \frac{\partial T}{\partial \theta} + w \frac{\partial T}{\partial z} \right) = \frac{1}{r} \frac{\partial}{\partial r} \left(\lambda r \frac{\partial T}{\partial r} \right) + \frac{1}{r^2} \frac{\partial}{\partial \theta} \left(\lambda \frac{\partial T}{\partial \theta} \right) + \frac{\partial}{\partial z} \left(\lambda \frac{\partial T}{\partial z} \right)$$
$$- \left(\frac{1}{r} \frac{\partial (r q_{rr}'')}{\partial r} + \frac{1}{r} \frac{\partial q_{r\theta}''}{\partial \theta} + \frac{\partial q_{rz}''}{\partial z} \right) + q''' + \beta T \left(\frac{\partial p}{\partial t} + u \frac{\partial p}{\partial r} + \frac{v}{r} \frac{\partial p}{\partial \theta} + w \frac{\partial p}{\partial z} \right) + \phi \quad , \quad \text{(D.54)}$$

where q_{rr}'', $q_{r\theta}''$, and q_{rz}'' are components of the radiation heat flux vector $\mathbf{q}_r''$ in the cylindrical coordinate system.

D.6 SPECIAL FUNCTIONS

Some useful special functions and relationships containing these function are given in this section.

D.6.1 BESSEL FUNCTIONS

Derivatives of Bessel functions:

$$\frac{dJ_0(x)}{dx} = -J_1(x),$$

$$\frac{dI_0(x)}{dx} = I_1(x),$$

$$\frac{dY_0(x)}{dx} = -Y_1(x),$$

$$\frac{dK_0(x)}{dx} = -K_1(x).$$

Integrals of Bessel functions:

$$\int J_1(x) dx = -J_0(x) + C,$$

$$\int I_1(x) dx = I_0(x) + C,$$

$$\int Y_1(x) dx = -Y_0(x) + C,$$

$$\int K_1(x) dx = -K_0(x) + C.$$

Roots of $J_0(x_n) = 0$: $x_1 = 2.40483...$, $x_2 = 5.52008...$, $x_3 = 8.65373...$, $x_4 = 11.79153...$, $x_5 = 14.93092...$
Roots of $J_1(x_n) = 0$: $x_0 = 0$, $x_1 = 3.83171...$, $x_2 = 7.01559...$, $x_3 = 10.17347...$, $x_4 = 13.32369...$
Roots of $Y_0(x_n) = 0$: $x_1 = 0.8936...$, $x_2 = 3.9577...$, $x_3 = 7.00861...$, $x_4 = 10.2223...$
Roots of $Y_1(x_n) = 0$: $x_1 = 2.1971...$, $x_2 = 5.4297...$, $x_3 = 8.5960...$, $x_4 = 11.7492...$

D.6.2 GAMMA FUNCTION

The Gamma function Γ is defined by the integral

$$\Gamma(z) = \int_0^\infty t^{z-1} e^{-t} dt, \quad \text{(D.55)}$$

which converges for any complex number z with real part greater than zero. When $z = n$ and n is a positive integer we have

$$\Gamma(n) = (n-1)! \qquad\qquad (D.56)$$

For the argument much large than one, a useful estimate of the factorial $n!$ is given by Stirling's approximation,

$$n! = \sqrt{2\pi n}\left(\frac{n}{e}\right)^n, \qquad\qquad (D.57)$$

where e is the base of the natural logarithm.
For $x > 2$

$$\Gamma(x) = (x-1)\Gamma(x-1) = (x-1)(x-2)...(x-k)\Gamma(x-k) \quad \text{where} \quad 1 < x-k < 2. \qquad (D.58)$$

For example:

$$\Gamma(4.7) = (3.7)! = 3.7 \times 2.7 \times 1.7 \times \Gamma(1.7) = 15.4314... \qquad\qquad (D.59)$$

Useful values:

$$\Gamma\left(\frac{1}{8}\right) = 7.5339... \quad \Gamma\left(\frac{1}{6}\right) = 5.5663... \quad \Gamma\left(\frac{1}{5}\right) = 4.5908... \quad \Gamma\left(\frac{1}{4}\right) = 3.6256... \quad \Gamma\left(\frac{1}{3}\right) = 2.6789$$

$$\Gamma\left(\frac{1}{2}\right) = \sqrt{\pi} = 1.7725... \quad \Gamma\left(\frac{3}{8}\right) = 2.3704... \quad \Gamma\left(\frac{2}{5}\right) = 2.2182... \quad \Gamma\left(\frac{3}{5}\right) = 1.4892... \quad \Gamma\left(\frac{5}{8}\right) =$$

Minimum value of the function:

$$\Gamma(1.461632...) = (0.461632...)! = 0.885603...$$

E Units

E.1 SI UNITS

The International System of Units defines seven units of measures as a basic set from which all other SI units can be derived. The base units include measures of length, time, electric current, temperature, luminous intensity, and amount of substance. New definitions of basic units, introduced in May 2019, are provided in Subsection E.1.1. The names and symbols of SI base units are written in lower case, except for symbols derived from person names. For example, the meter has the symbol m, whereas the kelvin has the symbol K, since it is named after Lord Kelvin. In the same way the unit of the electric current is ampere with symbol A, as it is named after André-Marie Ampère. The derived SI units are given in Table E.2.

E.1.1 BASE SI UNITS

The names of the base SI units and the corresponding symbols are shown in Table E.1. Since 20 May 2019, some of the base units (kilogram, ampere and kelvin) received new definitions. The new SI and the dependence of base unit definitions on physical constants with fixed numerical values (see Table E.1) are illustrated in Fig. E.1.

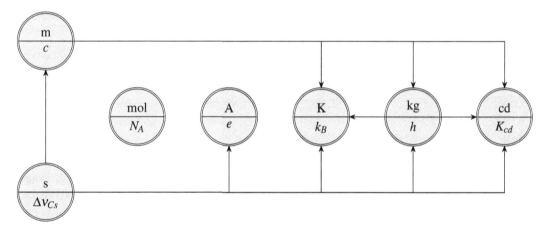

Figure E.1 Base unit relations in the new SI (since May 2019). For each base unit, the lower symbol represents the defining universal constant and arrows show dependency on other base units.

Table E.1
Base SI Units

Physical Quantity	Unit Name	Symbol
Length	meter	m
Mass	kilogram	kg
Time	second	s
Electric current	ampere	A
Thermodynamic temperature	kelvin	K
Luminous intensity	candela	cd
Amount of substance	mole	mol

DOI: 10.1201/9781003036982-E

Second

The second (s) is the unit of time and it is defined by taking the fixed numerical value of the caesium frequency Δv_{Cs}, the unperturbed ground-state hyperfine transition frequency of the caesium-133 atom, to be 9 192 631 770 when expressed in the unit Hz, which is equal to s^{-1}.

Meter

The meter (m) is the SI unit of length. It is defined by taking the fixed numerical value of the speed of light in vacuum c to be 299 792 458 when expressed in the unit $m \cdot s^{-1}$, where the second is defined as described above.

Kilogram

The kilogram (kg) is the SI unit of mass. It is defined by taking the fixed numerical value of the Planck constant h to be $6.626\ 070\ 15 \times 10^{-34}$ when expressed in the unit J·s, which is equal to $kg \cdot m^2 \cdot s^{-1}$, where the meter and the second are defined as described above.

Ampere

The ampere (A) is the SI unit of electric current. It is defined by taking the fixed numerical value of the elementary charge e to be $1.602\ 176\ 634 \times 10^{-19}$ when expressed in the unit C, which is equal to A·s, where the second is defined as described above.

Kelvin

The kelvin (K) is the SI unit of thermodynamic temperature. It is defined by taking the fixed numerical value of the Boltzmann constant k to be $1.380\ 649 \times 10^{-23}$ when expressed in the unit $J \cdot K^{-1}$, which is equal to $kg \cdot m^2 \cdot s^{-2} \cdot K^{-1}$, where the kilogram, meter, and second are defined as described above.

Mole

The mole (mol) is the SI unit of amount of substance. One mole contains exactly $6.022\ 140\ 76 \times 10^{23}$ elementary entities. This number is the fixed numerical value of the Avogadro constant, N_A, when expressed in the unit mol^{-1} and is called the Avogadro number. The amount of substance, symbol n, of a system is a measure of the number of specified elementary entities. An elementary entity may be an atom, a molecule, an ion, an electron, or any other particle or specified group of particles.

Candela

The candela (cd) is the SI unit of luminous intensity in a given direction. It is defined by taking the fixed numerical value of the luminous efficacy (a measure of how well a light source produces visible light) of monochromatic radiation of frequency 540×10^{12} Hz, K_{cd}, to be 683, when expressed in the unit $lm \cdot W^{-1}$, which is equal to $cd \cdot sr \cdot W^{-1}$, or $cd \cdot sr \cdot kg^{-1} \cdot m^{-2} \cdot s^3$, where the kilogram, meter, and second are defined as described above.

E.1.2   DERIVED, SUPPLEMENTARY, AND TEMPORARY SI UNITS

Units of measure which are defined in terms of the seven base units specified by the International System of Units are called the derived SI units. The SI has special names for 22 of these derived units. Selected derived SI units are shown in Table E.2. Selected accepted non-SI units are shown in Table E.3.

Table E.2
Physical Quantities Expressed with Derived SI Units

Physical Quantity	Unit Name	Symbol	In SI Units
Planar angle	radian	rad	m/m
Solid angle	steradian	sr	m^2/m^2
Frequency	hertz	Hz	s^{-1}
Force	newton	N	$kg\ m\ s^{-2}$
Pressure, stress	pascal	$Pa = N\ m^{-2}$	$kg\ m^{-1}\ s^{-2}$
Work, energy, quantity of heat	joule	$J = N\ m$	$kg\ m^2\ s^{-2}$
Power	watt	$W = J\ s^{-1}$	$kg\ m^2\ s^{-3}$
Electric potential difference	volt	$V = W\ A^{-1}$	$kg\ m^2\ s^{-3}\ A^{-1}$
Electric charge	coulomb	C	A s
Capacitance	farad	$F = C\ V^{-1}$	$kg^{-1}\ m^{-2}\ s^4\ A^2$
Electric resistance	ohm	$\Omega = V\ A^{-1}$	$kg\ m^2\ s^{-3}\ A^{-2}$
Electric conductance	siemens	$S = \Omega^{-1}$	$kg^{-1}\ m^{-2}\ s^3\ A^2$
Magnetic flux	weber	$Wb = V\ s$	V s
Magnetic flux density	tesla	$T = Wb\ m^{-2}$	$kg\ s^{-2}\ A^{-1}$
Inductance	henry	$H = Wb\ A^{-1}$	$kg\ m^2\ s^{-2}\ A^{-2}$
Temperature (with zero at 273.15 K)	degree Celsius	°C	K
Luminous flux	lumen	lm	cd sr
Illuminance	lux	$lx = lm\ m^{-2}$	$cd\ m^{-2}$
Radioactive decay rate	beckerel	Bq	s^{-1}
Absorbed dose	gray	$Gr = J\ kg^{-1}$	$m^2\ s^{-2}$
Equivalent dose	sievert	$Sv = J\ kg^{-1}$	$m^2\ s^{-2}$

E.2 SI PREFIXES AND CONVERSION FACTORS

To express very small and very large quantities, the SI units are scaled by using the prefixes given in Table E.4. Most common conversion factors for various units of length, area, volume, mass, energy and power are given in Tables E.5 through E.14. Length unit conversion factors are given in Tables E.5 and E.6 for micro-to-meso and meso-to macro scales, respectively. Area unit conversion factors are given in Tables E.7 and E.8 for micro-to-meso and meso-to macro scales, respectively. Volume unit conversion factors are given in Table E.9 and mass unit conversion factors are given in Table E.10. There are several units that are used to express amounts of energy and power levels. The unit conversion factors for energy in micro and meso-scales are given in Table E.11, whereas Table E.12 shows the factors for energy in mesoscales to describe energy content in various fuels and energy carriers. Macroscale energy conversion factors, useful to describe global energy amounts, are given in Table E.13. The most common power conversion factors are provided in Table E.14. For conversion of other units, very comprehensive unit conversion tools can be found on the Internet[1].

---

[1] www.unit-conversion.info

Table E.3
Physical Quantities Expressed with Accepted Non-SI Units

Physical Quantity	Unit Name	Symbol	Value in SI Units
Time	minute	min	1 min = 60 s
	hour	h	1 h = 60 min = 3 600 s
	day	d	1 d = 24 h = 86 400 s
Length	astronomical unit[a]	au	1 au = 149 597 870 700 m
Volume	liter	l, L	1 l = 1 L = 10^{-3} m^3
Mass	tonne (metric ton)	t	1 t = 1 000 kg
	amu[b]	u	1 u = 1.660 539 066 60(50) $\times 10^{-27}$ kg
Energy	electron volt	eV	1 eV = 1.602 176 634 $\times 10^{-19}$ J
Temporary Units[c]			
Length	nautical mile		1 nautical mile = 1852 m
	ångström	Å	1 Å = 0.1 nm = 10^{-10} m
Area	are	a	1 a = 100 m^2
	hectare	ha	1 ha = 10^4 m^2
	barn	b	1 b = 10^{-28} m^2
Pressure	bar	bar	1 bar = 0.1 MPa = 10^5 Pa
Radioactive decay rate	curie	Ci	1 Ci = 3.7×10^{10} Bq
Radiation exposure	röntgen	R	1 R = 2.58×10^{-4} C kg^{-1}
Absorbed radiation dose	rad	rad	1 rad = 10^{-2} Gy
Radiation dose equivalent	rem	rem	1 rem = 10^{-2} Sv

[a] The astronomical unit of length was redefined by the XXVIII General Assembly of the International Astronomical Union (Resolution B2, 2012) and has an exact value as given in the table.
[b] The unified atomic mass unit (amu) is equal to 1/12 of the mass of an unbound atom of the nuclide ^{12}C, at rest and in its ground state. The value must be obtained experimentally and is thus not known exactly.
[c] Other units outside the SI that are currently accepted for use with the SI, subject to further review.

Table E.4
SI Prefixes

Factor	Prefix	Symbol	Factor	Prefix	Symbol
10^{24}	yotta	Y	10^{-1}	deci	d
10^{21}	zetta	Z	10^{-2}	centi	c
10^{18}	exa	E	10^{-3}	milli	m
10^{15}	peta	P	10^{-6}	micro	μ
10^{12}	tera	T	10^{-9}	nano	n
10^{9}	giga	G	10^{-12}	pico	p
10^{6}	mega	M	10^{-15}	femto	f
10^{3}	kilo	k	10^{-18}	atto	a
10^{2}	hecto	h	10^{-21}	zepto	z
10^{1}	deca	da	10^{-24}	yocto	y

Note: Specific rules for using prefixes should be followed, such as:

- Never use double prefixes, such as $\mu\mu$g; use pg.
- Put prefix in numerator; thus km/s, not m/ms.
- When spelling prefixes with unit names that begin with vowel, suppress the ending vowel on the prefix; thus megohm and kilohm, not megaohm and kiloohm.
- Never use a hyphen with a prefix; thus microgram, not micro-gram.

Table E.5
Length Conversion Factors at Micro- and Mesoscale

$\downarrow \times \searrow = \rightarrow^a$	m	yd	ft	in	Å
meter (m)	1	1.0936	3.2808	39.37	1×10^{10}
yard (yd)	0.9144	1	3	36	9.144×10^{9}
foot (ft)	0.3048	0.3333	1	12	3.048×10^{9}
inch (in)	0.0254	0.02778	0.08333	1	2.54×10^{8}
ångström (Å)	1×10^{-10}	1.0936×10^{-10}	3.281×10^{-10}	3.937×10^{-9}	1

[a] These symbols mean: "take a quantity in any unit shown in the first column, multiply by a factor in the same row in the table and obtain the quantity in units shown in the first row of the corresponding column".

Table E.6
Length Conversion Factors at Macroscale

$\downarrow \times \searrow = \rightarrow^a$	m	mi	naut. mile	au	parsec
meter (m)	1	6.214×10^{-4}	5.400×10^{-4}	6.685×10^{-12}	3.241×10^{-17}
mile (mi)	1609	1	0.8690	1.076×10^{-8}	5.216×10^{-14}
nautical mile[b]	1852	1.151	1	1.238×10^{-8}	6.002×10^{-14}
astronomical unit[c] (au)	1.496×10^{11}	9.296×10^{7}	8.078×10^{7}	1	4.848×10^{-6}
parsec[d]	3.086×10^{16}	1.917×10^{13}	1.666×10^{13}	2.063×10^{5}	1

[a] See Table E.5 for explanation.
[b] Defined as exactly 1852 m. Historically it was defined as one minute of latitude along any line of longitude.
[c] Defined as the average distance from the Earth to the Sun.
[d] Defined as the distance at which one astronomical unit subtends an angle of one arcsecond, i.e. $1/3600^{th}$ of a degree.

Table E.7
Area Conversion Factors at Micro- and Mesoscale

$\downarrow \times \searrow = \rightarrow^a$	m^2	yd^2	ft^2	in^2	b
sq. meter (m^2)	1	1.196	10.76	1550	1×10^{28}
sq. yard (yd^2)	0.8361	1	9	1296	8.361×10^{27}
sq. foot (ft^2)	0.0929	0.1111	1	144	9.290×10^{26}
sq. inch (in^2)	6.452×10^{-4}	7.716×10^{-4}	6.944×10^{-3}	1	6.452×10^{24}
barn (b)^b	1×10^{-28}	1.196×10^{-28}	1.0764×10^{-27}	1.550×10^{-25}	1

^{a} See Table E.5 for explanation.
^{b} A unit of the microscopic cross section.

Table E.8
Area Conversion Factors at Macroscale

$\downarrow \times \searrow = \rightarrow^a$	m^2	are	acre	ha	sq. mile
sq. meter (m^2)	1	0.01	2.471×10^{-4}	0.0001	3.863×10^{-7}
are	100	1	0.02471	0.01	3.863×10^{-5}
acre	4047	40.47	1	0.4047	1.563×10^{-3}
hectare (ha)	10000	100	2.471	1	3.863×10^{-3}
square mile	2.589×10^6	2.589×10^4	640	258.9	1

^{a} See Table E.5 for explanation.

Table E.9
Volume Conversion Factors

$\downarrow \times \searrow = \rightarrow^a$	m^3	fl oz	gal	ft^3	bbl
cubic meter (m^3)	1	3.381×10^4	264.2	35.31	6.2898
fluid ounce (fl oz)	2.957×10^{-5}	1	7.813×10^{-3}	1.044×10^{-3}	1.860×10^{-4}
US gallon (gal)	3.785×10^{-3}	128	1	0.1337	2.381×10^{-2}
cubic foot (ft^3)	2.832×10^{-2}	957.5	7.481	1	0.1781
barrel petroleum (bbl)	0.1590	5376	42	5.615	1

^{a} See Table E.5 for explanation.

Table E.10
Mass Conversion Factors

$\downarrow \times \searrow = \rightarrow^a$	kg	t	lt	st	lb
kilogram (kg)	1	0.001	9.842×10^{-4}	1.102×10^{-3}	2.205
tonne (t)	1000	1	0.9842	1.102	2.205×10^3
long ton (lt)	1016	1.016	1	1.120	2.240×10^3
short ton (st)	907.2	0.9072	0.8929	1	2000
pound (lb)	0.4536	4.536×10^{-4}	4.464×10^{-4}	5×10^{-4}	1

^{a} See Table E.5 for explanation.

Table E.11
Energy Conversion Factors at Micro- and Mesoscale

$\downarrow \times \searrow = \rightarrow^a$	J	eV	cal_{th}	cal_{IT}	erg
J	1	6.24150×10^{18}	0.2390	0.2388	1×10^7
eV[b]	1.602×10^{-19}	1	3.829×10^{-18}	3.826×10^{-18}	1.602×10^{-12}
cal_{th}[c]	4.184	2.6114×10^{19}	1	0.9993	4.184×10^7
cal_{IT}[d]	4.1868	2.6132×10^{19}	1.0007	1	4.1868×10^7
erg[e]	1×10^{-7}	6.2415×10^{11}	2.390×10^{-8}	2.388×10^{-8}	1

[a] See Table E.5 for explanation.
[b] An electron volt (eV) is defined as $1\ eV = (e/C)J = 1.602\ 176\ 634 \times 10^{-19}$ J (exact).
[c] A thermochemical calorie (cal_{th}) is defined as the amount of energy exactly equal to 4.184 J.
[d] An International Steam Table calorie (cal_{IT}) is defined as exactly 1.163 mW·h = 4.1868 J.
[e] An erg is defined as the amount of work done by a force of one dyne exerted for a distance of one centimeter.

Table E.12
Energy Conversion Factors at Mesoscale

$\downarrow \times \searrow = \rightarrow^a$	kJ	BTU	kWh	toe	tce
kJ[b]	1	0.9478	2.778×10^{-4}	2.388×10^{-8}	3.4123×10^{-8}
BTU[c]	1.055	1	2.931×10^{-4}	2.519×10^{-8}	3.600×10^{-8}
kWh[d]	3600	3.412×10^3	1	8.597×10^{-5}	1.228×10^{-4}
toe[e]	4.1868×10^7	3.9682×10^7	1.163×10^4	1	1.429
tce[f]	2.93076×10^7	2.7780×10^7	8.141×10^3	0.7000	1

[a] See Table E.5 for explanation.
[b] A kilojoule (kJ) is equal to 1000 J.
[c] A British Thermal Unit (BTU) was originally defined as the amount of heat required to raise the temperature of 1 pound of liquid water by 1 degree Fahrenheit at a constant pressure of one atmosphere.
[d] A kilowatt-hour (kWh) is equal to 3.6 GJ.
[e] A ton of oil equivalent (toe) is defined by the International Energy Agency to be equal to 41.868 GJ.
[f] A ton of coal equivalent (tce) is defined by the International Energy Agency to be equal to 29.3076 GJ.

Table E.13
Energy Conversion Factors at Macroscale

$\downarrow \times \searrow = \rightarrow^a$	TWh	EJ	Mtoe	quad	Q
TWh[b]	1	3.600×10^{-3}	8.598×10^{-2}	3.412×10^{-3}	3.412×10^{-6}
EJ[c]	277.8	1	23.89	0.9479	9.479×10^{-4}
Mtoe[d]	11.63	4.187×10^{-2}	1	0.03969	3.969×10^{-5}
quad[e]	293.1	1.055	25.20	1	0.001
Q[f]	2.931×10^5	1055	2.520×10^4	1000	1

[a] See Table E.5 for explanation.
[b] A terrawatt-hour (TWh) is equal to 10^{12} watthours.
[c] An exajoule (EJ) is equal to 10^{18} joules.
[d] 1 Mtoe is equal to 10^6 toe.
[e] A quad is defined as equal to 10^{15} BTU.
[f] Q is defined as equal to 10^{18} BTU.

Table E.14
Power Conversion Factors

$\downarrow \times \searrow = \rightarrow^a$	W	kW	hp	BTU/s	ft·lbf/s
watt (W)	1	0.001	1.341×10^{-3}	9.478×10^{-4}	0.7376
kilowatt (kW)	1000	1	1.341	0.9478	737.6
horsepower (hp)	745.7	0.7457	1	0.7068	550
BTU/second	1055	1.055	1.415	1	778.2
ft·lbf/s	1.356	1.356×10^{-3}	1.818×10^{-3}	1.285×10^{-3}	1

^{a} See Table E.5 for explanation.

References

1. A.S. Adams and D.W. Keith. Are global wind power resource estimates overstated? *Environmental Research Letters*, 8:9pp, 2013. Online at stacks.iop.org/ERL/8/015021.
2. C. Angulo et al. A compilation of charged-particle induced thermonuclear reaction rates. *Nuclear Physics A*, 656:3, 1999.
3. C.L. Archer and M.Z. Jacobson. Evaluation of global wind power. *Journal of Geophysical Research*, 110:D12110, 2005.
4. A.C. Baker. *Tidal Power*. London: Peter Peregrinus, 1991.
5. R.G. Barry and R.J. Chorley. *Atmosphere, Weather and Climate*. New York, NY: Routledge, 9th edition, 2010.
6. N. Basu, G.R. Warrier, and V.K. Dhir. Onset of nucleate boiling and active nucleation site density during subcooled flow boiling. *Journal of Heat Transfer*, 124:717–728, 2002.
7. A. Bejan. *Convection Heat Transfer*. John Wiley & Sons, Inc., 605 Third Avenue, New York, NY, 2nd edition, 1995.
8. D.J. Bennet and J. R. Thomson. *The Elements of Nuclear Power*. Hoboken, NJ: Wiley, 1989.
9. P.J. Berenson. Transition boiling heat transfer from a horizontal surface. Technical Report 17, Massachusetts Institute of Technology, 1960.
10. A.I. Blokhin et al. New version of neutron evaluated data library BROND-3.1 (in Russian). *Nuclear and Reactor Constants Series*, 2:62–93, 2016.
11. British Petroleum. *BP Statistical Review of World Energy 2019*. BP p.l.c. 1 St James's Square London SW1Y4PD UK, 68th edition, 2019.
12. D.A. Brown et al. ENDF/B-VIII.0: The 8th major release of the nuclear reaction data library with CIELO-project cross sections, new standards and thermal scattering data. *Nuclear Data Sheets*, 148:1–142, 2018.
13. K. Caldeira and M.E. Wickett. Anthropogenic carbon and ocean pH. *Nature*, 425:365–365, 2003.
14. V.P. Carey. *Liquid-Vapour Phase-Change Phenomena: An Introduction to Thermophysics of Vaporization and Condensation Processes in Heat Transfer Equipment*. Hemisphere Publishing Corporation, 1992.
15. H.S. Carslaw and J.C. Jaeger. *Conduction of Heat in Solids*. University Press, Oxford, UK, 2nd edition, 1959.
16. J.C. Chen. A correlation for boiling heat transfer to saturated fluids in convective flow. *Industrial & Engineering Chemistry, Process Design and Development*, 5(3):322–339, 1966.
17. A.P. Colburn. A method of correlating forced convection heat transfer data and a comparision with fluid friction. *Transactions of the American Institute of Chemical Engineers*, 29:174–210, 1933.
18. C.F. Colebrook. Turbulent flow in pipes, with particular reference to the transition region between the smooth and rough pipe laws. *Journal of the Institution of Civil Engineering*, 11(4):133–156, 1939.
19. J.G. Collier and J.R. Thome. *Convective Boiling and Condensation*. McGraw-Oxford Univ. Press, London & New York, 3rd edition, 1994.
20. M. Cozzens and F.S. Roberts, editors. *Mathematical and Statistical Challenges for Sustainability*, May 2011. Report of a Workshop held November 15–17, 2010.
21. CWIF. Summary of wind turbine accident data to 31 march 2021. Available at www.caithnesswindfarms.co.uk/accidents.pdf (accessed April 4, 2021).
22. E.J. Davis and G.H. Anderson. The incipience of nucleate boiling in forced convection flow. *AIChE J.*, 12:774–780, 1966.
23. C. de Castro, M. Mediavilla, L.J. Miguel, and F. Frechoso. Global wind power potential: Physical and technological limits. *Energy Policy*, 39:6677–6682, 2011.
24. Y. Demirel. *Energy: Production, Conversion, Storage, Conservation, and Coupling*. Springer International Publishing, Switzerland, 2nd edition, 2016.
25. P.W. Dittus and L.M.K. Boelter. Heat transfer in automobile radiators of the tubuler type. *Univ. California Pub. Eng.*, 2(13):443–461, 1930.
26. J.J. Duderstadt and L.J. Hamilton. *Nuclear Reactor Analysis*. John Wiley & Sons, Hoboken, NJ, 1976.
27. J.A. Duffie and W.A. Beckman. *Solar Engineering of Solar Processes*. Hoboken, NJ: Wiley & Sons, Inc., 2013.

28. M. Emilio, J.R. Kuhn, R.I. Bush, and I.F. Scholl. Measuring the solar radius from space during the 2003 and 2006 mercury transit. *The Astrophysical Journal*, 750(2), 2012.

29. EU-OSHA. Occupational safety and health in the wind energy sector. Technical Report ISSN: 1831-9343, European Agency for Safety and Health at Work, 2013.

30. H.K. Forster and N. Zuber. Dynamics of vapor bubbles and boiling heat transfer. *AIChE J.*, 1(4):531–535, 1955.

31. GBD 2019 Risk Factors Collaborators. Global burden of 87 risk factors in 204 countries and territories, 1990–2019: a systematic analysis for the Global Burden of Disease Study 2019. *Lancet*, 396:1223–1249, 2020.

32. Z. Ge et al. CENDL-3.2: The new version of Chinese general purpose evaluated nuclear data library. *EPJ Web of Conferences*, 239:09001, 2020.

33. D. Giurco, E. Dominish, N. Florin, T. Watari, and B. McLellan. *Achieving the Paris Climate Agreement Goals: Global and Regional 100% Renewable Energy Scenarios with Non-energy GHG Pathways for +1.5°C and +2°C*, chapter Requirements for Minerals and Metals for 100% Renewable Scenarios. Springer International Publishing, Gewerbestrasse 11, 6330 Cham, Switzerland, 2019.

34. V. Gnielinski. New equations for heat and mass transfer for turbulent pipe and channel flow. *International Chemical Engineering*, 16:359–368, 1976.

35. M.A. Green, E.D. Dunlop, J. Hohl-Ebinger, M. Yoshita, N. Kopidakis, and X. Hao. Solar cell efficiency tables (version 56). *Progress in Photovoltaics: Research and Applications*, 28:629–638, 2020.

36. D.C. Groeneveld. Post-dryout heat transfer at reactor operating conditions. Technical Report 4513, AECL, 1973.

37. K. Gunn and C. Stock-Williams. Quantifying the global wave power resource. *Renewable Energy*, 44:296, 2012.

38. S.E. Haaland. Simple and explicit formulas for the friction factor in turbulent flow. *Journal of Fluids Engineering*, 105(1):89–90, 1983.

39. M.O.L. Hansen, J.N. Sørensen, S. Voutsinas, N. Sørensen, and H.Aa. Madsen. State of the art in wind turbine aerodynamics and aeroelasticity. *Progress in Aerospace Sciences*, 42:285–330, 2006.

40. D.L. Hartmann. *Global Physical Climatology*. Cambridge, MA: Academic Press, 1994.

41. W.M. Haynes. *CRC Handbook of Chemistry and Physics (97th ed.)*. Taylor & Francis, 2016.

42. H.C. Honeck. ENDF/B - specifications for an evaluated nuclear data file for reactor applications. Technical Report BNL-50066, Brookhaven National Laboratory, 1966.

43. Y. Y. Hsu. On the size range of active nucleation cavities in a heating surface. *ASME Journal of Heat Transfer*, 84(3):207–216, 1962.

44. Y.Y. Hsu and R.W. Graham. *Transport Processes in Boiling and Two-Phase Systems*. American Nuclear Society, La Grange Park, IL (USA), 1986.

45. IAEA. Classification of uranium reserves/resources. Technical Report IAEA-TECDOC-1035, International Atomic Energy Agency, 1998.

46. IAEA. Deterministic safety analysis for nuclear power plants. Specific safety guide. Technical Report SSG-2 (Rev. 1), International Atomic Energy Agency, 2019.

47. IEA. *Key World Energy Statistics*. International Energy Agency, 2018.

48. IEA/NEA. Projected costs of generating electricity. 2020 edition. Technical report, OECD Publishing, 2020.

49. H. Inhaber. Is solar power more dangerous than nuclear? *IAEA Bulletin*, 21(1):11–17, February 1979.

50. IRENA. Renewable energy and jobs. Annual review 2019. Technical Report, International Renewable Energy Agency, 2019.

51. J.G. Irwin and M.L. Williams. Acid rain: chemistry and transport. *Environmental Pollution*, 50(1–2):29–59, 1988.

52. M. Ishii and T. Hibiki. *Thermo-Fluid Dynamics of Two-Phase Flow*. Springer International Publishing, Switzerland, 2nd edition, 2011.

53. R.L. Jaffe and W. Taylor. *The Physics of Energy*. Cambridge University Press, 2018.

54. J. Jia, P. Punys, and J. Ma. Hydropower. *Handbook of Climate Change Mitigation*, pages 1355–1401. Springer, New York, 2012.

55. F. Kasten and A.T. Young. Revised optical mass tables and approximation formula. *Applied Optics*, 28:4735–4738, 1989.

56. G. Kopp and J.L. Lean. A new, lower value of total solar irradiance: Evidence and climate significance. *Geophysical Research Letters*, 38(1), 2011.

57. S.S. Kutateladze. Hydrodynamic model of heat transfer crisis in boiling liquid during free convection (in Russian). *Journal of Technical Physics*, 20(11):1389–1392, 1950.

58. J.C. Lee and N.J. McCormick. *Risk and Safety Analysis of Nuclear Systems*. Wiley, Hoboken, NJ, 2011.

59. M. Lenzen and U. Wachsmann. Wind turbines in Brazil and Germany: an example of geographical variability in life-cycle assessment. *Applied Energy*, 77:119–130, 2004.

60. L.L. Levitan and F.P. Lantsman. Investigating burnout with flow of a steam-water in a round tube. *Thermal Eng. (USSR)*, 22(1):101–105, 1975.

61. J. Lilley. *Nuclear Physics: Principles and Application*. Wiley, Hoboken, NJ, 2001.

62. J. Malone, A. Totemeier, N. Shapiro, and S. Vaidyanathan. Lightbridge corporations advanced metallic fuel for light water reactors. *Nuclear Technology*, 180:437–442, 12 2012.

63. A. Markandya and P. Wilkinson. Electricity generation and health. *The Lancet*, 370:979–990, 2007.

64. B.R. Martin. *Nuclear and Particle Physics: An Introduction*. John Wiley & Sons, New York, NY, 2nd edition, 2006.

65. E. Masanet, Y. Chang, A.R. Gopal, P. Larsen, W.R. Morrow III, R. Sathre, A. Shehabi, and P. Zhai. Life-cycle assessment of electric power systems. *Annual Review of Environment and Resources*, 38:107–136, 2013.

66. A.B. Meinel and P.B. Meinel. *Applied Solar Energy*. Addison Wesley Publishing Co., Michigan, 1976.

67. NEA-IAEA. Uranium 2018: Resources, production and demand. Technical Report NEA No. 7413, Nuclear Energy Agency and International Atomic Energy Agency, 2018.

68. M.E. Pasyanos. Lithospheric thickness modeled from long period surface wave dispersion. *Tectonophysics*, 481(1):38–50, January 2010.

69. A.J.M. Plompen et al. The joint evaluated fission and fusion nuclear data library, JEFF-3.3. *The European Physical Journal A*, 56(181):1–108, 2020.

70. N.C. Rasmussen et al. Reactor safety study: An assessment of accident risk in U.S. commercial nuclear power plants. Technical Report WASH-1400 (NUREG-75/014), U.S. NRC, 1975.

71. S.T. Revankar. *Storage and Hybridization of Nuclear Energy*, chapter Chemical Energy Storage, pages 177–227. Academic Press, Cambridge, Massachusetts, 2019.

72. W.M. Rohsenow. A method of correlating heat transfer data for surface boiling of liquids. *Trans. ASME*, 84:969, 1962.

73. W.M. Rohsenow. *Handbook of Multiphase Systems*, chapter Nucleate Boiling. McGraw-Hill Book Company, New York, 1982.

74. S. Ruhle. Tabulated values of the shockley-queisser limit for single junction solar cells. *Solar Energy*, 130:139–147, 2016.

75. K. von Schuckmann et al. Heat stored in the Earth system: where does the energy go? *Earth System Science Data*, 12(3):2013–2041, 2020.

76. K. Shibata et al. JENDL-4.0: New library for nuclear science and engineering. *Journal of Nuclear Science and Technology*, 48(1):1–30, 2011.

77. I. Shiklomanov. World's fresh water resources. *Water in Crisis: A Guide to the World's Fresh Water Resources*. Oxford University Press, New York, 1993.

78. W. Shockley and H.J. Queisser. Detailed balance limit of efficiency of *p-n* junction solar cells. *Journal of Applied Physics*, 32:510, 1961.

79. E.N. Sieder and G.E. Tate. Heat transfer and pressure drop of liquids in tubes. *Industrial & Engineering Chemistry Research*, 28:1429–1436, 1936.

80. B.K. Sovacool. The costs of failure: a preliminary assessment of major energy accidents, 1907 to 2007. *Energy Policy*, 36(5):1802–1820, 2008.

81. B.K. Sovacool, R. Andersen, S. Sorensen, K. Sorensen, V. Tienda, A. Vainorius, O.M. Schirach, and F.B. Thygesen. Balancing safety with sustainability: assessing the risk of accidents for modern low-carbon energy systems. *Journal of Cleaner Production*, 112:3952–3965, 2016.

82. W.M. Stacey. *Nuclear Reactor Physics*. Wiley-VCH, 2004.

83. T.F. Stocker et al., editors. *IPCC, 2013: Climate Change 2013. Contribution of Working Group I to the Fifth Assessment Report of the Intergovernmental Panel on Climate Change*. Cambridge: Cambridge, United Kingdom and New York, NY, USA, 2013.

84. L.S. Tong and Y.S. Tang. *Boiling Heat Transfer and Two-Phase Flow*. Taylor and Francis, Washington, D.C., 1997.

85. R. Turconi, A. Boldrin, and Th. Astrup. Life cycle assessment (LCA) of electricity generation technologies: Overview, comparability and limitations. *Renewable and Sustainable Energy Reviews*, 28:555–565, 2013.

86. U.S. National Academy of Sciences. *Energy in Transition 1985–2010. Final Report of the Committee on Nuclear and Alternative Energy Systems*. W.H. Freeman and Company, San Francisco, 1980.

87. G.B. Wallis. *One-Dimensional Two-Phase Flow*. McGraw-Hill, New York, 1969.

88. L.C. Weiss, L. Pötter, A. Stieger, S. Kruppert, U. Frost, and R. Tollrian. Rising pCO_2 in freshwater ecosystems has the potential to negatively affect predator-induced defenses in Daphnia. *Current Biology*, 28(2):327–332, 2018.

89. WMO. State of the global climate 2020: Provisional report. Technical report, World Meteorological Organization, 2020.

90. N. Zuber. *Hydrodynamic Aspects of Boiling Heat Transfer*. PhD thesis, University of California, Los Angeles, 1959.

91. N. Zuber and J. Findlay. Average volumetric concentration in two-phase flow systems. *ASME Journal of Heat Transfer*, 87(C):453–468, 1965.

Index

Printed in the United States
by Baker & Taylor Publisher Services